International agri-food chains and networks

International agri-food chains and networks

Management and organization

edited by:
J. Bijman
S.W.F. Omta
J.H. Trienekens
J.H.M. Wijnands
E.F.M. Wubben

Wageningen Academic
Publishers

ISBN-10: 90-76998-95-7
ISBN-13: 978-90-76998-95-4

First published, 2006

Wageningen Academic Publishers
The Netherlands, 2006

Preface

This book covers a broad range of issues in the field of management in international agri-food chains and networks. The 23 papers in this book were selected from a total of 175 papers submitted to the 7th International Conference on Management in Agri-food Chains and Networks. This conference, held in The Netherlands, the 1st and 2nd of June 2006, brought together more than 200 academics and business practitioners from all over the world, presenting and discussing the latest advancements in theory and practise of management in chains and networks. In addition to the academic papers a number of business cases were presented by managers from international agri-food companies. This mix of scientific and business papers lead to a fruitful interaction between two worlds that do not mingle too often.

Besides providing new insights into chain and network management, the papers in this book present empirical trends and cases from different parts of the world. Empirical research has been conducted in Germany, The Netherlands, United Kingdom, Italy, Hungary, Poland, Costa Rica, Brazil, Bolivia, Kenya, Australia, and China. In addition, various parts of the agri-food industry have been investigated. The reader will find information on the chains producing, processing and trading the following food products: dairy (5x), fruits and vegetables (4x), pork, beef and poultry (4x), fish, and quinoa.

The papers in this volume have been selected after a double blind review processes, by reviewers from several universities and research organizations. We would like to thank them for their time-consuming and thoughtful reviewing, which has really contributed to the academic quality of this book.

The editors

Table of contents

Preface 7

Part 1. Introduction

Management and organization in international agri-food chains and networks 15
J. Bijman, S.W.F. Omta, J.H. Trienekens, J.H.M. Wijnands and E.M.F. Wubben

Part 2. Governance

Does guanxi matter for vegetable supply chains in China? A case study approach 31
H. Lu, J.H. Trienekens and S.W.F. Omta

Is more vertical integration the future of food supply chains? Empirical evidence and theoretical considerations from German pork production 49
B. Schulze, A. Spiller and L. Theuvsen

Contracts in agribusiness: A survey in the Hungarian dairy sector 65
G.G. Szabó and K. Bárdos

Part 3. Horizontal collaboration

Producer groups as a form of entrepreneurship and a tool for strengthening the competitive position of private farms in Poland 83
M. Adamowicz and M. Lemanowicz

The role of producer organizations in the process of developing an integrated supply chain; experiences from Quinoa chain development in Bolivia 97
G. Ton and J. Bijman

Horizontal cooperation among small and medium-sized supermarkets as a tool for strengthening competitivenes 113
F.A. Ghisi, D.P. Martinelli and T. Kristensen

Part 4. Supply Chain Management

Logistics outsourcing in food supply chain networks: Theory and practices 135
H.I. Hsiao, J.G.A.J. van der Vorst and S.W.F. Omta

About the preparedness of dairy companies for ECR 151
T. Obersojer and H. Weindlmaier

Using data mining to improve operations management in food supply networks 163
Y. Li, M.R. Kramer, A.J.M. Beulens and J.G.A.J. van der Vorst

Tailored improvement of supply chain processes in agri-food networks 179
M. Fritz and T. Hausen

Part 5. Power

A power-based approach to the analysis of the food system 199
V. Sodano

Chain management challenges and market power in the Chinese fresh produce sector 217
B.J. Revell and X. Liu

Bargaining power in mango supply chains: An experimental gaming approach 231
G. Zúñiga-Arias, S. Meijer, R. Ruben and G.J. Hofstede

Part 6. Performance

Modelling the recall costs of consumption milk in relation to time 259
A.G.J. Velthuis, C. van Erve and M.P.M. Meuwissen

Alliances between wholesalers and retailers in the fresh vegetable industry 269
J.H.M. Wijnands, W.B. Koot and S. Rosenkranz

Part 7. Quality, safety and sustainability

Hybrid forms in the supply of safe food 287
G. Martino and C. Perugini

Quality assurance and vertical alliances: Case studies in the UK and Brazilian beef chains 303
M.D. de Barcellos, G.C. Ferreira and L.M. Vieira

Vertical coordination, quality management and firm performance of the pork processing industry in China 319
J. Han, J.H. Trienekens, T. Tan, S.W.F. Omta and K. Wang

Improving quality and ecological sustainability of natural resources in international supply chains: The role of market-based incentives 333
E. Kambewa, P. Ingenbleek, A. van Tilburg and I. van der Lans

Assessing and stimulating environmental pro-activeness in food supply chains 343
D.J. Haverkamp, H.J. Bremmers and S.W.F. Omta

Part 8. Case studies

Mapping and quantification of the milk chain in São Paulo State, Brazil 359
M.F. Neves, M.A. Consoli, F.F. Lopes, E.M. Campos and H.A.B. Paiva

Impacts of the rapid development of supermarkets on China's dairy industry 373
D. Hu, F. Fuller and T. Reardon

Organizational learning in the context of an agri-food chain 387
A.J. Dunne

Authors and affiliations 403

Part 1. Introduction

Management and organization in international agri-food chains and networks

J. Bijman, S.W.F. Omta, J.H. Trienekens, J.H.M. Wijnands and E.M.F. Wubben

1. Introduction

The papers in this book present theoretical and practical insights about various management issues in agri-food chains and networks. Topics covered include classical supply chain management issues like logistics, information exchange (*e.g.*, tracking and tracing), quality control, safety assurance, and chain performance. Other timely issues covered are joint innovation and shared responsibility for the sustainability in agri-food supply chains. Special attention is given to the issues of governance in chains and networks, such as the role of power and the interaction between horizontal collaboration (for instance in producer cooperatives) and vertical collaboration in international supply chains.

The study of chains and networks has grown rapidly as an area of academic interest from the early 1990s onwards. Economists, sociologists, and management scholars acknowledged that studying firms individually could not sufficiently explain real life phenomena, and did not lead to useful recommendations for practitioners. Inter-organizational relationships became a whole new area of research, emphasising the collaboration and coordination between firms. Interestingly, from both a social science perspective, looking at human and organizational behaviour and performance, and a technical perspective, studying the serially linked processes and their interfaces, scholars have greatly contributed to this emerging field of chain and network studies.

The study of (agri-food) chains and networks can be divided into three streams (Trienekens *et al.*, 2003). One stream is the Supply Chain Management (SCM) literature, which has its roots in logistics and operations research. SCM seeks to optimize the processes within and between the firms participating in the chain, such as logistic processes, (electronic) information exchange and quality control. SCM studies the whole chain and not just one dyadic part of it. In this line of research the agri-food product is followed on its journey through the chain from farm to fork, and, if needed, the food product traced back from consumer to farmer. SCM literature is based on operations research and logistics models, combined with more inductive models based on best business practices. The second stream is the economic organization approach, *e.g.* Transaction Cost Economics. This approach, with its roots in both economics and organization theory, seeks to analyse transactions and to design the most efficient governance structure for a particular transaction. In this line of research, the actors in the chain are also object of study. Performance is determined by the actors' investments decisions, which are influenced by the bounded rationality of the investor and the uncertainty about the behaviour of its collaboration partners. Most of this literature has a dyadic instead of a full chain perspective, studying supplier-buyer relationships. The third stream is the network approach, emphasising multiple relationships among firms. This approach investigates the network instead of the chain as firms are often participating in multiple chains, and have relationships with companies that are not part of the supply chain (such as providers of equipment, labour, advisory services, accountancy, etc). Network studies go beyond the functional perspective (as in SCM) or the purely economic perspective (as in TCE), by incorporating the embeddedness of a firm in a (social) network (*e.g.*, Powell, 1990; Uzzi, 1997). Recently the second and third stream are becoming more integrated, as scholars have combined the economic and social perspectives into studies of relational or network governance (Borgatti *et al.*, 2003).

It should not come as a surprise to learn that the study of chains and networks is well developed for the disciplines that take the agri-food industry as their object of study. In recent years, the various stages of production, processing, and trade have strengthened their collaboration, for several reasons. First, the rise of food safety as one of the most important issues of public and private concern has made the different actors in the chain aware that assuring safety of the final food product requires proper alignment of the activities of all chain participants. Second, there are few industries where the final consumer good so closely resembles the original 'raw material', while it has gone through many hands and crossed many borders. Consumers increasingly value traditional and/or regional characteristics of the food product, strengthening the link between final food product and farming conditions. The latter are in some cases even used as a marketing tool, emphasising that the food product comes straight from the farm, where it was produced in a traditional way with respect for animals, natural environment and human labour conditions. Third, the perishability of agricultural and food products requires efficient logistic processes, to move the product through the chain as rapidly as possible and to maintain valuable quality and safety characteristics. The acknowledgement that all stages of the agri-food chain are closely related, and that improving the performance in the final stage depends on the effort in all preceding stages has lead to a large and still growing body of literature on agri-food chains and networks.

In this introductory chapter we refer to chains and networks instead of using the term supply chain. There has been quite some debate on the proper name of the supply chain; Is it the same as a channel? Is it also a network? In the marketing literature, it is common to use the term marketing channel (Stern *et al.*, 1996). Supply chain management (SCM) studies interfirm processes both at the input and the output side of the firm, and emphasises the link between the supply and marketing side of the focal company. More recently, SCM literature also acknowledges that (focal) companies participate in multiple supply chains, which together form a supply chain network (Lambert and Cooper, 2000). While in SCM the firm participates in multiple supply chains, the recently introduced netchain perspective (Lazzarini *et al.*, 2001) emphasises that firms are part of multiple networks that are sequentially arranged based on the vertical ties between firms in different layers.

The 23 papers have been clustered into seven groups. Each group of papers starts with a general introduction to the subject, after which we give a brief introduction to the individual papers. We start with the three groups of papers that focus on inter-organizational relationships. The papers in Section 2 deal with governance issues, those in Section 3 with horizontal collaboration, while Section 4 presents several supply chain management topics. Following are two sections that focus more on the role of the individual company within the chain or network: Section 5 concentrates on power issues, and Section 6 on performance. Section 7 is devoted to issues of quality, safety and sustainability. The five papers in this section all conclude that quality, safety and sustainability can be enhanced by strengthening vertical collaboration. We conclude this book, in Section 8, with three cases studies: one on the dairy chain in Brazil, one on the rise of supermarkets and their impact on the dairy industry in China, and one on organizational learning in chains and networks.

2. Governance

New institutional economics has been one of the main bodies of literature applied in the study of the organization of agri-food chains and networks. New institutional economics uses a transaction costs perspective. The focus is on the transaction (understood as transfer of rights) because taking advantage of the division of labour requires complex devices at the micro level (modes of organising the transfer) as well as at the macro level (institutions facilitating and enforcing this transfer). The emphasis on transaction costs is explained by the costs of those devices: comparing these costs is crucial for understanding how (and what) institutions and organizations allow benefiting from specialization. This economic organization perspective on transactions enables to study the efficiency of the various organizational structures that can be found in the agri-food industry. As Ménard and Valceschini (2005: 426) have put it, "a fundamental lesson of the new institutional approach to the organization of agricultural activities is that costs of alternative arrangements of transactions at the micro level as well as costs imposed on transactions by alternative policies at the macro level must be assessed in order to understand how specific solutions are selected and why some work better than others".

Agri-food chains and networks are typically hybrid governance structures (Ménard, 1996). According to Ménard (2004), hybrids are characterised by: (1) coordinating partners maintaining distinct legal property rights while pooling particular resources; (2) the use of contractual mechanisms of coordination; and (3) persisting competition among partners to hybrid arrangements as well as between hybrids and alternative modes of organization. As Williamson (1991) has posed, each governance structure has at least three functions: enabling coordination (or alignment) of activities carried out by the collaborating parties; safeguarding the investments that parties make specifically for the transaction; and leaving room for *ex post* adaptation of the transaction in case changes in markets, policies and technologies require such adjustments. Transactions in agri-food chains and networks are characterized by frequent and time-critical production and logistic processes, high uncertainty particularly in upstream parts of the chain, a large number of small-scale primary producers, and high asset specificity in several stages of the chain. As the specific coordination, safeguarding and adaptation requirements of agri-food transactions make it almost impossible to only rely on command or prices, hybrid governance structure are dominant in the agri-food industry.

An interesting development in studying governance structure choice is the issue of using particular governance mechanisms (Bijman, 2006). While institutional economics traditionally focuses on formal arrangements (such as property rights, contracts, authority), sociological theory has emphasised the role of informal institutions such as norms and social ties in governing a transaction (*e.g.*, Granovetter, 1985; Powell, 1990). Repeated exchanges provide the opportunity for social relationships to grow. These social processes promote norms of flexibility, solidarity, and information exchange. Through these social processes and the resulting norms, relational governance may function to mitigate the same exchange hazards that formal institutions address (Jones *et al.*, 1997). For long, the new institutional economics and the sociology approaches have developed separately, with some authors even claiming that formal and informal institutions should be considered as substitutes (Ghoshal and Moran, 1996; Dyer and Singh, 1998). Recently, economic organization theorists began to acknowledge that all inter-firm transactions are carried out within a specific set of formal and informal institutions (Zenger *et al.*, 2002). Several studies have shown that formal contracts and relational governance function as complements (Poppo and Zenger, 2002; Lazzarini *et al.*, 2004). Most recently, authors have focussed on the dynamic interaction between formal and informal mechanisms of governance (Klein Woolthuis *et al.*,

2005; Lazzarini and Zenger, 2002), concluding that over time a differential mix of formal and informal mechanisms may lead to the most efficient outcome.

The paper by Lu, Trienekens and Omta shows how making use of informal ties, individuals and firms can strengthen collaboration in the chain. Lu *et al.* have studied the function of a person's social network in buyer-seller relationships in the vegetable chains in Chinese province of Jiangsu. This social network is called a guanxi and can be defined as a set of personal relationships that can be used to secure resources or to draw benefits when doing business (besides playing a role in social life). A guanxi can consist of family, friend and business relationships. As people have guanxi with various other people, a network of relationships exists: the guanxi network. The authors have found that guanxi networks improve trust between business partners, and support investments in specific assets, for instance needed to comply with quality and safety requirements of customers. Using the guanxi network influences marketing behaviour, and may even improve business performance.

Schulze, Spiller and Theuvsen present a critical review on the ubiquitous argument that agri-food chains need to strengthen vertical coordination through formal mechanisms like contracts and vertical integration. They claim that improved supplier relationship management is a good alternative to more formal governance mechanisms. This argument is based on theoretical and empirical analyses of the German pork chain. The theoretical part of their study uses the transaction costs perspective. They argue that new institutions like quality certification and new automated sorting technologies in slaughterhouses reduce transaction costs and thereby reduce the need for contractual integration of farmers into food supply chains. In other words, these institutional and technical advances permit more market type of governance structures in the farmer-processor part of the pork chain. In addition to developing these theoretical considerations, the authors have carried out a survey among German pork producers, to find out about farmers' attitudes towards contracts and vertical integration. This empirical study focussed on farmers' general evaluation of contracts, perceived advantages of contracts, and the relationship between contract behaviour and trust and commitment. The authors conclude that the limited level of trust in the meat supply chain indicates unused potential in improving vertical coordination through informal governance mechanisms.

The paper by Szabó and Bárdos presents the results of an empirical study into the contractual relationships between producers and processors in the Hungarian dairy industry. The authors have built their theoretical framework on the basis of contract theory and Transaction Cost Economics. The main purpose of the study was to analyse the effectiveness of coordination mechanisms and to explain the latent dimensions of contract motivation in the EU market environment. The findings show that contracting arrangements can be segmented according to asset specificity, bargaining power, and price determination. It was also found that the volume of milk sold had a positive, significant effect on the bargaining power of the producers. On the basis of this finding, the authors conclude that government policy to support the establishment of cooperatives or producer associations would be advisable.

3. Horizontal collaboration

One of the organizational issues in chains and networks of particular importance in the agri-food industry is how (small) primary producers are organised in a horizontal way to strengthen their position in vertical relationships. Producers often have established collective organizations, to benefit from bargaining power, risk sharing and/or economies of scale. These so-called economic producer organizations (POs) play a major role in national and international agri-food supply chains, as they are an intermediary between an often large number of individual farming households and a small number of customers (or suppliers). POs are also an important instrument for rural development, because of their ability to support both economic development and social cohesion. POs have several economic functions, such as collecting, processing and marketing agricultural products, implementing quality assurance programs, and giving advise and training to their members. By exploiting economies of scale and scope as well as by reducing transaction costs, POs can improve the efficiency and efficacy of agri-food supply chains.

In developed countries, POs are experiencing restructuring processes, due to their changing role and function in (international) supply chains and networks. Changes in competitive conditions, consumer demands and public policies have forced the producer organization to shift from a producer (or supply) orientation to a customer (or demand) orientation. POs as well as their members have increased their effort in marketing, innovation, and internationalization. As the PO - customer relationship has become relatively more important compared to the producer - PO relationship, we have witnessed shifts in investment patterns, in coordination mechanisms, and in control relationships. Many theoretical and empirical studies have addressed the restructuring of traditional agricultural cooperatives (*e.g.*, Cook, 1995; Bijman and Hendrikse, 2003; Chaddad and Cook, 2004). Most of these studies apply organizational economics theory (notably transaction cost, agency, property rights/incomplete contracting approaches) to analyse the structural changes taking place within cooperatives. The restructuring processes among POs in developing and transition countries are similar to those in developed countries, although the institutional context may be quite different. What is similar is the need to redesign the role of the organization in the supply chain, trying to strengthen the link between producer and (final) customer, as part of the overall aim to improve the market orientation of agricultural production. This volume includes two papers on POs, one from a transition country (Poland), and one from a developing country (Bolivia). In addition, a paper is presented on horizontal collaboration among small and medium-sized retailers in Brazil. While being active in a different stage of the agri-food chain, the question of how and why to collaborate is not fundamentally different for retailers compared to farmers.

Adamowicz and Lemanowicz analyse the performance of newly established economic producer organizations in Poland. With the fundamental changes in economic conditions after the 1990 introduction of the free market economy, individual farmers experienced the disadvantages of low bargaining power vis-à-vis their wholesale and processor customers. Having memories of state-controlled cooperatives in mind, they are reluctant to establish collective organizations. Still, horizontal collaboration in some kind of jointly owned and controlled organization may be the only viable way to strengthen the bargaining position of farmers. Adamowicz and Lemanowicz show, in an empirical study among 62 economic producer organizations, that horizontal cooperation can be beneficial to individual farmers. This study is particularly valuable as empirical studies of the economic advantages for farmers joining a producer organization are rather scarce.

The paper by Ton and Bijman uses a network perspective in explaining the special position that economic producer organizations have in the supply chain. In developing countries, the process of supply chain development is often facilitated by state agencies and international development NGOs. These facilitators, however, do not always appreciate the role of the PO in the supply chain, as they do not reckon with the special characteristics of a PO. As a PO bases part of its organizational efficiency on social relations, and has both an economic and an interest representation function, it cannot be considered as just another commercial chain participant. The authors present a case study of a chain development process for quinoa in Bolivia, showing how chain facilitators struggle with the special features of POs. The paper suggests that externally facilitated chain development processes will be more successful when the special organizational features of a PO are taken into account.

Ghisi, Martinelli and Kristensen use network theory to study horizontal collaboration among small and medium-sized supermarkets in Brazil. For supermarkets that are not part of large retail chain, horizontal cooperation is one of the few tools available for strengthening their competitive position (and thereby economic sustainability) as it allows for strengthening purchasing power and obtaining economies of scale, without giving up independence. Ghisi *et al.* have conducted a large study, using both qualitative and quantitative methods, to analyse (positive and negative) experiences of small retail companies participating in a buying group. They provide a number of propositions on the advantages and disadvantages of participating in a buying group. Many of the buying associations have actually developed into more comprehensive business groups. As organizational culture matures, new value added activities are incorporated in the network. For example, the partners begin to exchange key information; develop joint training programmes; develop collective marketing activities such as private labels; share technology; professionalize business management; make improvements in innovation; and modernize their stores. The authors found that commitment is, without doubt, one of the most critical aspects of the success of a horizontal network. Interestingly, commitment is also considered to be of crucial importance for the viability and efficiency of agricultural cooperatives, another form of horizontal collaboration.

4. Supply Chain Management

With the growing emphasis in business literature on the role of interfirm collaboration as a source of added-value and competitive strength, the field of supply chain management (SCM) has grown tremendously. While traditionally SCM was considered equivalent to integrating logistics across the supply chain, the current understanding focuses on integrating and managing (all) key business processes across the supply chain. An often used conceptual framework in the SCM literature is the one developed by Lambert and Cooper (2000). Their SCM framework consists of three closely interrelated elements: the supply chain network structure, the supply chain business processes, and the supply chain management components. This framework can be used to map and analyse the supply chain for any focal company. As this framework makes clear, SCM is a broad field of study, seeking combinations of various theoretical approaches, studying various real life phenomena, and applying various methodologies. Several papers in this volume apply SCM concepts, such as logistics, efficient consumer response, ICT support systems, and tracking and tracing.

Hsiao, Van der Vorst and Omta present an overview of recent theoretical and practical developments in logistics outsourcing in food supply chain and networks. The recent trend towards more integrated supply chain management in the agri-food industry has resulted in the rise of a new type of logistics service providers offering advanced services to food manufacturers. One of the key supply chain management questions is what logistics activities can be outsourced, to which type of logistics service providers, under what specific conditions. Combining theoretical arguments from Transaction Costs Economics, Resource-Based View and Supply Chain Management, as well as results from six case studies, the authors build a conceptual framework for decision-making on the type of logistics activities that should be outsourced, namely execution, value-adding and planning activities.

The paper by Obersojer and Weindlmaier presents a study on the preparedness of food companies for Efficient Consumer Response, a supply chain strategy for the retail industry. The paper provides an overview of the ECR concept and its requirements for successful implementation. ECR concepts combine three focal areas: enabling technologies (the technological platform), category management (the demand side), and management of replenishment (the supply side). The paper investigates to what extent dairy companies in Germany, Austria and Switzerland are ready to implement the ECR concept and which instruments are preferably applied. The authors found that few of the essential requirements for successful implementation are met, and that ECR is often only tailored to IT concerns. Distrust in the relationship between dairy companies and retailers and a lack of strategic orientation are the main obstacles for a consequent implementation.

The paper by Li, Kramer, Beulens and Van der Vorst describes a method to identify factors that cause food quality problems by employing data mining techniques. Operations management in food supply networks is especially complex due to the variability of primary inputs and the perishability of products. To prevent problems in food quality and operations, early warning and proactive control systems may be used. Data mining is the process of extracting valid, previously unknown, comprehensive and actionable information from large databases and using it to make crucial business decisions. The authors investigated the applicability of data mining by applying three data mining techniques in the chicken chain of a large Dutch food company. This case study has illustrated that data mining techniques can help to efficiently find relations between performance indicators and variables related to operational processes.

Fritz and Hausen present a five-stage procedure model for the analysis, tailored improvement, development, and testing of a solution for a specific supply network. The paper shows results from the application of the procedure model to the agri-food production network. The emphasis in the model is on customized electronic supply chain management, to improve efficiency and coordination in complex supply networks. Several efficiency improvements can be realized by a tailored and embedded supply chain management platform. Empirical evidence was collected by laboratory experiments and from expert evaluations. Complexity could be reduced, process efficiency improved, and change efforts could be reduced.

5. Power

A special issue covered in this volume, and one that is receiving growing attention from chain and network scholars, is the use and distribution of power in the chain (or network). Hitherto, the concept of power has rarely been discussed in supply chain management, except to deny its importance, or to argue that power should be not exploited in chain management because lean approaches should be based on equity, trust and openness (Cox, 1999). However, it is increasingly acknowledged that studying the effect of bargaining or market power is important to understand the mechanisms and dynamics of business exchanges. In agri-food chains, this interest in power relationships has also been fuelled by consolidations at the retail stage of the chain. But also the shift in the agri-food system from competitive (*i.e.*, spot markets) to imperfectly competitive environments justifies attention to the power issue, because in situations of incomplete contracts actors in the chain may use asymmetries in information and control to further their private interests.

The issue of power has also been raised in the context of international commodity chains. Several authors have developed the concept of 'global commodity chains' (Gereffi, 1994; Daviron and

Gibbon, 2002; Gereffi *et al.*, 2005). They use this concept to address the issue of who controls global trade and industry, how they do so, and with what consequences for producers in developing countries. Using a political economy perspective on international (food) chains, these authors claim that power in chains and networks is associated with system coordination. Daviron and Gibbon (2002: 140) have stated that "... functional coordination only comes about through the emergence of a chain 'driver' – the expression of a relation of power in a commodity chain." Vertical coordination occurs in forms that allow 'driving' agents to institute measures that simultaneously increase sales, reduce costs and risks, and increase speed and reliability of supply. Gereffi (1994) has characterized such forms of vertical coordination as "buyer-driven' chains.

The paper by Sodano presents an overview of the literature on power and economic theory. The author argues that in order to understand the current restructuring processes in the agri-food system it is necessary to take explicitly into account the role of power as a driving organizational force in the system. The paper presents and comments the following economic and social perspectives on power: standard (Walrasian) economics, contested exchange approach, influence cost theory, structural approaches, and new institutional economics. As conclusion to the literature review, the author claims that although new institutional economics is grounded in the same epistemological assumptions as the standard model, the organizational issues it raises draw a bridge between the latter and new theoretical strands of radical economics and economic sociology. The paper concludes with indicating three issues in the agri-food system that merit special attention from a power perspective: welfare assessment of vertical contractual arrangements, understanding recent processes of networking, and the multiform power of large firms.

Revell and Liu establish a link between power and supply chain management. The authors analyse how the possibilities for chain management and logistics are linked to the distribution of market power within the supply chain for fresh produce in China. The analysis starts from the assumption that each supply chain participant tries to appropriate as much value as possible. Appropriating a certain share of the chain value is done on the basis of owning or controlling resources that are both unique (or difficult to imitate) and vital to customers or suppliers. These 'critical supply chain assets' are the foundations of supply chain and market power. Revell and Liu show the power structure for each tier of the chain as well as for the whole chain. On the basis of the power structure of the whole chain the authors derive possible chain management solutions. For instance, positive pro-active chain management is only enforceable or likely to emerge where there is consistent direction in dominance or interdependence between buyers and sellers throughout the chain. The authors conclude that in the Chinese fresh produce industry two chains are likely to apply chain management practices that may overcome current logistics problems. One is the supermarket - general wholesaler - individual farmers chain, with overall buyer dominance and giving suppliers no option but to comply with buyer-driven practices. The other is the supermarket - specialist fresh produce wholesaler - farming company, with interdependencies at both buyer-seller dyads, lending itself to pro-active non-adversarial collaborative management approaches in developing chain quality standards, in sharing value and in investment.

Zuniga-Arias, Meijer, Ruben and Hofstede analyse (perceived) bargaining power in the mango supply chain of Costa Rica. Transactions between producers, traders, retailers and consumers in the mango supply chain are characterised by contractual arrangements concerning outlet choice, price, volume, quality and frequency. Contract choice is influenced by the anticipation of the behaviour and response of potential trading partners. The authors used a gaming simulation approach to identify the intrinsic agent characteristics that typically result in specific contracts and assess the underlying differences in perceived bargaining power. The game design closely mimics the negotiation conditions in the Costa Rican mango supply chain. The game results confirm the important role of trust and information exchange for reaching mutually acceptable contracts.

6. Performance

The study of chains and networks requires performance indicators to assess the effectiveness and efficiency of chain organization and management. Performance indicators measure the value added of the chain, evaluate how work is done, and can help managers to direct the activities. Performance has been since long recognized as important (Beamon, 1999, Gunasekaran, 2004). As the performance of chains and networks depends on the effort of several independent actors together, performance measurement is more complicated than measuring performance of an individual firm (Aramyan *et al.*, 2006). Van der Vorst (2000) distinguishes three levels at which performance can be measured: the chain or network level, the organization or firm level, and the process level. Indicators are, for instance, product availability, quality, responsiveness, reliability, throughput time, and costs or profits. A method to measure these multiple indicators, specifically designed for chains, is the Supply-Chain Operations Reference SCOR®) (Supply-Chain Council, 2004). This method provides most of the metrics needed for chain performance, but does not attempt to address every chain process. Further research is needed to design multiple performance indicators, which recognize each firm separately as well as the full chain or network (Aramyan *et al.*, 2006).

The paper of Velthuis, Van Erve and Meuwissen addresses the performance of processes. Their case is the recall of milk as contamination is traced. They provide a modelling approach to investigate the relation between the time at which the recall of consumption milk takes place and the recall costs. A stochastic (Monte Carlo) model simulates the flow of one batch consumption milk of 150,000 kilograms in the chain from farm to consumer. The model predicts the amount of milk in each stage of the milk- supply chain at a certain moment in time and calculates corresponding direct recall costs. From the modelling results it can be concluded that the direct recall costs are less than € 100,000 in the first 16 hours after the milk being collected from the farms. The costs increase rapidly to more than €200,000 with a recall moment later than 16 hours, due to the costly media announcement. This model provides useful information to the supply chain manager (*i.e.*, usually the dairy company) about the costs of a possible recall in relation to the recall moment. This information is helpful in developing a strategy to monitor the quality of the milk and detect a contamination.

Wijnands, Koot and Rosenkranz focus on the possibilities of improving the chain performance by vertical collaboration between wholesalers and retailers in the Dutch fresh vegetable chain. They have characterized transactions between wholesale and retail by a high level of competition and rivalry between the actors; relatively low switching costs; abundant availability of substitutes, and wholesalers having low bargaining power compared to the retailers. Using a micro-economic model, the authors show that collaboration offers opportunities for cost control in the chain by reduction of waste levels. However, the benefits from collaboration are unequally distributed between retailer and wholesaler. The conventional wisdom that collaboration increases the profit of each player is not obvious. Even when total chain profit increases, some actors may experience a decrease in profit. Redistribution should be bargained in order to have an incentive for each chain actor to contribute to the waste reduction.

7. Quality, safety and sustainability

As consumers are becoming more demanding and legislation is becoming more stringent, firms in the supply chain are strengthening their effort to improve, maintain and control the quality and safety of agri-food products. One expression of this effort is the rapid growth in the number and scope of quality assurance systems, which are established to enable the application and verification of quality control measures. Quality assurance systems (including those for sustainable agriculture) have been set up by public authorities at regional or national level, by private companies, or by collectives of private companies. Most of these systems have multiple objectives, such as compliance to legal requirements, risk reduction, marketing, and reduction of (foreign) competition. These assurance systems often include a tracking and tracking system, as required by legislation or as a private strategy to reduce costs and strengthen the reputation among consumers (Trienekens and Van der Vorst, 2006). While the implementation of a traceability system is mostly a technical issue, which is greatly supported by the rapid advances in information technology, the introduction of a quality assurance system is much more an issue of strategy and competition. Individual companies, horizontal groups of companies, supply chain networks, and even (regional) governments compete to establish a reputation of being a provider of high quality (high safety) products. Much of current research on quality and safety in agri-food chains focuses on how particular organizational structures (or governance structures in the terminology of Transaction Cost Economics) can support the implementation of quality assurance systems.

Another important topic for research on quality and safety in (international) agri-food chains concerns the use of grades and standards. Not only are quality requirements rapidly becoming more strict, there is a shift from public to private standards. The use of standards for product quality as well as for sustainability of natural resources faces challenges in the upstream parts of supply chains, particularly chains originating from a developing country. Implementing the requirements stated in codes of conduct for quality and safety often requires non-trivial investments that go beyond what producers and other chain participants in developing countries can afford (Henson and Loader, 2001). In addition, codes of conduct for conserving natural resources face non-compliance problems, often because primary producers in developing countries encounter institutional, technological and socio-economic constraints.

Martino and Perugini apply the transaction costs approach in studying the relationship between food safety and organizational form. They have built their analytical framework on the basis of the studies by Raynaud *et al.* (2005), who argue that governance structure choice is a function of the (firm or chain) strategy for guaranteeing quality (including food safety). Martino and Perugini have applied this framework in a case study of a large Italian food distribution company. Uncertainty and asset specificity in the supplier-food company relationship turn out to be important factors determining the type of contractual relationship.

The paper by De Barcellos, Ferreira and Vieira, on quality assurance and vertical alliances, presents case studies of beef chains from the United Kingdom and Brazil. The focus of the paper is on strategic alliances between beef producers, abattoirs and retailers. These complementary strategic alliances have been established to produce, process and sell a differentiated beef product. In order to guarantee final consumers a premium quality beef product, the alliance partners have set up a quality assurance scheme involving contractual arrangements. The paper particularly reports on the producers' perception of the advantages and disadvantages of the alliance. The results indicate that producers in both the UK and Brazil consider the alliance to be successful when producers are paid premium prices for the quality beef they supply to the slaughterhouses. Joining a strategic alliance with other chain partners with the goal to develop, produce and supply quality guaranteed beef seems a good strategy for producers to improve their overall competitiveness.

Han, Trienekens, Tan, Omta and Wang explore the link between vertical coordination, quality management and firm performance, in an empirical study of the Chinese pork industry. With the rise of supermarkets (see also the paper by Hu *et al.* in this volume) and the advent of higher incomes in China, some pork processors have started to develop more close relationships with their suppliers and customers, in order to be able to produce high quality pork products. Thus, strengthening vertical coordination with its suppliers and customers is part of the quality management strategy of the processing companies. The results of the empirical study, which focussed on the supplier-processor transaction, show that firms applying a quality management strategy more often use contracts to govern the relationship with their suppliers. In addition, the study shows a positive relationship between applying a quality management strategy and firm performance.

The paper by Kambewa, Ingenbleek, Van Tilburg and Van de Lans, establishes a bridge between improving food quality and strengthening ecological sustainability. It does so by studying an international supply chain for a developing country product. The study explores whether market-based incentives may induce fishermen of fresh Nile perch in Lake Victoria, East-Africa, to implement practices that improve quality as well as conserve natural resources. The incentives included in the study are access to special investments (such as fishing gear and quality management tools), price information, selection of transacting partners (processors or middlemen), and sanctions for non-compliance. Data were collected through a conjoint design from 275 fishers who rated their preferences to enter into contracts to implement better practices. The results show that access to specific resources was the most important incentive, followed by price information. Selection of a trading partner and enforcement were third and last in importance respectively. The results can help chain participants downstream, such as traders and retailers, to identify appropriate incentives to promote both quality and sustainability in the upstream parts of the chain.

Haverkamp, Bremmers and Omta study the relative importance of various factors that stimulate environmental pro-activeness among food companies. One of the factors included in the study is the influence of supply chain and network relationships with suppliers and buyers. The authors hypothesised that supply chain and network relationships would result in beyond-compliance, which is a particular form of environmental pro-activeness. Beyond-compliance aims at cooperation with buyers and/or suppliers in order to reduce the environmental impact beyond the individual production location. Beyond-compliance is obtained by applying a chain-oriented environmental management system (EMS). Because environmental impacts go beyond the boundaries of the individual firm (or plant), a chain-oriented EMS may be more efficient and effective in improving sustainability. However, governmental policies for environmental protection are mainly process-oriented and do not promote particular chain or network measures and procedures. In an empirical study among Dutch food companies, the above stated hypothesis was confirmed. Relationships with private stakeholders in the chain or network of a food company affect beyond-compliance environmental behaviour, such as the implementation of a chain-oriented EMS. On the basis of this result, the authors recommend public policy makers to develop and implement chain-oriented environmental policies, for instance by linking governmental permits to products instead of production units.

8. Special case studies

The book concludes with three special case studies. The paper by Dunne is on organizational learning in the context of an agri-food chain. As innovation has become more important for competition and meeting shifting consumer demand, learning processes beyond firm boundaries is receiving growing attention in the agri-food chain and network literature. The papers by Neves *et al.* and Hu *et al.* present studies from two countries that show fascinating developments in the field of agri-food chains and networks: Brazil and China.

Brazil has become a world power in agricultural production and trade. Globally, Brazil belongs to the three largest producers and exporters of coffee, citrus, beef, poultry, soybeans and sugar. The agricultural sector has grown with more than 5 percent per year since 1999, and the country still has an enormous potential to expand production and improve productivity. Export opportunities, domestic deregulation and technological developments have lead to restructuring of the agri-food industry, leading to internationalization and concentration. Also improved methods of chain coordination as well as the application of new logistic systems have supported growth and restructuring processes. Several papers in the volume study chains and networks in the Brazilian agri-food industry. De Barcellos et al, focus on the beef industry, Ghisi *et al.* Have investigated the food retail industry, and Neves *et al.* provide a case study of the dairy sector in the State of Sao Paulo.

China is a fascinating country for scholars of agri-food chains and networks. Not only for its sheer size, but particularly for the large economic growth (10 percent per annum since 1992) and the rapid transformation that are both cause and effect of this growth. A number of papers in this volume deal with chain and network issues in China. Lu *et al.* study the importance of social networks in the fresh produce industry. Revell and Liu have also investigated the fresh produce industry, to analyse power relationships among the chain actors. The paper by Han *et al.* analyses quality management in the pork industry.

The paper by Neves, Consoli, Lopes, Campos and Paiva presents the results of a University of Sao Paulo (USP-PENSA) project on "Mapping and Quantification of the Milk Chain in Brazil". While Sao Paulo State is the largest consumer of dairy products, with 7.6 billion litres per year, it only produces 1.7 billion litres. Over the years, milk production in Sao Paulo has decreased, leading to larger import from other states. The study was set up to analyse the competitiveness of the dairy industry in Sao Paulo State. The paper is a nice example of how to describe chain and network relationships, both qualitatively and quantitatively. Also the methodology for the mapping exercise is clearly presented.

Hu, Fuller and Reardon present a case study on the effect of the rapid development of supermarkets on China's dairy industry. In the past decade the food retail sector has gone through rapid restructuring, not only in China, but also in Latin America and other parts of Asia (Reardon *et al.*, 2003). As the supermarket sector develops, leading stores are rapidly adopting technological and governance changes in their product procurement systems, resulting in changing supply conditions both for primary producers and for processors of farm products. This study on the impact of supermarket development on the Chinese dairy industry shows that supermarkets have encouraged the consolidation of dairy firms into large-scale dairy processing enterprises with centralized management. Along with an improvement of dairy product distribution, supermarkets have created opportunities for consumers in remote townships and rural areas to gain access to a greater variety of dairy products. The gradual increase in the number of rural children consuming dairy products may lead to tremendous growth in dairy product demand by the next generation. China, with a population of 1.3 billion, may one day become the largest consumer of dairy products in the world.

Dunne reports on the outcomes of the first stage of a longitudinal study that focused on the transformational change process being undertaken within the supply chain and operations area of a major Australian food manufacturing company. Organizational learning is an essential prerequisite for any successful change process. An organization's ability to learn is dependent on the existence of an environment within the organization that nurtures learning and the presence of key enablers that facilitate the learning process. An organization's capacity to learn can be enhanced through its ability to form and sustain collaborative relationships with its chain partners. The results of the study show that an environment that supports organizational learning is being developed through consultative leadership and the empowerment of individuals within a culture that supports innovation and cross-functional teamwork but demands responsibility and accountability. The case study company has involved its supply chain partners in a series of mutually beneficial projects that have improved communication and built trust thereby laying the foundations for more collaborative chain relationships.

References

Aramyan, L. C. Ondersteijn, O. van Kooten and A. Oude Lansink, 2006. Performance indicators in AGRI-Food production chains. In: Ondersteijn, C.J.M., J.H.M. Wijnands, R.B.M. Huirne and O. van Kooten. Quantifying the supply chain. Springer, Heidelberg, 47-64.

Beamon, B., 1999. Measuring supply chain performance. International journal of operations and production management, 19(3/4), 275-292.

Bijman, J., 2006. Governance structures in the Dutch fresh produce industry. In: C.J.M. Ondersteijn, J.H.M. Wijnands, R.B.M. Huirne and O. van Kooten. Quantifying the supply chain. Springer, Heidelberg, pp. 205-221.

Bijman, J. and G. Hendrikse, 2003. Co-operatives in chains: institutional restructuring in the Dutch fruit and vegetables industry. Journal on Chains and Network Science, 3(2): 95-107.

Borgatti, S.P. and P. C. Foster, 2003. The Network Paradigm in Organizational Research: A Review and Typology. Journal of Management, 29(6): 991-1013.

Chaddad, F.R. and M.L. Cook, 2004. Understanding new cooperative models: An ownership-control rights typology. Review of Agricultural Economics, 26(3): 348-360.

Cook, M.L., 1995. The Future of U.S. Agricultural Cooperatives: A Neo-Institutional Approach. American Journal of Agricultural Economics, 77(December): 1153-1159.

Cox, A., 1999. Power, value and supply chain management. Supply Chain Management, 4(4): 167-175.

Daviron, B. and P. Gibbon, 2002. Global Commodity Chains and African Export Agriculture. Journal of Agrarian Change, 2(2): 137-161.

Dyer, J.H. and H. Singh, 1998. The relational view: Cooperative strategy and sources of interorganizational competitive advantage. Academy of Management Review, 23(4): 660-679.

Gereffi, G., 1994. The Organization of Buyer-Driven Commodity Chains: How US Retailers Shape Overseas Production Networks. In: G. Gereffi and M. Korzeniewicz (eds.), Commodity Chains and Global Capitalism, Westport: Preager, pp. 95-122.

Gereffi, G., J. Humphrey and T. Sturgeon, 2005. The governance of global value chains. Review of International Political Economics, 12(1), 78-104.

Ghoshal, S. and P. Moran, 1996. Bad for practice. A critique of the transaction cost theory. Academy of Management Review, 21(1): 13-47.

Granovetter, M., 1985. Economic-Action and Social-Structure - the Problem of Embeddedness. American Journal of Sociology, 91(3): 481-510.

Gunasekaran, A., C. Patel and R.E. McGaughey, 2004. A framework for supply chain performance measurement. International journal of Production Economics, 87, 333-347.

Henson, S. and R. Loader, 2001. Barriers to Agricultural Exports from Developing Countries: The Role of Sanitary and Phytosanitary Requirements. World Development, 29(1): 85-102.

Jones, C., W.S. Hesterly, S.P. Borgatti, 1997. A general theory of network governance: exchange conditions and social mechanisms. Academy of Management Review, 22(4): 911-945.

Klein Woolthuis, R., B. Hillebrand, B. Nooteboom, 2005. Trust, Contract and Relatonship Development. Organization Studies, 26(6): 813-840.

Lambert, D.M. and M.C. Cooper, 2000. Issues in Supply Chain Management. Industrial Marketing Management, 29: 65-83.

Lazzarini, S.G., F.R. Chaddad, M.L. Cook, 2001. Integrating supply chain and network analyses: The study of netchains. Journal on Chain and Network Science, 1(1): 7-22.

Lazzarini, S.G. and T.R. Zenger, 2002. The Strength of Churning Ties: A Dynamic Theory of Interorganizational Relationships. Working Document.

Lazzarini, S.G., G.J. Miller, T.R. Zenger, 2004. Order with Some Law: Complementarity versus Substitution of Formal and Informal Arrangements. The Journal of Law, Economics, & Organization, 20(2): 261-298.

Menard, C., 1996. On clusters, hybrids and other strange forms: the case of the French poultry industry. JITE, 152(1): 154-183.

Ménard, C., 2004. The economics of hybrid organizations. JITE, 160(1): 1-32.

Menard, C. and E. Valceschini, 2005. New institutions for governing the agri-food industry. European Review Of Agricultural Economics, 32(3): 421-440.

Ménard, C. and M.M. Shirley (eds.), 2005. Handbook of New Institutional Economics. Dordrecht, etc.: Springer.

Poppo, L. and T. Zenger, 2002. Do formal contracts and relational governance function as substitutes or complements? Strategic Management Journal, 23(8): 707-725.

Powell, W.W., 1990. Neither Market nor hierarchy: Network forms of organization. In: B. M. Staw and L. L. Cummings. Research in Organizational Behavior. Greenwich, CT and London, UK, JAI Press, 12: 295-336.

Raynaud, E., L. Sauvée and E. Valceschini, 2005. Alignmnet between quality enforcement devices and governance structures in the agro-food vertical chains. Journal of Management and Governance, 9(1): 47-77.

Reardon, Th., C.P. Timmer, Ch.B. Barrett and J. Berdegué, 2003. The rise of supermarkets in Africa, Asia and Latin America. American Journal of Agricultural Economics, 85(5): 1140-1146.

Stern, L.W., A.I. El-Ansari and A.T. Coughlan, 1996. Marketing channels (5th ed.), Upper Saddle River : Prentice Hall.

Supply-Chain Council, 2004. Avaliable at : www.supply-chain.org/index

Trienekens, J.H., A.J.M. Beulens, J.M. Hagen and S.W.F. Omta, 2003. Innovation through international supply chain development: a research agenda. International Food and Agribusiness Management Review, 6(1).

Trienekens, J. and J. van der Vorst, 2006. Traceability in food supply chains. In; P.A. Luning, F. Devlieghere and R. Verhé (eds.), Safety in the agri-food chain. Wageningen: Wageningen Academic Publishers, pp. 439-470.

Uzzi, B., 1997. Social structure and competition in interfirm networks: The paradox of embeddedness. Administrative Science Quarterly, 42(1): 35-67.

Van der Vorst, J.G.A.J., 2000. Effective food supply chains; generating, modelling and evaluating supply chain scenarios, PhD-thesis Wageningen University, Wageningen.

Williamson, O. E., 1991. Comparative Economic Organization: The Analysis of Discrete Structural Alternatives. Administrative Science Quarterly, 36(June): 269-296.

Zenger, T. R., S. G. Lazzarini, L. Poppo, 2002. Informal and formal organization in new institutional economics. In: P. Ingram and B. S. Silverman. New Institutionalism in Strategic Management. Amsterdam, JAI-Elsevier Science. 19: 277-305.

Part 2. Governance

Does guanxi matter for vegetable supply chains in China? A case study approach

H. Lu, J.H. Trienekens and S.W.F. Omta

Abstract

A case study approach was applied to investigate the impacts and contributions of the Chinese cultural and historical embedded concept of guanxi networks on buyer-seller relationships in vegetable supply chains in Jiangsu China. 8 case companies, 5 typical supply chains and 9 buyer-seller relationships were thoroughly studied. Results demonstrate that guanxi networks have diversified impacts on different types of relationships involved in various marketing chains both in domestic and international markets. Guanxi networks stimulate investment behaviour, improve trust in counterparts, influence decision making in terms of marketing channel and transaction governance choices. Guanxi networks may increases the efficiency of buyer-seller relationships and improve quality performance in management practices.

Keywords: guanxi network, buyer-seller relationship, vegetable, performance, China

1. Introduction

Since the start of 1990s, relationship marketing has attracted much attention from academics and practitioners, especially in the business-to-business marketing (Dwyer *et al.*, 1987; Ganesan, 1994; Morgan and Hunt, 1994; Sharma and Pillai, 2003; Sheth and Shah, 2003). Guanxi, the social networks of personal relationships in China, has become a growing research interest in recent years (Ambler, 1994; Davies *et al.*, 1995; Lovett *et al.*, 1999; Yau *et al.*, 2000; Wang, 2005).

Guanxi is translated as personal connections or relationships on which an individual can build to secure resources or to draw benefits when doing business as well as in the course of social life (Davies, 1995). Guanxi can be generally classified into three categories: family, friend and business (Fan, 2002). Family guanxi is generally a relatively permanent and stable social relationship. It occurs mostly among members of such primary groups as family and it is governed by the need rule for social exchange and resource distribution within a family. Friend guanxi is rather stable and long term relationship as a means or an instrument with other people outside the family to attain material and/or mutual goals. Friend guanxi follows the reciprocity rule. Business guanxi is defined as the process of finding business (rather than personal) solutions through personal connections (Fan, 2002). Business guanxi is governed by equity rule. Because guanxi is transferable from person to person, the development of individual guanxi ultimately leads to a network of relationships: guanxi network.

Some theoretical literatures suggests that guanxi is crucial to the attainment of long-term business success in China (Yeung and Tung, 1996; Lou, 1997; Vanhonacker, 1997). Guanxi network also shows potential benefits and direct impacts on market expansion and sales growth (Kao, 1993). Guanxi network enhances firm performance (Luo and Chen, 1996; Peng and Luo, 2000); shows transaction cost advantages (Standifird and Marshall, 2000); help to secure off-farm job (Zhang and Li, 2003).

The previous studies on guanxi first appeared in the West in the 1980s in popular business writings that advised about cultural factors affecting doing business in China (Pye, 1982; Butterfiled, 1983; Alston, 1989). Later studies mainly focus on the comparison of the western concept of relationship and Chinese concept of guanxi (Arias, 1998; Wong and Chan, 1999; Ramasamy *et al.*, 2006). Though, some researchers already applied relationship (guanxi) marketing theory in China to air industry (Yang and Liu, 2003), direct selling (Merrilees and Miller, 1999), industrial marketing (Fang, 2001), and nonfarm employment (Zhang and Li, 2003). Literatures less focused on the agri-food sector (Cunningham, 2001). So we would like to fill this gap by raising and answering the question: does guanxi matter for vegetable supply chains in China?

In doing this, we are going to study the importance and the impacts of guanxi network on buyer-seller relationships and performances in vegetable supply chains in China (Figure 1). The supply chain is more than the sum of its individual members since the members maintain relationships with each other. Thus, how the supply chain is managed depends on the way relationships are organized. From marketing channel perspective, trust is an important concept to make buyer-seller relationships distinguishable channels (Dwyer *et al.*, 1987; Heide, 1994). The need for trust between business partners has been identified as an essential element of buyer-supplier relationships (Anderson and Narus, 1990; Geyskens *et al.*, 1998; Rousseau *et al.*, 1998). Transaction specific investment is an important mechanism for achieving closeness in a buyer-seller relationship. Creating specific assets is known as creating credible commitments (Heide and John, 1988) or pledges (Anderson and Weitz, 1989). Governance may take various forms within a supply chain (Slack *et al.*, 2001) which may influenced by guanxi networks. Thus we focus on four basic concepts for buyer-seller relationships in this study: trust, transaction specific investments, and channel and governance choices.

The remainder of this paper is structured as follows. The literature review and hypotheses are shown in Section 2. Section 3 describes the research method. The basic information of case companies and case results are shown in Section 4 and 5 respectively. Conclusions and limitations are drawn at the end of this paper.

2. Literature review and hypotheses

2.1 Guanxi network and trust

Empirical research showed that guanxi network promotes interpersonal trust (Granovetter, 1985; Farh *et al.*, 1998). There are a lot of aspects attached to trust, such as individual personal contacts, prior business experience, communicating market information and reputation. In a guanxi network, the cost of opportunism is the potential loss of exchange opportunities with all members of the network. So guanxi network provides some assurance of exchange partner behaviour (Standifird and Marshall, 2000). Strong personal relationships lead to the generation of relationship-sustaining factors such as trust and commitment which reduces the exchange

Figure 1. Conceptual research model.

partner's opportunistic behaviour. Thus we expect a positive relation between guanxi network and trust.

2.2 Guanxi network and transaction specific investments

The flexible and socially-based nature of guanxi permits members of a guanxi network to deal with unforeseen contingencies arising after the agreements are reached. Guanxi possesses the capacity to reduce transaction costs associated with environmental and behavioural uncertainties. The well-developed guanxi network can handle an increased level of asset specificity (Standifird and Marshall, 2000).

Transaction specific investments made by sellers stimulate cement long-term relationships with buyers. Such investments promote relational exchanges and increase the commitment between partners (Blau, 1964; Cook and Emerson, 1978). Guanxi network thus safeguard transactions by reduce the opportunistic behaviour (Standifird and Marshall, 2000). We then expect that the guanxi network has a positive impact on the willingness to invest in transaction specific investments.

2.3 Guanxi network and channel and governance choices

From a transaction cost perspective, guanxi-based business practices offer certain advantages over existing structural alternatives (Standifird and Marshall, 2000). Contractual arrangements function as the basic arrangements for transactions, incorporating a relatively high degree of flexibility through relational arrangements established via the guanxi network (Standifird and Marshall, 2000). With the dismantling of the Chinese legal and judicial system after 1949, and the consequent difficulty to enforce contracts, relations are an economically better mode of governance of transactions than contracts (Davies, 1995). As uncertainty rise in the chain, we expect that relational arrangements will be used to reinforce the safeguards of contingencies that can not be specified in a contract.

Economically, companies are more willing to engage in high quality category to pursue a high profit margin which always company with high risk and uncertainty. With the environmental uncertainty and behaviour uncertainty becoming lower within the well-developed guanxi networks, company becomes more successful in international markets and newly developed domestic markets, supermarkets as an example.

2.4 Guanxi network and performance

Transaction cost and quality satisfaction are important performance attributes in this research. The level of transaction costs incurred by the companies and their buyers indicates the level of chain efficiency for each combination of channel and governance. Transaction costs may arise in the form of direct or opportunity costs (Malone, 1987; Masten *et al.*, 1991); these costs are directly related to asset specificity, environmental uncertainty, and behavioural uncertainty. Direct transaction costs would include costs of crafting safeguard; communication, negotiation and co-ordination costs; screening and selection costs (*ex ante*) and measurement costs (*ex post*) (Rindfleisch and Heide, 1997).

In a well developed guanxi network, the transaction costs associated with opportunistic behaviour are low, so the information and negotiation costs are also low for the trusted counterparts. Thus we expected that the chain participants can achieve effective business relationships with well developed guanxi network.

Guanxi serves as a safeguarding mechanism for quality and safety control in China. Guanxi also increases the problem tolerance and speeds up the problem solving. So we propose that in a well developed guanxi network, the quality satisfaction for both sides will be higher than in a less developed guanxi network.

3. Research methods

Case study approach was applied to this research. Case study research is useful to explore an area in which the researcher has little or no control over events and the focus is on an existing event in a real life context (Eisenhardt, 1989, 1991; Yin, 2003). To conduct case studies in vegetable marketing in Jiangsu Province, we had asked for the perception of the case company interviewees about the concepts in our model related to guanxi network, buyer-seller relationships and the performance.

Data for this study were collected from May to October, 2005 based on personal interviews. We carefully selected our cases from different areas (developed and less developed economic zone) in Jiangsu Province for different type of the companies (producer, processor, exporter, and supermarket). The interview topics were based on pre-designed case protocols. Each interview lasted between 50 to 120 minutes. The interviewees are those who have full knowledge about the company's marketing strategy and administrations.

4. The basic information of the companies participating in the case studies

All the companies that participated in this case study interviews have been involved in vegetable industry for several years and functioned at different stages in the vegetable supply chain. Basically we distinguish two different relationships for companies: the relationships with their sellers and buyers separately.

Table 1 provides basic information and marketing strategies relating to the buying and selling behaviours of the interviewed companies. Results showed that all the interviewed companies are rather new and small in terms of their establishment and sales. Vegetable producer company, processing companies and cooperative fully engaged in vegetable business, while exporting companies and retailers diversify their business in different sectors. They all showed significant differences for their sourcing and selling strategies. 5 typical vegetable supply chains with different buyer-seller relationships will be detail discussed below.

Vegetable producer company case: This vegetable producer company owns an advanced greenhouse imported from the Netherlands to produce limited varieties (pepper, Holland cucumber and tomato). This company promises to produce high quality vegetables with high productivity level. All vegetables from this company are delivered to local supermarkets (relationship 1 in Figure 2). This company has well built long term relationships with different supermarkets in Jiangsu. The company delivers all vegetables to supermarket outlets directly using a cooling truck. Annually contracts are signed to regulate their transactions. Payments, quality requirements and delivery

Table 1. Basic information of case companies.

Cases / Characteristics	Producer company	Processing company 1	Processing company 2	Exporting company 1	Exporting company 2	Cooperative	Supermarket 1	Supermarket 2
Percentage of vegetable business	100	100	100	40	30	100	<5	10
Number of fixed employees	36	35	40	7	5	28	10	15
Sales scale (10^6 ¥)	1-5	4-5	2	5-10	5-10	5-10	>10	>10
Applied vegetable quality standards*	PFF	PFF	Specific for buyers	Specific for buyers	Specific for buyers	PFF	PFF & GF	PFF
Number of source channels	n.a	1[a]	2[a, b]	1[b]	1[b]	1[a]	3[a, c, d]	3[a, d, e]
Longest duration of source channel	n.a.	6	2	7	6	4	10	6
Number of destination channels	1[1]	2[2]	2[3]	1[4]	1[5]	3[6]	1[7]	1[7]
Longest duration of destination channel	4	6	2	7	6	4	15	6
Company establishment	2001	1999	2003	1998	1999	2001	1990	1999
Persons interviewed	General manager	General manager	General manager	CEO & General manager	General manager	General manager	Deputy general manager	Manager for fresh sector

* There are three quality standards in China, named as Pollution Free Food (PFF), Green Food (Green food), and Organic Food (OF). No organic food was found in case companies during the interview.

[a] Vegetable farmer; [b] Vegetable company; [c] Produce base; [d] Wholesale market; [e] Distribution centre

[1] Supermarket; [2] America, Japan, Canada, Hong Kong; [3] Australia, Netherlands, America, Middle-east, Supermarket in China; [4)] America, German, Japan; [5] Japan, Korea, Singapore; [6] Wholesale market, supermarket, vegetable company; [7] Consumer

Source: Field survey of quality management for vegetable supply chain in Jiangsu China, May-Oct. 2005.

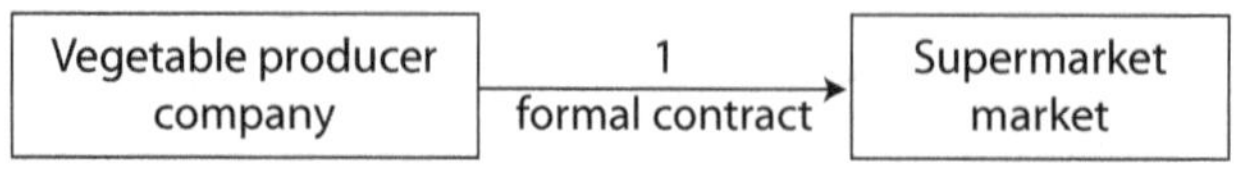

Figure 2. Vegetable flow of producer company chain.

condition are fixed in the contracts. Prices are monthly determined based on market price, while the daily delivery volume is based on daily order. Pollution Free Food quality standards are applied in this producer-supermarket chain.

Vegetable processing company case: The two processing companies have similar sourcing and destination channels (Figure 3). They buy fresh vegetables mostly from local vegetable farmers (relationship 2), and only partly buy processed vegetables from domestic food companies during the supply shortage (left dash line). The international markets are the major destinations for their processed products (frozen vegetables, cooling vegetables *etc.*) to pursue high profits with large quantities (relationship 3). They also sell part of their produces, mostly second class quality products, either to domestic food companies or supermarkets in domestic markets (right dash line box).

During the interviews, we found that guanxi networks were more important for processing companies in international markets than in domestic markets. Formal contracts are often used for purchasing vegetables from domestic food companies and selling final products to either domestic or international markets. There are no formal contracts for transactions with farmers, only in certain cases, informal (oral) agreements occur. It is important to mention that guanxi is an important factor to enhance and lubricate transactions in international markets although they have contracts with their buyers. Vegetable quality is based on each buyer in international markets. Pollution Free Food, Green Food in some cases, is required by international buyers.

Vegetable exporting company case: The two interviewed exporting companies use similar vegetable sourcing and destination markets. Both of them buy processed vegetables (dried, frozen or cooling vegetables) from domestic food companies and sell to international markets (Figure 4). As trading companies, they did not invest for their own produce and processing facilities, but they put lots of efforts to acquire produce and marketing information. They also train employees to seeking right products effectively and sell the products efficiently and profitably. Thus exporting companies incur high level of training costs and information costs.

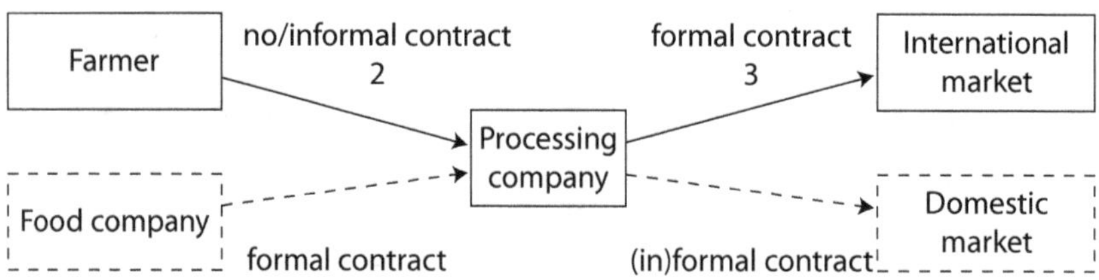

Figure 3. Vegetable flow of the processing company chain.

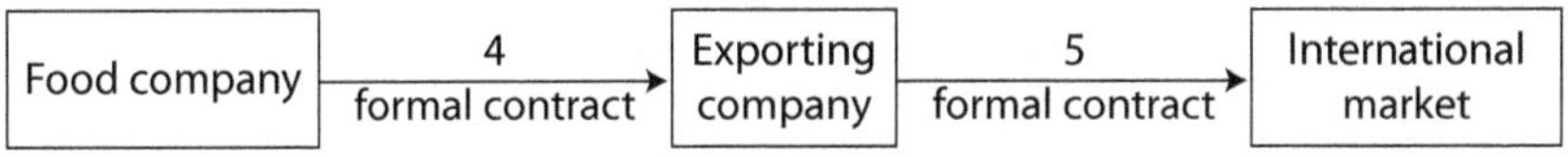

Figure 4. Vegetable flow of the exporting company chain.

Formal contracts are widely used in their purchasing and selling activities (relationship 4 and 5). Similar as processing companies, guanxi is also an important influencer for the transactions in international markets. Pollution Free Food or Green Food is required by the international buyers.

Vegetable cooperative case: The vegetable cooperative in this research has single vegetable source and diversified destinations in domestic markets (Figure 5).

The cooperative is fully responsible for its members' vegetable sale. Formal contracts and the cooperative membership indicate their buyer-seller relationships (relationship 6). Pollution free vegetables are required by the cooperative. The major markets for the cooperative's products are domestic wholesale market (relationship 7). The cooperative also sells part of vegetables to supermarkets and domestic vegetable companies. Contracts are used in the transactions with supermarkets and vegetable companies while cash payments are popular in the wholesale market.

Supermarket case: The two interviewed supermarkets have different strategies for their vegetable business (Figure 6). Supermarket 1 had a long time experience but still showed lack of knowledge for vegetable sector operations. It started to purchase vegetables from state-owned vegetable company, but it failed after several years' of cooperation. Then it turned to purchase vegetables from different sources. Leaf vegetables were mainly bought from local farmers or farmer associations (relationship 8) and local specified vegetable production sites (farmer based) to promise freshness and high quality. Part of leaf vegetables were also bought from the wholesale market to guarantee large quantities demand. Supermarket 1 also purchased long shelf products from long distance production sites. This diversified sourcing strategy showed positive results for supermarket 1 so far. Supermarket 2, however, had another way to arrange its vegetables. It started with central purchasing strategy. All the vegetables were purchased from a vegetable company based in Shanghai (relationship 9). Recently, supermarket 2 also bought some specific varieties from local farmers and wholesale markets with small volumes. Both supermarkets use

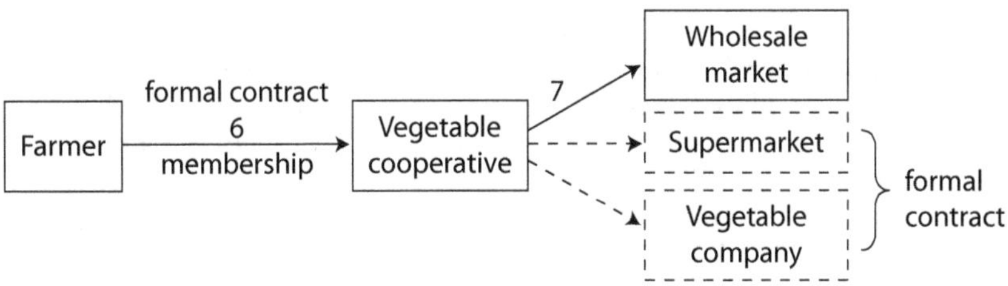

Figure 5. Vegetable flow of the cooperative chain.

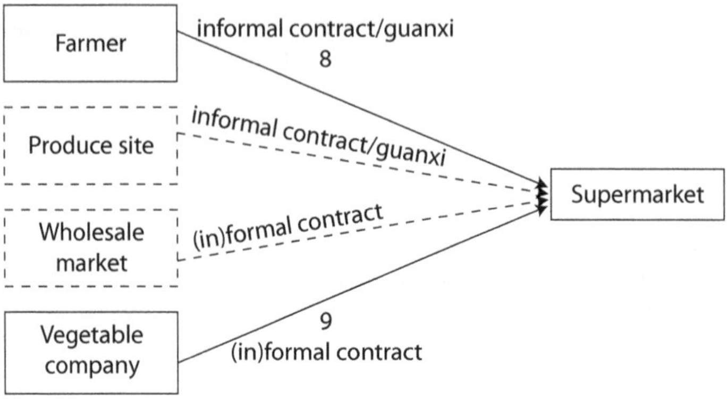

Figure 6. Vegetable flow of supermarket chain.

contracts to regulate their transactions. Results showed that both sourcing strategies are rather successful so far.

5. Case study results

5.1 Guanxi network and vegetable supply chain

The case studies showed that the vegetable related companies use various guanxi to help them build up business relationships. Their family guanxi and friend guanxi did show positive effects on their business development. Family members may occupy an important position in the company. This is because the private company normally started with joint investments from family members and/or with close friends. So the owners feel confident and trusted to work with their family members.

Friend guanxi also matters for company business. The friends here can be own friends, friends of the friends, classmates, *etc.* Friendship is very important for the companies' business. Marketing information, managing knowledge and capital investment are the major help from friend networks. "We do business with friends". Chinese people trust friends while keeping distance with strangers.

> Box 1.
> *"My nephew is the general manager of my company and I trust him. He also showed a good performance for the last several years in developing markets and in general administration. His inspiration and innovation ideas help my company to grow fast and to be the leader in vegetable industry in Jiangsu province. I trust him not just because he is my nephew, but also he is young and creative".*
> Interview with exporting company 1

The most important guanxi for the companies' business was so called business guanxi. Nearly all the companies said that they will carefully and sincerely deal with their business partners. The international customers are so important to a company because of the difficulties of finding and maintaining business relationships with them. Buyers are the guarantee of the business success.

Box 2.
"My business guanxi networks are my most important treasures; I have to rely on them to continue my company. But they are so difficult to fully rely on them same as my family member or close friends, we still should keep distance with them, but for sure we can not lose them".
Interview with processing company 2

The guanxi network has significant impacts on companies' business, especially for their marketing development. Nearly all the companies agreed that their guanxi network helps them to start their business in terms of searching market information, finding new buyers or accessing new markets *etc.* Once the contact person moved to another company, they still can keep relationships with him, so that they can extend their business relationships with a new company, maybe even move to a new market.

5.2 Trust

Processing and exporting cases showed that guanxi networks increase interpersonal trust. This may because of both the processing and exporting companies have long term relationships with their buyers, and their previous business experiences develop mutual trust. Telephone contact and personal visits also increase trust. Companies often mentioned that trusting their customers is the most important factor to maintain and coherent their business relationships and to improve their performance.

Trust increases with a good business experience. Most of the companies said that it is difficult for them to fully trust a customer for the first transaction. They get to know the reputation of this customer via their guanxi network before the formal business negotiation starts. A small quantity trial transaction is another common solution for most of the companies which allows a constructive relationship to develop future deals is very important for the international business both for sellers and buyers. Based on successful prior experiences, reputation was built up.

Good reputation increases trust. The companies are willing to keep their reputation to each customer because of the transferability and transparency of the guanxi network. Any opportunistic behaviour to any customer may lead to the reputation damage within the whole guanxi network. So people like to trust those who with good reputations.

Trust changes the way of business transactions. Trust is the most important factor influence companies' behaviour. Due to the uncertainty and high risk in the international business, nearly all the transactions are conducted under Letters of Credit (L/C) which is issued by the authorized bank with a high commission charge. So L/C incurs a high transaction costs to buyers and sellers and it takes a long period to complete a transaction. But it is rather safe both for buyers and sellers. With the increasing of trust, a new, faster and cheaper method, Telegraphic Transfer (T.T.), was preferred in transactions. T.T. allows buyers to pay after they receive the goods. So it implies higher risk for sellers compared to L/C. But T.T. is a very convenient way to conduct transactions with low costs.

Box 3.
"We are not going to do wrong thing to treat our customers, we have to keep our reputation to them, otherwise, we will not only lose this customer, but also lose the entire market in this region."
"We trust our customers because we had a happy transaction experience; we believe they will be honest to us in the future as well".
Interview with exporting company 1 & 2

5.3 Transaction specific investments

Transaction specific investments are required for the company to be involved in the vegetable industry. Both physical and human investments are required. Case interview showed that guanxi network can influence companies' behaviours in terms of transaction specific investments. Most of the companies agreed that they got requests form their buyers regarding investment to improve food quality and safety. The decision will be made depend on the expectation of the future business chance with this buyer.

Investment makes buyer-seller relationships closer. The physical investments are the equipments and machineries for processing and packaging which is crucial to improve product quality. Equipments (such as cooling vehicle) to transport products also facilitate deliveries from the suppliers' location to the warehouse of the company or directly to the retailers' shops. Companies perceived transaction specific investments as improving the relationship with buyers and to enhance the transactions in the future.

Box 4.
"The greenhouse is a huge investment for us. But it helps us to negotiate with supermarkets. They show confidence with our product quality and we can easily make a deal and access more markets."
Interview with the producer company

The internet construction is also another important physical investment for the company. It is a good and reliable media to introduce the product, management, experts, company culture to their customers. *"We launched a French version website this year when we had a French girl do internship in our company. It is a good way for us to keep close contact with our French customers especially."*
Interview with processing company 2

Human specific investments are mainly focused on knowledge training and transfer. It is import to train sale representatives to get familiar with international rules and the commercial skills to deal with foreign customers. It is also important to know the significant different cultures in different markets. Interviews showed that knowledge training for employees is an important investment for gaining efficiencies. The buyers should have knowledge about where they can find the right product with a competitive price. The way of contact with foreign customers is totally different from the way of communication with domestic customers. Nearly all the companies highlighted the importance of human investments to decrease conflicts with customers and to benefit of relationship building.

5.4 Channel and governance choices

Companies have significant different marketing strategies regarding their channel and governance choices. For their buying behaviour, most of the companies try to concentrate on preferred suppliers to acquire consistent product quality. Contracts are used for transactions.

Companies try to keep regular contacts and long term relationships with their customers. They mostly keep several preferred customers for their entire business. The businesses relationships will not be stop if there is no serious problem. Nearly all the case companies kept their longest relationships since they set up their company. Companies also showed tendency to diversify their sale markets. They try to keep rather more buyers compared to suppliers. But several important buyers are still preferred to guarantee minimum transactions. But they will never stop searching new buyers to achieve business development.

Contracts are the basic governance mechanism for all international transactions due to security reasons while their guanxi networks make a role to enhance and lubricate these transactions. Companies' guanxi network showed positive contribution to international business. Guanxi network increases problem tolerance and improves problem solving. Guanxi network also increases rebuy possibilities and extends business durations.

Payment method is very important as indicator for guanxi and trust. Security is the major concerns for international transactions. The sellers not only evaluate the reputation of the international buyers, they also use less risky methods to conduct transactions, such as L/C. But it will be replaced by a more risk taking methods (T.T.) once they build a good relationship and trust each other. However, transaction with cash is more often used in domestic markets. In doing this, buyers can get a lower price to acquire a high margin. Companies also contact traders/middlemen to reduce collecting costs in domestic markets. Contacting with farmer associations is a common way. General contracts are used to regulate the transactions. Quality, delivery time and place, and price are the most important items in the contracts. Stable delivery and consistent quality are the important judging criterions.

5.5 Performance

Case company interviews showed that building strong guanxi relations with the right person is important to achieve business success in China. The guanxi network determines companies' marketing strategies in terms of marketing channel and governance choices; as a result, it also influences relationship performance regarding transaction costs and quality satisfaction.

Transaction cost

Relations reduce the costs for transactions. Guanxi network in research area showed advantages to reduce transaction costs. Good guanxi network helps companies find valuable information, to enter a new market, to transfer reputation to potential buyers, *etc.* So less time needed for companies for negotiation process. They also can use more flexible and fast ways to conduct transactions with low risk.

However, maintain a guanxi network is also a costly investment. Companies have to contact regularly with their guanxi partner to show their concerns. Personal visits are also common to keep contact. All these activities make cost since reception and gifts are necessary to build up and to improve their guanxi network.

Marketing channels and governance arrangements also showed certain differences regarding transaction costs. The negotiation and enforcement cost are higher in domestic markets due to highly opportunism and lack of trust. Written contract reduce the risks involved in transactions and relationships showed complementary function to contracts.

Quality satisfaction

Quality satisfaction is the results of the delivered production quality and quality expectation. International buyers require higher quality products than in domestic markets according to their own quality standards. Pesticides residuals are major safety concerns for buyers. HACCP and ISO certification implies quality guarantees to convince buyers. Companies have to offer their best quality products to international markets due to higher quality requirements, and deliver the second class quality products to domestic markets. This diversified quality management practices lead to different quality images. Case studies showed that the quality image is higher in international markets than in domestic markets. Guanxi network shows positive impacts on the quality perception. Good guanxi improves quality management practice and reduces quality problem both in buying and selling stages which results a high level of satisfaction.

6. Conclusions and limitations for further research

6.1 Conclusions

The aim of this research is to study if the Chinese cultural and social embedded guanxi networks are important in business practices and how they influence buyer-seller relationships and performances in vegetable supply chains in China. We interviewed 8 case companies to explore their guanxi network, trust, transaction specific investments, channel and governance choices and performance attributes of transaction costs and quality satisfaction. Summaries of the 9 selected buyer-seller relationships (relationship 1-9 in Figure 2-6) for each case company are listed in Table 2.

This case study demonstrated that guanxi network has diversified contributions to buyer-seller relationships involved in various vegetable supply chains in Jiangsu Province. Guanxi network stimulates companies to make investment; improves trust between counterparts; influences marketing behaviours in terms of sourcing and destinations of their products and institutional arrangements of transactions. Also company business performances can be improved through well developed guanxi networks in practices.

Guanxi networks: Guanxi networks are important for company based transactions while they are less important for individual (farmer) related transactions. So it is reasonable for companies using guanxi network to enhance business relationships in international markets and in newly developed domestic markets, such as supermarkets.

Trust: Interpersonal trust is important for transactions in international markets and in supermarkets. Companies are not likely to show trust in the relationships with farmers and traders in the wholesale markets. Trust is also low in the relationships between domestic food company and exporting company. This may because of less monitoring efforts needed for cash

Table 2. Summary of the concepts for different buyer-seller relationships.

Relationship	Producer company–supermarket (1)	Farmer–processor (2)	Processor–international market (3)	Food company–exporter (4)	Exporter–international market (5)	Farmer–cooperative (6)	Cooperative–wholesale market (7)	Farmer–supermarket (8)	Vegetable company–supermarket (9)
Concept									
Guanxi networks	important	limited	important	important	important	limited	limited	limited	important
Trust	high	low	high	low	high	low	low	low	high
Transaction specific investments	high	low	high	low	high	low	high	low	high
Governance	formal contract	no/ informal contract	formal contract	formal contract	formal contract	contract/ membership	(in)formal contract	informal contract	formal contract
Transaction costs	low	high	low	Low	low	high	high	high	low
Quality satisfaction	high	low	high	high	high	low	low	low	high

and carry transactions or less feasible to build trust with individual counterparts due to high opportunistic behaviour.

Transaction specific investments: Similar to trust, transaction specific investments are also high in international and supermarket chains due to long term business relationships and high quality and safety requirements. Guanxi network in the relationships with international and supermarket buyers improves the willingness to invest to comply with the quality and safety requirements in practice.

Governance choices: Contracts are used to safeguard transactions and to reduce risks in business. Formal contracts exist in situations of transactions with companies both in domestic and international markets for large volumes and reoccurrence of transactions. Informal contracts more present in domestic markets and transactions with farmers.

Transaction costs: Transaction costs close related to trust and reputation in buyer-seller relationships. Guanxi accelerates problem solving and increase problem tolerance in relationships with international customers which enhance long term business success. Thus transaction costs in international markets are lower than in domestic markets for higher level of trust and creditability of reputations which lead to less information and enforcement costs.

Quality satisfaction: The required and delivered product quality is much higher in international markets than in domestic markets which leads to a high quality image and high level of quality compliance in international markets. Compare to the wholesale market, the quality satisfaction is higher in supermarkets in domestic markets. This may also due to the restricted market access for quality requirements for supermarkets.

6.2 Limitations for further research

We also found certain limitations for this case study which pave the ways for further research.

First of all, the sample size of this study is limited. Practically it is difficult to use a rather small group of 8 case companies to sketch the picture of the entire vegetable chain. So the results showed here are more based on the position of individual companies and on buyer-seller relationships. More cases study is needed to see how guanxi networks work in the scope of the entire vegetable chain.

Secondly, this case study focuses on one direction impacts. We studied the impacts of guanxi networks on buyer-seller relationships and on quality performance. We ignored the interrelations among the buyer-seller relationships and performance indicators. We also excluded the loop effects of performance on developing and the quality of guanxi network. It is reasonable to take these interrelations and loop effects into account for further research.

Thirdly, this case study gives qualitative results. So we are not confident to say that all these impacts or relationships will be valid in a more general sense. So quantitative analysis is necessary to further validate these relationships. Therefore a survey with a large sample base is necessary. Fortunately, the necessary data have been collected in a wide range of company types with different marketing channels. The findings of this case study will be further tested through econometric methods in the follow-up research.

7. Managerial implications

The results of this study suggested that the guanxi network showed positive contribution to buyer-seller relationships and may substantially enhance the quality performance in vegetable sector in China. Managers then may use this study and its empirical evidence as a check on the adequacy of their existing guanxi networks and type of benefits their networks might provide. Guanxi networks increase the success possibility to access new markets and to maintain long term relationships. So it is important for the companies putting more efforts to build up strong guanxi networks to expand their markets and to develop their business. Companies should then increase face-to-face encounters, frequent contracts, information sharing and showing honesty and sincerity to each person in their guanxi networks. But case study also implied that they also should be aware of the contingency of the costs to build and maintain such guanxi networks. If managers either under- or overestimate the negative/positive effects of guanxi networks, their efforts would be misguided with which would eventually lead to satisfaction decrease. But the economic effects of guanxi networks at the individual and organizational level should be evaluated with further study.

Vegetable farmers also can use this study to improve their marketing positions. Case study showed that farmers' association and close relations with buyers improve the accessibility to newly developed markets, such as supermarket and international markets, with better prices and more stable transactions. Farmers can improve their negotiation power and get better marketing chances to achieve better income in situations of being organized. Besides cooperative, Farmers' Professional Associations, farmer volunteered or initiated by government, should be developed with efforts.

Acknowledgements

We wish to acknowledge financial support for the implementation of this research from WOTRO (grant file number W01.65.2001.010). We would like to thank Dr. Jos Bijman for his valuable comments on the earlier draft of this paper.

Reference

Alston, J., 1989. Wa, Guanxi, and Inwa: Managerial principles in Japan, China, and Korea. Business Horizon (March), 26–31.

Ambler, T., 1994. Marketing's third paradigm: Guanxi. Business Strategy Review 5, 69- 80.

Anderson, E. and B. Weitz, 1989. Determinants of continuity in conventional industrial channel dyads. Market Science 8, 310-323.

Anderson, J.C. and J.A. Narus, 1990. A model of distributor firm and manufacturer firm working partnerships. Journal of Marketing 54, 42-58.

Arias, J.T.G., 1998. A relationship marketing approach to guanxi. European Journal of Marketing 32, 145-156.

Blau, P.M., 1964. Exchange and power in social life. New Brunswick, NJ., Transaction Publishers.

Butterfiled, F., 1983. China: Alive in Bitter Sea, Coronet Books, N.Y.

Cook, K.S. and R.M. Emerson, 1978. Power, equity, and commitment in exchange networks. American Sociological Review 43, 721-739.

Cunningham, D.C., 2001. The distribution and extent of agrifood chain management research in the public domain. Supply Chain Management 6, 212-215.

Davies, H., 1995. Interpreting Guanxi: the role of personal connections in a high context transitional economy. China Business: Context and Issues, Longman, Hong Kong, pp. 155-69.

Davies, H., T.K.P. Leung, S.T.K. Luk and Y.-H. Wong, 1995. The benefits of "Guanxi": The value of relationships in developing the Chinese market. Industrial Marketing Management 24, 207-214.

Dwyer, F.R., P.H. Schurr and S. Oh, 1987. Developing buyer-seller relationships. Journal of Marketing 51, 11-27.

Eisenhardt, K.M., 1989. Building theories from case study research. Academy of Management Review 14, 532-550.

Eisenhardt, K.M., 1991. Better stories and better constructs: The case for rigor and comparative logic. Academy of Management Review 16, 620-627.

Fan, Y., 2002. Questioning guanxi: definition, classification and implications. International Business Review 11, 543–561.

Fang, T., 2001. Culture as a driving force for interfirm adaptation: A Chinese case. Industrial Marketing Management 30, 51–63.

Farh, J.L., A.S. Tsui, K. Xin and B.S. Cheng, 1998. The influence of relational demography and guanxi: The Chinese case. Organization Science 9, 471-488.

Ganesan, S., 1994. Determinants of long-term orientation in buyer-seller relationships. Journal of Marketing 58, 1-19.

Geyskens, I., J.-B.E.M. Steenkamp and N. Kumar, 1998. Generalizations about trust in marketing channel relationships using meta-analysis. International Journal of Research in Marketing 15, 223-248.

Granovetter, M., 1985. Economic action and social structure: The problem of embeddedness. American Journal of Sociology 91, 481-510.

Heide, J.B., 1994. Interorganizational governance in marketing channels. Journal of Marketing January 58, 71-85.

Heide, J.B. and G. John, 1988. The role of dependence balancing in safeguarding transaction-specific assets in conventional channels. Journal of Marketing 52, 20-35.

Kao, J., 1993. The worldwide web of Chinese business. Harvard Business Review 71, 24-36.

Lou, Y., 1997. Guanxi and performance of foreign-invested enterprises in China. Management International Review 37, 51-70.

Lovett, S., L.C. Simmons and R. Kali, 1999. Guanxi versus the market: Ethics and efficiency. Journal of International Business Studies 20, 231-248.

Luo, Y. and M. Chen, 1996. Managerial implications of guanxi-based business strategies. Journal of International Management 2, 193-316.

Malone, T.W., 1987. Modeling coordination in organizations and markets. Management Science 33, 1317-1332.

Masten, S.E., J.W. Maeehan and E.A. Snyder, 1991. The costs of organization. Journal of Law, Economics and Organization 7, 1-25.

Merrilees, B. and D. Miller, 1999. Direct Selling in the West and East: The Relative Roles of Product and Relationship (Guanxi) Drivers. Journal of Business Research 45, 267-273.

Morgan, R.M. and S.D. Hunt, 1994. The commitment-trust theory of relationship marketing. Journal of Marketing 58, 20-38.

Peng, M. and Y. Luo, 2000. Managerial ties and firm performance in a transition economy: The nature of a micro-macro link. Academy of management Journal 43, 486–501.

Pye, L., 1982. Chinese commercial negotiating style. In: Advances in Consumer Research, Is relationship marketing culturally bound: a look at guanxi in China? L.C. Simmons and J.M. Munch (eds.), Cambridge: Oelgeschlager. Gunnand Hain Inc. 23, pp. 92–96.

Ramasamy, B., K.W. Goh and M.C.H. Yeung, 2006. Is Guanxi (relationship) a bridge to knowledge transfer? Journal of Business Research 59, 130-139.

Rindfleisch, A. and J.B. Heide, 1997. Transaction cost analysis: Past, present, and future applications. Journal of Marketing 61, 30-54.

Rousseau, D.M., S.B. Sitkin, R.S. Burt and C. Camerer, 1998. Not so different after all: A cross-discipline view of trust. Academy of Management Review 23, 393-404.

Sharma, A. and K.G. Pillai, 2003. The impact of transactional and relational strategies in business market: An agenda for inquiry. Industrial Marketing Management 32, 623-625.

Sheth, J.N. and R.H. Shah, 2003. Till death do us part.but not always: Six antecedents to a customer's relational preference in buyer- seller exchanges. Industrial Marketing Management 32, 627- 631.

Slack, N., S. Chambers and R. Johnston, 2001. Operations management, Prentice Hall: New York.

Standifird, S.S. and R.S. Marshall, 2000. The transaction cost advantage of Guanxi-based business practice. Journal of World Business 35, 21-42.

Vanhonacker, W., 1997. Entering China: an unconventional approach. Harvard Business Review (March-April), 130-140.

Wang, C., 2005. Guanxi vs. relationship marketing: Exploring underlying differences. Industrial Marketing Management (in press).

Wong, Y.H. and R.Y.-K. Chan, 1999. Relationship marketing in china: guanxi, favouritism and adaptation. Journal of Business Ethics 22, 107-118.

Yang, J.Y. and A. Liu, 2003. Frequent flyer program: a case study of China airline's marketing initiative - Dynasty Flyer Program. Tourism Management 24, 587-595.

Yau, O.H.M., J.S.Y. Lee, R.P.M. Chow, L.Y.M. Sin and A.C.B. Tse, 2000. Relationship marketing the Chinese way. Business Horizons (January-February), 16-24.

Yeung, I.Y.M. and R.L. Tung, 1996. Achieving business success in Confucian societies: The importance of guanxi (connections). Organizational Dynamics 25, 54-65.

Yin, R.K., 2003. Case study research: design and methods. (3rd Edition), Sage Publications, Thousand Oaks, London, New Delhi.

Zhang, X. and G. Li, 2003. Does guanxi matter to nonfarm employment? Journal of Comparative Economics 31, 315-331.

Is more vertical integration the future of food supply chains? Empirical evidence and theoretical considerations from German pork production

B. Schulze, A. Spiller and L. Theuvsen

Abstract

In the literature it is often hypothesized that contracts and vertical integration are paramount for the future competitiveness of meat supply chains. In this paper we review the empirical evidence and theoretical considerations underlying this argumentation. We confront the contract and integration hypothesis with transaction cost considerations taking into account current developments at the slaughterhouse level and in quality assurance systems and results from a recent survey in German pork production. Cluster analysis was performed to classify farmers with different attitudes towards contract farming. It is concluded that improved supplier relationship management may be a viable alternative to more contracts and vertical integration in meat supply chains.

Keywords: pork production, supplier relationship management, transaction cost economics, vertical coordination

1. Introduction

In recent years contracts and vertical integration in meat supply chains were among the most vividly discussed topics in agriculture and the food industry. This discussion is fuelled by several developments. Some authors identify increasing requirements of large retailers and fast-food companies concerning product quality and traceability as important drivers towards more integrated food supply chains (Den Ouden *et al.*, 1996; Lawrence *et al.*, 2001). Others highlight the important role contracts and vertical integration may have played in the emergence of the United States and Denmark as leading pork producers and exporters (Windhorst, 2004).

It is also argued that buyers' appreciation of improved food traceability is often lower than expected and can in many cases easily be met without redesigning supply chains (Theuvsen and Hollmann-Hespos, 2005). Furthermore, quality assurance schemes such as the Dutch IKB or the German Quality and Safety System have reduced quality uncertainties and, thus, favour spot market transactions over alternative ways of organizing meat supply chains (Schramm and Spiller, 2003).

In this paper we critically review the arguments and propose improved supplier relationship management as an alternative to more contracts and vertical integration in meat supply chains. Our recommendation is based on a large-scale survey in German pork production as well as an analysis of the effects of recent structural and technological changes in slaughterhouses on transaction costs.

2. Vertical coordination in meat supply chains

Vertical coordination describes the way relationships between producers and processors are organized in food supply chains which is not a yes-or-no decision. In fact, there is a broad spectrum of alternatives farmers and processors can choose from when designing their business relationships (Peterson *et al.*, 2001). The classification of distinct hybrid coordination forms between market and hierarchy is not really consensus. Peterson *et al.* (2001: 151), for instance, describe specification contracts, relation-based alliances and equity-based alliances. Figure 1 presents some of the most important alternatives of organizing livestock production.

The typology refers to coordination intensity resulting from time horizon, centralization of decisions and ownership. It integrates non-contractual relationships as well as some specific arrangements in pork production.

In *spot market* exchanges, pork producers and slaughterhouses negotiate only short-term contracts and are ready to change their selling respectively buying behaviour very quickly; *"control is fully located at the separate stages (...) and coordinated solely by market prices"* (Den Ouden *et al.*, 1996: 281). The transactions are governed by classical contract law: *"sharp in by clear agreement, sharp out by clear performance"* (Williamson, 1991: 271).

Although not bound by contracts, many farmers and abattoirs actually prefer *long-term relationships* with a limited number of livestock dealers or slaughterhouses to sell their finished pork. Stable business connections allow more reliability and lower search costs without losing switchover facilities.

Marketing contracts establish selling and buying obligations, *i.e.* farmers have to sell their finished pork to the contracting abattoir which on the other hand binds itself to buy all pigs from the contracting farmer. Besides secured supply for the processor and market access for the farmer, marketing contracts often shift price risks from farmers towards the processor, *e.g.* through window contracts or feed-price formula contracts (Roe *et al.*, 2004). The underlying assumption is that farmers are more risk averse.

Production contracts do not only establish selling and buying obligations but do also prescribe some details of the production process such as feed, breed or animal welfare aspects. Especially in markets with high information asymmetries, the contractor tries to control a number of inputs in order to reduce the scope of moral hazard problems and adverse selection (Goodhue, 1999: 617).

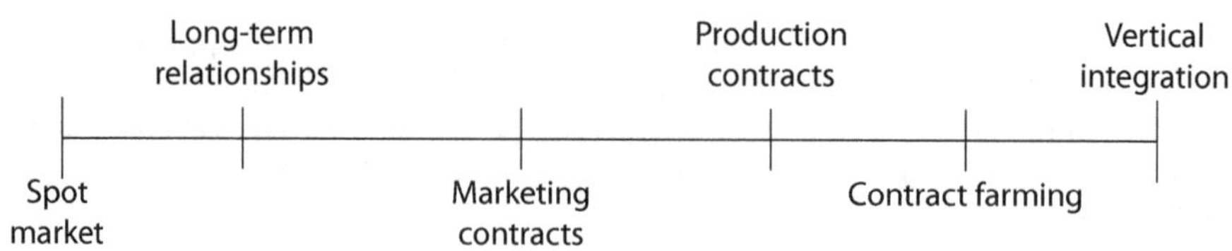

Figure 1. Vertical coordination of meat supply chains (Spiller et al.*, 2005: 398).*

Contract farming schemes represent an even more integrated relationship between farmers and abattoirs. Farmers remain legally autonomous but are heavily dependent on a contractor, for instance an abattoir, who provides all critical resources (*e.g.* piglets, feed, managerial oversight) (Mighell and Jones, 1963). Contract farming is characterized through centralized decision making and long-term contracts relying on the processor's ability to manage the whole supply chain.

In *vertically integrated* pork production chains, there is joint ownership of resources on the farm and the slaughterhouse level, be it that abattoirs have invested in pork production facilities or that farmers have control over slaughtering due to, for instance, the prevalence of direct marketing or cooperative slaughterhouses. However, the classification of coops in this coordination scheme is not easy. On the one hand the slaughterhouse is in the ownership of the farmers, on the other hand the farmers are losing influence on modern cooperatives acting more or less similar to stock companies (Cook, 1995).

In the terminology of transaction cost economics, vertical integration represents hierarchical governance mechanisms, whereas the various forms of contracts as well as repeated transactions are hybrid governance structures (Williamson, 1985). Throughout the world, very diverse pork production systems have evolved. In this respect, important pork producing countries can be roughly divided into two groups. In most of the European countries, like Germany, The Netherlands, Belgium, and France spot markets, long-term relationships and marketing contracts are dominating (Traupe, 2002; Boston *et al.*, 2004; Spiller *et al.*, 2005). In other countries, *e.g.* the United States and Denmark, partially also in Brazil and Spain, production contracts, contract farming, and vertical integration have largely replaced less integrated forms of pork production. Today more than 70% of U.S. pork production is produced in stricter coordinated meat supply chains (Figure 2). This parallels a trend in U.S. agriculture according to which nowadays about 36% of agricultural products are produced under contracts and in vertical integration (Martinez, 2002a, b; MacDonald *et al.*, 2004).

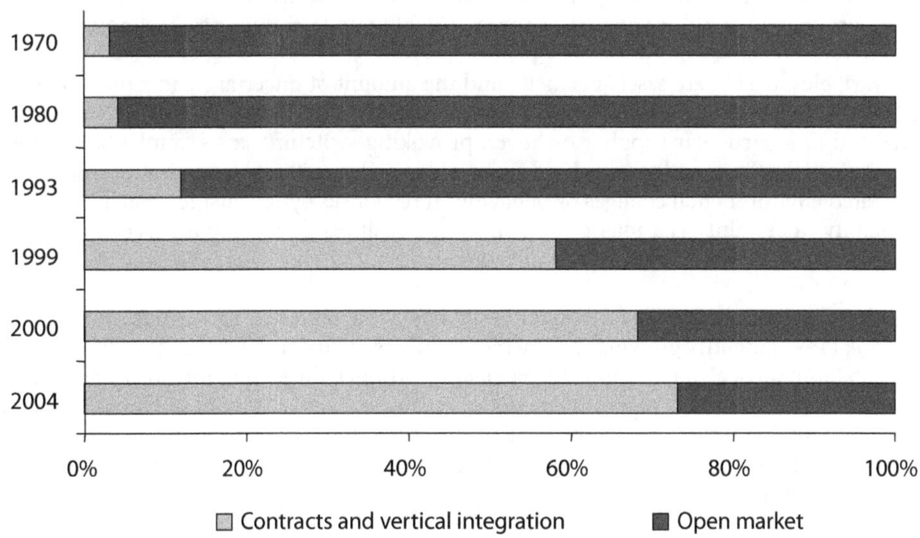

Figure 2. Contracts and vertical integration in US pork production (Martinez, 2002a, b; Haley, 2004).

3. Transaction cost savings through contracts and vertical integration: some critical remarks

The continued existence of two different types of pork production chains throughout the world – less integrated systems in, for instance, many Western European countries and stricter coordinated systems based on contracts and vertical integration in countries such as the U.S. and Denmark – is rarely discussed in the agribusiness literature. Most authors favour a trend towards more intense vertical coordination and consider the loose relationships in some old European markets as an obsolescent model (Windhorst, 2004; Bhuyan, 2005). It is linked with a view of a well organized supply chain coordinated by focal chain leadership. Past studies have revealed that the poultry industry was the first strictly integrated branch and that this trend has been spreading into other industries since the 1980s (Bhuyan 2005: 429). Even in developing and transition economies contract farming is seen as a means to integrate small scale farmers into modernized agri-food chains. Boger (2001) for example discusses the use of production contracts as an instrument in the Polish pork market in order to establish high quality markets without widely used grading systems for pigs. Other authors (Guo *et al.*, 2005) analyze the danger that contract farming bypasses small scale producers therefore deepening the gap between local producers and the requirements of export orientated sectors.

In the following sections we will discuss the assumed prevalence of contract farming and vertical integration from two perspectives. First, the theoretical foundation of vertical coordination is questioned by referring to some new arguments contesting the traditional transaction cost view. The second argument is an empirical one, stemming from a farmer survey.

Transaction Cost Economics (TCE) provides the most common theoretical framework for contracts and vertical integration in livestock production. Contracts may reduce moral hazard problems through centralized decisions about input factors (feed, genetic, *etc.*) and production standards. The problem of adverse selection in case of unobservable quality characteristics (credence attributes) is decreased by contract systems with inherent monitoring approaches. A long-term orientation could enhance the processor's ability to introduce new technologies.

The key variables in TCE are asset specificity and the amount of uncertainty in a market (Klein *et al.*, 1978; Williamson, 1985). Asset specificity represents the degree to which an investment is specialized to a particular supplier or buyer, provoking switching costs. Sunk costs enforce opportunistic behaviour and create hold-up problems. Uncertainty contains the degree of unanticipated environmental changes or behavioural responses by the business partner. Besides prices, quality uncertainty is a major concern in agribusiness due to the different food crises especially in the meat industry (Hobbs *et al.*, 2002).

Empirical evidence supporting TCE hypotheses has been obtained in various studies. Den Ouden *et al.* (1996) identify growing quality requirements of customers as a major driving force of contracts and vertical integration. In particular, product differentiation in order to meet changing consumer demands regarding credence attributes such as animal welfare, food safety and environmental issues is considered a main driver of closer ties in the meat supply chain. Transmitting the changing demands to the farm stages is considered more transaction cost efficient under contracts and in vertically integrated systems.

Lawrence *et al.* (1997) offer a similar explanation. According to their empirical studies, both market sides, *i.e.* farmers and slaughterhouses, save transaction costs through contracts and vertical integration. For abattoirs high quality of pork and consistency of supply with adequate quantities are paramount. Lawrence *et al.* (1997: 28) argue that under these circumstances long-term contracts allow transaction cost savings compared to traditional marketing channels. Farmers may also save transaction costs through long-term contracts, *e.g.* by settling a premium for higher quality with a one-time negotiation. Since this advantage is less relevant for small farmers, mainly large farmers enter contractual arrangements.

Hornibrook and Fearne (2005) found that retailers who have introduced private labels or use their company's name as a brand put great emphasis on product safety and quality. In these cases, specific investments in brands require safeguards against food crises and a subsequent loss of reputation. Therefore, retailers have strengthened their influence on the meat supply chain and largely refrain from spot market transactions.

Serving special market segments may also require specific investments and/or a high collaborative adaptiveness of the supply chain. The Japanese pork market, for instance, is famous for its specific consumer demands concerning taste, marbling, cutting and so on (Makise, 2002). Since vertically disintegrated chains do not provide the required quality, the Japanese market has mainly been served by contractually bound or vertically integrated pork producers from Denmark or the United States.

The underlying assumption behind the different arguments is that with higher market segmentation the need for processors to define stricter governance structures grows. Asset specificity and market uncertainty are not a premise but variables in TCE which may change due to new technologies and institutional innovations. In the following part two main recent developments are described as driving forces for lower vertical coordination in the pork industry: introduction of new technologies of pre-slaughter and on-line sorting as well as comprehensive quality management metasystems. Both aspects reduce the necessity for contractual arrangements between pork producers and buyers.

The Danish pork production is the most widely cited example of successful meat packing supplying uniform animals to a slaughterhouse. Special pig genetic is restricted geographically (Bogetoft and Olesen, 2002: 190). Farmers close to a certain abattoir are forced by contracts to produce a special pig in order to supply homogeneous animals. The respective slaughterhouse is specialized on different market segments, *e.g.* the British bacon market, the Japanese market, *etc.* The Danish market leader, the coop Danske Slagterier, uses contract farming to produce homogeneous pork products in large quantities. During the 1980s and 1990s this production systems worked quite well and contributed to Denmark's emergence as the leading pork exporter in the high quality segment. "*The trend towards more uniform production processes is a central feature of agricultural industrialization*" (Hennessy, 2005: 601). This emphasis on uniformity in pig phenotypes, feed, medication, and sometimes animal welfare seems to be a crucial factor of market orientation.

However, supplying uniform animals by contract farming is only one option to produce homogeneous products in high quantities. During the last years, new sorting technologies in combination with the enormous growth of slaughterhouses have been allowing the same output with pre-slaughter sorting instead of vertical coordination. In Germany a successful example of this new strategy is Toennies. Toennies, market leader in packed pork in Germany, created over 70 different internal classification categories into which the animals are sorted using an automatic classifying technology. In the next step the different batches are divided by automated sorting technologies to produce about 1,000 different products tailor-made for special market destinations. Processing capacities of 20,000 pigs a day enable the company to produce sufficient quantities of uniform meat without defining homogeneous input factors. Quite contrary to that,

Toennies strongly prefers diverse animals to be sure that suitable meat is available for every market segment. A core competence of Toennies is the ability to develop new markets for the huge number of different products. The necessary basic quality is guaranteed by a usual grading system (payment grid) and the participation in the German Quality and Safety System, a quality certification in the German meat business.

All in all, new sorting technologies in combination with large-scale slaughtering lower the degree of asset specificity between pork producers and processors. In the Danish example, farmers and their local abattoir are heavily dependent on each other by the use of special genetic. Toennies reaches a similar output without contracts on a spot-market basis or, at best, long-term relationships with their suppliers.

Besides the advantages of new sorting technologies, the example of Toennies also demonstrates the impact of quality management metasystems on governance structures. Quality certification systems enforce spot markets. They build standards for the whole industry thus reducing the imperative for company specific quality approaches on a contract basis. *"Certification is the (voluntary) assessment and approval by an (accredited) party on an (accredited) standard"* (Meuwissen *et al.*, 2003). A key feature of a certification system is that inspections are carried out by independent bodies (third-party audits) beholden to standards laid down by external organizations (Luning *et al.*, 2002). By means of regular control and – whenever necessary – additional sampling, neutral inspection institutions monitor the entire supply chain. Once having been awarded the requisite certificate, companies are entitled to make use of the quality label for marketing purposes. The supplier provides a certificate serving as a quality signal, which is issued by a neutral certifier based on the quality and certification standard laid down by the standard owner. Most certification schemes are privately organized and are used predominantly as a business-to-business (B-to-B) marketing tool. Well-known examples of certification are the EurepGap standard, covering agricultural producers, the BRC (British Retail Consortium) Global Standard or its German (and since 2003 also French) equivalent IFS (International Food Standard), directed towards the manufacturers of private labels. Most certifications are based on the retailers' efforts to control the suppliers. Nevertheless, as a countervailing power there are also certification systems initiated by suppliers such as the Assured Farm Standard (AFS) in British agriculture or the IKB standard of the Dutch meat industry or chain-wide certification standards such as the German Quality and Safety System.

Reliable certification systems in a market work in the same way as public quality regulation. They establish substitutes for company specific monitoring and control in the supply chain (Schramm and Spiller, 2003; Giraud-Héraud *et al.*, 2005). In most cases certification systems formulate a baseline assuring minimum standards like compliance with legal requirements. Therefore, they are quite inexpensive. The most important advantage is the comprehensive acceptance of a successful scheme. If all processors in a market recognize a certificate as a minimum standard farmers are free to choose the suitable buyer for every transaction. Hence, the introduction of comprehensive quality management metasystems like IKB in the Netherlands, Quality and Safety in Germany or EurepGap reduces buyers' quality uncertainty and, hence, search and quality control costs. According to TCE, this enforces arm's length solutions in food supply chains (Figure 3).

Today, more than 54,000 farmers are members of the Quality and Safety System in Germany producing, among other things, about 120,000 pigs for slaughter every day. In The Netherlands, IKB integrates more than 90% of all farmers. In 2004 QS, IKB and Danske Slagterier agreed on a common quality approach called European Meat Alliance. Through a benchmarking process other countries can join this association. The same benchmarking is possible for the retailer-

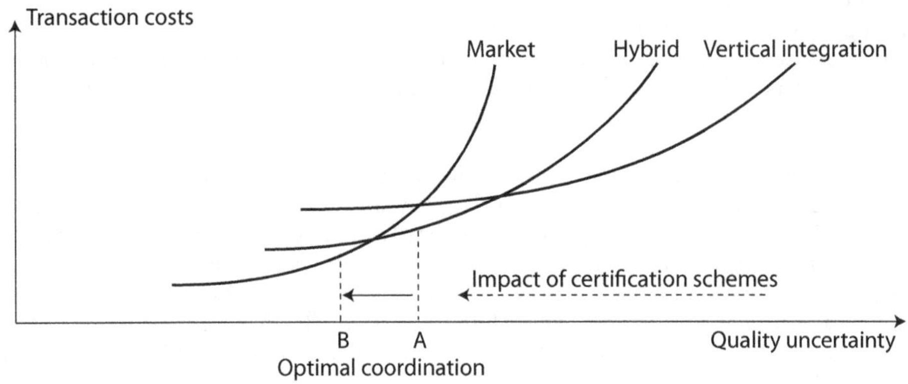

Figure 3. Transaction-cost saving effects of certification schemes.

driven EurepGap system, which is widely used in international fruit and vegetables production and now trying to become a common certification standard on a whole farm approach.

All in all the introduction of new institutions like quality certification and new automated sorting technologies in larger slaughterhouses ceteris paribus reduces the requirements for the contractual integration of farmers into food supply chains and supports more flexible market structures. Researchers as well as practitioners often underestimate the value of spot markets, *i.e.* greater flexibility and stronger pricing pressure. However, TCE is a comparative static model which only allows simple ceteris paribus analyses so that the exact position on the vertical coordination continuum (see Figure 1) remains somewhat vague. In addition there are other factors besides TCE-arguments prejudicing the governance structure, *e.g.* monopoly motives, core competencies, or behavioural variables such as motivation, involvement and attitudes.

4. Farmers' acceptance of more vertically integrated meat supply chains

The previous section provided some theoretical support for the model of a less integrated pork supply chain, but TCE as a comparative static model is not sharp in defining governance decisions. There remain some arguments for better supply chain coordination which leads us to the second step of our analysis. An important but often neglected factor of choosing the right governance structure is farmers' acceptance of stricter coordinated chains, which is crucial at least for short-term changes. In Germany, the concentration ratio at the processor level is growing but there are still enough alternative slaughterhouses. Farmers have the choice between abattoirs requiring contracts (*e.g.* the leading German co-operative Westfleisch *e.g.*), and others preferring spot markets like the market leader for packed pork, Toennies. Farmers' freedom of choosing alternative marketing channels is leading us to the assumption that their attitude has to be taken into account.

Farmers' contracting behaviour has been researched by a number of economists. The contributions can be distinguished into econometric and behavioural approaches. The first category comprises studies modelling contract decisions as a function of structural variables, *e.g.* personal and farm characteristics, which are based on statistics about real contract behaviour (Katchova and Miranda, 2004; Key and McBride, 2003; Key, 2004). Key (2004), for example, states that contracting is highly correlated with size of farms. Other variables explaining the adoption of marketing contracts are education, the use of advisory services and the use of marketing plans and futures (Katchova and Miranda, 2004: 101). All in all, the authors resume that adoption of contracting in the real world stands a little bit behind the theoretical advice.

There are only a few studies that apply behavioural approaches, thus addressing farmers' attitudes towards contracting (Guo *et al.*, 2005) or contract attributes (Lajili *et al.*, 1997; Roe *et al.*, 2004). They work with stated preferences, typically in simulated experiments or surveys. Guo *et al.* (2005: 12-13) find evidence that Chinese farmers have a rather positive attitude towards contracts, as 21.2% out of 1,036 surveyed farmers already had contracts and another 76 % said that they would accept contracts if they were offered. According to these results, low participation in contract farming is mostly due to a lack of opportunity (52.2 %), interest of buyers (24.5 %) or obvious benefits (20.7 %).

Lajili *et al.* (1997) reveal that asset specificity and uncertainty influence farmers' preference for stricter contracts in line with TCE considerations. Boger (2001) discusses the use of production contracts as an instrument in the Polish pork market in order to establish high quality markets without developed grading systems.

The mainstream conclusion in agricultural economics is that contracts are a highly preferable option for farmers to reduce (price) risks and to safeguard specific investments. The main problem of farmers is not to be excluded from contractual relationships. From this viewpoint, the relevant question is whether contract farming bypasses small scale producers especially in developing countries.

Contrary, Roe *et al.* (2004: 123) reveal preferences for short term contracts, for contracts with coops, smaller minimum delivery requirements and for marginal increases in the price ceiling instead of marginal increases in the price floor. These U.S. results imply some barriers to the adoption of stricter forms of governance by farmers. Furesi *et al.* (2006) reveal similar results for the Italian poultry sector. For Germany, anecdotic evidence shows that many farmers still strongly reject more vertical coordination of supply chains. In the German pork industry, for instance, "free entrepreneurs don't need contracts" is still a very popular slogan (AgraEurope, 2004). Contracts in farming based on persuasion and pression is threatened by a low level of involvement and internal motivation on the farmers' side.

All in all, the results of behavioural approaches reveal some scepticism among farmers in developed markets regarding stricter vertical coordination, whereas in developing and transformational countries various forms of contracts provide farmers with price security and market access. Finally, we assume great attitudinal differences depending on the respective country.

As existing studies do not discuss in depth farmers' general attitudes towards contracts and vertical integration, an own survey was conducted among 357 large-scale pork producers in North-Western Germany, *i.e.* the Westfalen-Lippe and the Weser-Ems regions, the centres of German pork production where larger farms and slaughterhouses are sited (Table 1). We analyse farmers' general evaluation of contracts, perceived advantages of contracts and the relationship between contract behaviour and trust and commitment. The latter are included here since they are viewed as substitutes as well as complements of contracts (Poppo and Zenger, 2002).

Table 1. Sample description (mean values).

	Total	Westfalen-Lippe	Weser-Ems
Sample size	357	188	169
Farm size (ha)	93	97	89
Lease land (ha)	46	49	43
	(N = 329)	(N = 172)	(N = 157)
Herd size (feeder pigs)	1,413	1,162	1,681
Herd size (sows)	197	168	229
	(N = 101)	(N = 54)	(N = 47)
Age of respondents (years)	41	41	40

The attitude model mostly consisted of seven point-Likert scales ranging from "strongly disagree" (scale = -3) to "strongly agree" (scale = +3) to evaluate farmers' attitude towards contracts as well as the level of trust and commitment and their switching behaviour.

17.1% of the surveyed farmers have marketing contracts for their whole production, 1.7% have only partially contracted their production and the majority of 81.2% does not have any contractual linkages to a processor at all. Among the latter group of farmers, only 14.6% say that they often or very often switch between the different processors available in their region (score 60 and higher on a 100-point rating scale, where 0 represented "I supply my pigs to [abattoir X] for a long time already." and 100 "I switch from one buyer to the next very often."). Another 14.2% declare a medium level of switching frequency, and the great majority switches seldom (24.6%; score 21-40) or rarely (46.6%; score 0-20) their buyers. There is a highly significant but rather low correlation of switching behaviour and perceived number of alternative buyers ($r = 0{,}19^{***}$). To sum up, the degree of vertical coordination in the German pork production is quite low but with a strong focus on long-term relationships.

Answers of contract farmers and independent farmers differ significantly. For example, the statement "contractual arrangements are only favourable for slaughterhouses, farmers don't benefit at all" is rejected by contract farmers ($\mu = -0.67$, $\sigma = 1.55$) whilst farmers without contracts clearly agree ($\mu = 0.84$, $\sigma = 1.48$). Another question asks for the perceived benefits of contracts in terms of enhanced planning, which are recognized by contract farmers ($\mu = 0.96$, $\sigma = 1.44$) and rather neglected by the others ($\mu = -0.51$, $\sigma = 1.43$). High standard deviations show that there is no consensus amongst the farmers in either of the groups. Even contract farmers do not have a general positive view of contracts as one could assume. One reason might be that farmers feel dependent on one specific buyer because of the regional structure or historical linkages to their coop. All in all, attitudes towards contracts are much more complex and sceptical than has been recognized by scientists up to now.

Farmers without contractual arrangements were asked whether they think they will inevitably have to use contracts in the future. With an average of -1.26 ($\sigma = 1.53$), this statement is strongly rejected. A correlation with farm size, which was measured in hectares and herd size can only be identified for the statement "In my opinion it would be better if farmers would engage in long-term contracts with slaughterhouses" ($r = -0.13^{***}$). Nevertheless, there is a clear willingness to cooperate more closely with a buyer, if the latter turns out to be a good business partner, which is shown by 42.5 % of farmers (strongly) agreeing and another 27.5 % rather agreeing to the respective statement, resulting in an average of 0.96.

Due to the high standard deviations reported above, a cluster analysis was carried out based on 5 variables representing the most important statements towards contracting (*contr 1-5*). Euclidian Distance serves as proximity measure. The optimal number of clusters is first defined using Ward method. A four cluster solution is chosen based on scree test, dendrogram and plausibility considerations. In order to refine this solution in a second step, a K-means cluster analysis is conducted.

In Table 2 clusters are described by means and standard deviations of the active (cluster building) variables as well as important passive variables (statements see appendix). The four groups can be characterized as "inveterate antagonists", "indifferent farmers", "cooperation-oriented farmers" and "contract supporters". As show the F-values, the strongest differences between the clusters can be observed for the statements concerning the perceived necessity to contract with a slaughterhouse (contr 3 and 4). The first cluster ("antagonists") contains 103 farmers accentuating the strongest rejection, which account for about 30% of the sample. The answers to the statement "I don't want to give up my entrepreneurial freedom" (free) are consistent with these results and demonstrate that on the whole, more than two thirds of German pork producers are not willing to participate in contract farming.

A look at current contracting behaviour of the cluster members shows that cluster 4 – the "contract supporters" – consists of 53.4% contractees, whilst in the other clusters there are far less farmers with contracts, as could be expected. Put differently: 66% of all contract farmers in our sample are members of cluster 4. Another 15% belong to the group of the "cooperation-oriented farmers", 14% are in the group of "indifferent farmers", and 5% are "inveterate antagonists". The other way around, one third of contract farmers do not show positive attitudes towards contracting. It is

Table 2. Cluster analysis: attitudes towards contracting.

Cluster (N)		contr 1	contr 2	contr 3	contr 4	contr 5	free	trust 1	trust 2	com 1	com 2	swit
		active variables					passive variables					
1 (103)	μ	2.15	-1.75	-2.51	-2.07	0.81	2.34	0.86	60.92	-1.02	0.03	34.83
	σ	0.86	1.05	0.64	0.82	1.56	1.02	1.22	19.07	1.57	1.42	26.18
2 (85)	μ	0.59	-0.40	-0.55	-0.65	0.12	1.29	0.92	60.24	-0.71	0.00	37.44
	σ	1.00	0.89	1.09	1.01	1.11	1.12	1.06	16.54	1.41	1.45	24.09
3 (82)	μ	-0.04	0.50	-1.59	0.33	1.65	1.20	1.01	64.57	-0.21	0.36	24.61
	σ	1.20	1.18	0.83	1.03	0.76	1.28	1.23	21.48	1.69	1.45	23.34
4 (73)	μ	-0.88	1.23	1.73	1.38	1.95	-0.41	1.40	71.81	0.64	0.90	12.23
	σ	1.34	1.07	0.84	0.81	0.55	1.43	1.17	14.97	1.49	1.53	10.91
Total (343)	μ	0.59	-0.24	-0.90	-0.41	1.08	1.22	1.03	63.97	-0.39	0.29	28.10
	σ	1.57	1.54	1.77	1.58	1.31	1.55	1.19	18.74	1.66	1.49	24.30
F-value		121.9	133.0	375.2	223.2	44.7	74.4	3.4	6.5	17.6	6.5	20.1
p-value		0.000	0.000	0.000	0.000	0.000	0.000	0.018	0.000	0.000	0.000	0.000

Note: Seven point Likert scales ranging from "strongly disagree" (scale = -3) to "strongly agree" (scale = +3); exception: switching behaviour (swit) and trust 2 are measured on a 100-point rating scale.

interesting to note that cross tabulations do not show significant differences between the clusters in terms of farm size (finishing capacities and hectares) and significant differences in the age of the farmers do not reflect a linear relationship.

Cluster 4 is the only group which perceives some farmer benefits from marketing contracts, especially assured market access and planning reliability. Nevertheless, the group's positive attitude towards contracting is somewhat modest.

Clusters 2 and 3 are a little bit surprising. On the one hand, respondents in cluster 3 show a distinctive willingness to build closer relationships with their preferred customers; on the other hand, they perceive their market opportunities as sufficient with a lot of alternative slaughterhouses. Farmers in cluster 2 are quite indifferent but show the lowest degree of cooperative intentions. Furthermore, this group is characterized through a comparatively high switching behaviour. Altogether the amount of trust and especially of commitment in the pork chain is rather low in all segments. Contract farmers show significant higher values but regarding their long-term and ongoing relationships a weak approval is not convincing.

If we compare farmers' attitude towards contracting on the one hand and TCE considerations on the other hand congruent results can be observed. The degree of asset specificity and uncertainty is rather low in most market segments and a lot of farmers tend to protect their economic freedom which indicates a similar assessment of vertical coordination. Only 20% of all suppliers are willing to participate in marketing or production contracts. The majority prefers arm's length structures or stable, but non-constrained relationships. The rejection of contract farming is very manifest for the first cluster. Notwithstanding respondents in this group show a certain disposition to build stable relationships with one slaughterhouse. In contrast to this the second cluster is indifferent towards vertical contracts but with a low willingness to engage in closer business relationships. Both clusters are characterized by a low level of trust and commitment. Overall, these farmers are suitable to supply markets without higher specialities and production requirements.

Farmers belonging to cluster 4 are supporting the idea of contract farming but not as manifest as cluster 1 is rejecting it. Trust (trust 1 and 2) and commitment (comm1 and 2) are higher than in all other groups, but also to a limited extent. For these farmers we propose expanded endeavours to build up trust in the supply chain. With reservations the same conclusions also are advisable for cluster 3. But in this case, trust management should not imply marketing contracts.

All in all, we state strong negative attitudes towards contractual relationships, but nevertheless a broader willingness to cooperate more closely with processors if they in turn show cooperative behaviour. The limited level of trust in the meat supply chain indicates unused potential in this direction.

5. Discussion

The question of vertical coordination is more complex than the simple independent farmer-contract farmer dichotomy suggests. Both theoretical considerations and results of the farmer survey nourish doubts about an ongoing trend towards more contracts and vertical integration. In fact, for the German pork production we argue for a dominating sector of independent farming combined with an emphasis on more trust between transacting parties (*i.e.* farmers and abattoirs) and more commitment to long-term business relationships. Faced with the manifest rejection of contracts by most farmers, it seems to be appropriate to think more about non-contractual options of improved supplier relationship management. This is especially true in a country, characterized by family farming and few opportunities of building new big pork houses. To our minds, closer vertical relationships in German pork production are and will remain limited to smaller market segments with above-average quality requirements concerning, for instance, animal welfare or region-of-origin cues.

The currently low level of trust between farmers and abattoirs has led to a number of coordination problems in the German meat industry, *e.g.* failed attempts to successfully introduce salmonella monitoring or state-of-the-art carcass grading systems. These problems are not (only) due to various trade-offs between both sides but the consequence of information and communication conflicts in the supply chain. We therefore suggest the development of relationships characterized by stability, calculability, and reliability which offer the possibility of optimizing processes, enhancing chain-wide information flows and ameliorating product quality. As both transaction parties can draw advantages from this kind of relationship they are likely to sacrifice short-term advantages that arise from market changes. This option of long-term non-contractual relationships based on improved mutual trust and commitment is often overlooked in the discussion about alternative coordination systems in food supply chains.

The new requirements of meat quality (*e.g.* new European food hygiene regulations coming into force in 2006) enhance the necessity to introduce more reliability into the rather conflictual relationships between farmers and slaughterhouses. Better relationships cannot be developed without a minimum of trust (Galizzi and Venturini, 1999). Game theory provides insights into the "functioning" of collaboration. As long as the game is definitely limited to a certain number of transactions, none of the players will cooperate. Only if there is a perspective for infinite collaboration, a "tit for tat"-strategy becomes interesting for the players, where one player does in one round exactly what the other did in the round before. By this, confidence in the other's behaviour can be developed so that the game ends up as a non-zero sum game (Axelrod, 1984). Thus, if one party goes ahead with trust-building measures, and the business partner reacts favourably, stable relationships which are resistant towards incentives for short-term opportunistic behaviour can evolve over time.

Thus, building trust requires one party to make the first step. In our context, trust building instruments have to be taken by the slaughterhouses signalling the farmers that the enterprise is trustworthy and reliable. This requires the commitment of the whole enterprise, not only of those employees who directly communicate with farmers. Thus, internal changes such as introducing a "code of ethics" (Wieland, 1994, 1999) have to be considered in a business environment which is characterized so far by distrust and opportunism.

Trust communication is complex because it is always contested by the danger of being perceived as unbelievable public relations. For that we propose a comprehensive supplier relationship management concept which goes beyond trust communication as it implies new internal management tools (*e.g.* supplier selection and development programs) as well as instruments to enhance vertical transparency, personal bonds, participation, complaint management, *etc.* (Ryder and Fearne, 2003; Stoelzle and Heusler, 2003). Such a strategy is appropriate for those processors which aim at working closer together with their farmers. In particular, target groups for a supplier relationship management are cluster 3 ("cooperation-oriented farmers") and a part of cluster 4 ("contract supporters") which are interested in what has become known as co-opetition without structural bonds.

To resume: Reflections on transaction costs taking into account current developments in the slaughterhouse industry and in quality assurance schemes indicate relative advantages of pork markets with a low degree of vertical coordination over more integrated meat supply chains, which is supported by results of a recent survey in German pork production. We thus argue for more research in the area of non-contractual long-term relationships which are prevalent so far in Germany and some other European countries. Trust management can be a suitable instrument to gain some of the advantages which are usually attributed to contracts. It should be integrated in a broader approach of supplier relationship management.

References

AgraEurope, 2004. Freie Unternehmer brauchen keine Vertraege. 12/2004, 33-34.

Axelrod, R., 1984. The evolution of cooperation. Basic Books.

Bhuyan, S., 2005. An empirical evaluation of factors determining vertical integration in U.S. food manufacturing industries. Agribusiness 21, 429-446.

Boger, S., 2001. Quality and contractual choice: A transaction cost approach to the Polish pork market. Eur. Rev. Agric. Econ. 28, 241-261.

Bogetoft, P. and H.B. Olesen, 2002. Ten rules of thumb in contract design: Lessons from Danish agriculture. Eur. Rev. Agric. Econ. 29, 185-204.

Boston, C., C. Ondersteijn and G. Giesen, 2004. Using stakeholder views to develop strategies for the Dutch pork supply chain. Paper presented at the 14th Annual IAMA Conference, June 12-15, 2004, Montreux.

Cook, M.L., 1995. The future of U.S. agricultural cooperatives: A neo-institutional approach. Am. J. Agric. Econ. 77, 1153-1159.

Den Ouden, M., A.A. Dijkhuizen, R.B.M. Huirne and P.J.P. Zuurbier, 1996. Vertical cooperation in agricultural production-marketing chains, with special reference to product differentiation in pork. Agribusiness 12, 277-290.

Furesi, R., G. Martino and P. Pulina, 2006. Contractual choice and food safety strategy: some empirical findings in Italian poultry sector, 99th EAAE Seminar Bonn "Trust and Risk in Business Networks, February 8-10, 2006, Pre-Print, pp. 475-483.

Gallizi, G. and L. Venturini, 1999. Towards a theory of succesfull vertical cooperation in the food system. In: Gallizi, G. and L. Venturini (eds.): Vertical relationships and coordination in the food systems. Heidelberg, pp. 61-92.

Giraud-Héraud, E., H. Hammoudi and L.-G. Soler, 2005. Retailer-led regulation of food safety: back to spot markets, paper presented at 11th conference of the EAAE, Copenhagen, Denmark, August, pp. 24-27.

Goodhue, R.E., 1999. Input control in agricultural production contracts. Am. J. Agric. Econ. 81, 616-620.

Guo, H., R.W. Jolly and J. Zhu, 2005. Contract farming in China: Supply chain or ball and chain? Proceedings of the Minnesota International Economic Development Conference, University of Minnesota, April 29-30, 2005.

Haley, M.M., 2004. Market integration in the North American pork industries. USDA, Agricultural Economic Research Service, Washington, DC.

Hennessy, D.A., 2005: Slaughterhouse rules: Animal uniformity and regulation for food safety in meat packing. Am. J. Agric. Econ. 87, 600-609.

Hobbs, J.E., A. Fearne and J. Spriggs, 2002. Incentive structures for food safety and quality assurance: An International Comparison. Food Control 13, 77-81.

Hornibrook, S. and A. Fearne, 2005. Demand driven supply chains: Contractual relationships and the management of perceived risk. Paper presented at 2nd European Forum on Market-Driven Supply Chains. Politecnico di Milano, April 5 - 6, 2005.

Katchova, A.L. and M.J. Miranda, 2004. Two-step econometric estimation of farm characteristics affecting varketing contracts decisions. American Journal of Agricultural Economics, 86, 88-102.

Key, N., 2004. Agricultural contracting and the scale of production. Agric. Res. Econ. Rev. 33, 255-271.

Key, N. and W.D. McBride, 2003. Production contracts and productivity in the U.S. pork sector. Am. J. Agric. Econ. 85, 121-133.

Klein, B., R.G. Crawford and A.A. Alchian, 1978. Vertical integration, appropriable rents, and the competitive contracting process. J. Law Econ. 27, 297-326.

Lajili, K., P.J. Barry, S.T. Sonka and J.T. Mahoney, 1997. Farmers' preferences for crop contracts. J. Agric. Res. Econ. 22, 264-280.

Lawrence, J.D., V.J. Rhodes, G.A. Grimes and M.L. Hayenga, 1997. Vertical coordination in the US pork industry: Status, motivations, and expectations. Agribusiness 13, 21-31.

Lawrence, J.D., T.C. Schroeder and M.L. Hayenga, 2001. Evolving producer-packer-customer linkages in the beef and pork industries. Rev. Agric. Econ. 23, 370-385.

Luning, P.A., W.J. Marcelis and W.M.F. Jongen, 2002. Food quality management: A techno-managerial approach. Wageningen Pers.

MacDonald, J.M., J. Perry, M. Ahearn, D. Banker, W. Chambers, C. Dimitri, N. Key, K. Nelson and L. Southard, 2004. Contracts, markets, and prices: Organizing the production and use of agricultural commodities. USDA, Agricultural Economic Report No. 837. Washington, DC.

Makise, K., 2002. Creating special pork for the Japanese market. Advances in Pork Production 13, 233-237.

Martinez, S.W., 2002a. A comparison of vertical coordination in the U.S. poultry, egg and pork industry. USDA, Economic Research Service, Agriculture Information Bulletin No. 747-05. Washington, DC.

Martinez, S.W., 2002b. Vertical coordination of marketing systems: Lessons from the poultry, egg and pork industries. USDA, Economic Research Service, Agricultural Economic Report No. 807. Washington, DC.

Meuwissen, M.P.M., A.G.J. Velthuis, H. Porkeveen and R.B.M. Huirne, 2003. Technical and economic considerations about traceability and certification in livestock production chains. A.G.J. Velthuis, L.J. Unnevehr, H. porkeveen and R.B. . Huirne (eds.), New approaches to food safety economics, pp. 41-54. Kluwer Academic Publishers.

Mighell, R. L. and L.A. Jones, 1963. Vertical coordination in agriculture. USDA, Economic Research Service, Agricultural Economic Report No. 19. Washington, DC.

Peterson, H.C., A. Wysocki and S.B. Harsh, 2001: Strategic choice along the vertical coordination continuum. IFAMR 4, pp. 149-166.

Poppo, L. and T. Zenger, 2002. Do Formal Contracts and Relational Governance Function as Substitutes or Complements? Strat. Manag. J. 23, 707-726.

Roe, B., T.L. Sporleder and B. Belleville, 2004. pork producer preferences for marketing contract attributes. Am. J. Agric. Econ. 86, 115-123.

Ryder, R. and A. Fearne, 2003. Procurement best practice in the food industry: supplier clustering as a source of strategic competitive advantage. Supply Chain Management 8, 12-16.

Schramm, M. and A. Spiller, 2003. Farm-Audit und Farm-Advisory-System: Ein Beitrag zur Oekonomie von Qualitaetssicherungssystemen. Berichte ueber Landwirtschaft 81, 165-191.

Spiller, A., L. Theuvsen, G. Recke and B. Schulze, 2005. Sicherstellung der Wertschoepfung in der Schweineerzeugung: Perspektiven des Nordwestdeutschen Modells. Muenster.

Stoelzle, W. and F. Heusler, 2003. Supplier relationship management. In: R. Bogaschewsky, and U. Goetze (ed.). Management und Controlling von Einkauf und Logistik. Gernsbach, pp. 167-194.

Theuvsen, L. and T. Hollmann-Hespos, 2005. The economics of traceability: A model of investments in tracking and tracing systems in agriculture and the food industry. J. Boaventura Cunha and R. Morais (eds.), Proceedings of the EFITA/WCCA 2005 Joint Conference. Vila Real (Portugal), pp. 914-921.

Traupe, C., 2002. Schlachtschweinevermarktung in Niedersachsen - Stand, Defizite, Entwicklungsmoeglichkeiten. Cuvillier.

Wieland, J., 1994. Organisatorische Formen der Institutionalisierung von Moral in der Unternehmung: empirische Resultate, theoretische Reflexionen. Unternehmen und Organisationen, philosophische Begruendungen, individuelle und kollektive Rationalitaet 2, 11-35.

Wieland, J., 1999. Die Ethik der Governance. Metropolis-Verlag.

Williamson, O.E., 1985. The economic institutions of capitalism. Free Press.

Williamson, O.E., 1991. Comparative economic organization: The analysis of discrete structural alternatives. ASQ 36, 269-296.

Windhorst, H.-W., 2004. Qualitaetssicherung in der Lebensmittelkette – wo liegen die Herausforderungen? In: Dachverband Agrarforschung (ed.), Lebensmittelqualitaet und Qualitaetssicherungssysteme. VerlagsUnionAgrar, pp. 21-33.

Appendix 1. List of variables/statements.

Contr1	Contractual arrangements are only favourable for the slaughterhouses, farmers don't benefit from them at all.
Contr2	Contracts provide me with more planning security.
Contr3	In the long run I will have to sign contracts to finish pork profitably.
Contr4	In my opinion it would be better if farmers engaged in long-term contracts with slaughterhouses.
Contr5	I prefer cooperation with only one processor if he has turned out to be a good business partner.
Free	I don't want to give up my entrepreneurial freedom due to contractual arrangements.
Trust1	I can trust in XY.
Trust2	Please indicate your level of trust with regard to XY: 0 = "slaughterer pulls a fast one on me whenever possible."; 100 = "I can blindly trust in the slaughterhouse".
Com1	I feel committed to XY.
Com2	Even if something goes wrong I stick to XY.
Swit	Please indicate your switching-behaviour: 0 = "I have supplied my pork to XY for a long time already."; 100 = "I very often switch from one buyer to another.

Contracts in agribusiness: A survey in the Hungarian dairy sector

G.G. Szabó and K. Bárdos

Abstract

In some cases, spot markets failure to effectively govern the whole or a part of the marketing channel effectively and contractual relations are gaining more importance. It is especially true of agricultural markets, since these markets became more differentiated and market players are often vulnerable. The aim of our research is to present a theoretically structured framework of contracting arrangements of milk producers based on Transaction Cost Economics' (TCE) predictions and an empirical analysis of the key determinants of governance structure between farmers and dairy processors in Hungary. The source of the research is a theoretical argument based partly on a review of Hungarian and international literature on relevant market channels, economics of contracting and governance structures. This review provided the theoretical determinants of testable propositions. In the second quarter of 2005 we carried out a postal survey among milk producers. A total of 300 questionnaires containing closed and open questions were sent to milk producers; 68 returned questionnaires could be evaluated. The results were structured and electronically converted for applying SPSS-software. The data were analyzed by employing multivariate techniques (cluster analysis, linear regression, multidimensional scaling). We grouped our 7 hypotheses into three sections related to governance structure, contract features, and the bargaining power-cooperation field. We found that two of the hypotheses had to be rejected, the others were confirmed.

Keywords: contracts, dairy sector, governance structure, cooperation, transaction cost economics

1. Introduction: theoretical background and main research aims

Vertical coordination has been an important topic in the agricultural marketing literature since the beginning of the industrialization of agriculture. Recent literature has distinguished two extreme coordination mechanisms: spot markets (external coordination) and vertical integration (internal coordination). Instead of discrete governance structures, Peterson and Wysocki (1997) rank vertical coordination on a continuum that moves from external mechanisms to internal mechanisms with three transitional stages (contracts, strategic alliances, formal cooperation) between two extreme polar forms.

As agricultural markets become more differentiated, open market transactions do not always prove to be the most appropriate form for the exchange of goods. Contractual relations are gaining more importance.

Crises in Hungarian dairy sector can be traced back to the following coordination insufficiencies:

1. *Failure of public coordination means*: owing to permanent problems, the state is forced to intervene by constituting additional legal rules (e.g. decree on loss reducing, etc.), whereas market coordination should prevent failures.

2. *Unsatisfactory level of market coordination processes*:
  - Lack or partial presence of cooperatives, producers' groups and other interest-enforcing and bargaining organizations.
  - Problem of market structure: failures of competition due to dominant and growing bargaining power of retail chains.
  - Due to the effect of high consumer prices on producers' prices, the production is ineffective and the income from dairy production is uncertain. Therefore there is a lack of the necessary level of investments. These factors contribute to a further increase of costs which raises consumer prices, causing a lower level of consumption and so on.

Crises in the Hungarian dairy sector caused by failure of public coordination and unsatisfactory level of market coordination demonstrate how contractual systems work in this situation.

Our research aims are to present a theoretically structured framework of contracting arrangements of milk producers based on predictions of Transaction Cost Economics (TCE) and economics of contracting and to carry out an empirical analysis of the key determinants of governance structure among farmers and dairy processors in Hungary. The main purpose of the paper is to analyze the effectiveness of coordination mechanisms based on the importance of contracting practice and explain the latent dimensions of contract motivation in the EU market environment. Literature on relevant marketing channels, economics of contracting and governance structures gave the theoretical determinants of testable propositions. The following main research questions were asked: how are contracts arranged, what kinds of diversifications exist in contracting practice and what are the driving forces behind the chosen governance structure. Primary importance was given to developing a model framework based on multivariate analysis technique, which enabled us to prove or reject our hypotheses.

The paper is organized as follows: following the introduction, the second section briefly reviews the relevant current literature. The third section focuses on the dairy sector and its special features from the point of view of EU accession and market development are in the focus of third section, concentrating on transactions between milk producers and processors. The analytical framework is set up in section four for empirical analysis, illustrating survey methods, questionnaire design, sampling and responsiveness. Section five presents multivariate analysis with their methodological characteristics, containing descriptive statistical evaluation of the variables applied. This section also presents the main empirical findings. Finally we draw conclusions with their implications for practical applicability and outline the necessity of establishing bargaining power (cooperatives) on behalf of milk producers.

2. Selected literature review

2.1 Transaction cost theory in agricultural economics

Three branches of current literature are relevant as a theoretical background: transaction cost economics (TCE), vertical coordination, and contracting. One can differentiate between two approaches within the transaction cost theory: the governance approach and the measurement approach. The similarity between these two concepts is that both attempt to identify factors that influence the formation of distinct organizations resulting from the development of transaction costs. The literature dealing with the governance approach focuses on the characteristics of transactions, while the measurement approach literature concentrates on the costs of measuring product features.

Transaction cost theory is based on three behavioural assumptions: bounded rationality, opportunism, and risk neutrality. According to Williamson (1985) three contractual features influence the level of the costs accompanying transactions: (1) transaction-specific investments, (2) uncertainty accompanying the transaction, and (3) frequency of transactions. In the case of agricultural products/produce (like milk) the most distinctive product feature is perishability. This implies several contractual risks, for example the opportunistic behaviour of the contracting parties, or the so-called hold-up problem that *"comes up if one contracting party tries to exploit the other party's vulnerability connected to his asset specific investments"* (Royer, 1999: 49).

In the agricultural economy, TCE is basically used for analyzing two issues (*e.g.* Aust, 1997; Hobbs, 1997; Loader, 1997; Banker and Perry, 1999; Boger, 2001a, b): different forms of agricultural organization and the causes of vertical coordination. The second approach studies the different phases of vertical coordination, namely the relations between farmers, processors, as well as wholesalers and retailers that make up the total supply chain (Young and Hobbs, 2000). We dealt with farmers-processors contractual relations in our present research.

Lyons (1996) gives testable hypotheses connected to TCE and examines empirical relevance. As a conclusion of his paper, he states: *"...the formal law of contract is only limited relevance for many on-going business relationships, where reputation is a powerful enforcement mechanism"*. However, he encourages more empirical work *"...to distinguish the roles of expected future trade and social norms as forces supporting reputation."* (Lyons, 1996: 49). The vertical integration, or make-or-buy, decision has been the most extensively studied question in the empirical transaction cost literature (Shelanski and Klein, 1995; Crocker and Masten, 1996).

2.2 Empirical studies on vertical coordination and contracting in agriculture

The application of TCE is becoming more and more popular in empirical studies dealing with vertical coordination in agriculture (Frank and Henderson, 1992; Hobbs, 1997; Boger, 2001a, b; Fertő and Szabó, 2002; Szabó, 2002; Fertő *et al.*, 2004; Montaigne *et al.* 2005).

Hobbs (1997) in her seminal paper analyzes transaction costs variables that have a significant effect on the cattle-breeders' decision of whether to sell deadweight, direct-to-packer or live weight, through a live-ring auction. According to Bash and Davies (1998) foreign direct investments are influencing factors in agribusiness contract choice. With the help of multinomial logit model Boger (2001a, b) analyzes the Polish pig meat market's distribution channels and contracts and identifies the application of contractual forms: none (63.2%), relational contract (23.6%), and neoclassical contracts (13.2%). Fraser (2004) examines wine grape supply contracts used in the main grape growing regions of Australia. Lower-quality regions place a greater reliance on grape quality assessment, while higher-quality regions place greater emphasis on explicit winery involvement and direction vineyard management.

Among papers dealing with vertical coordination issues in Hungarian dairy supply chain, Szabó and Tóth (1998) examine market development and government policy in the milk/dairy sector, giving an insight analysis into the organization of the whole supply chain, including a detailed (marketing) channel mapping. Szabó (1999) analyzes vertical coordination of the dairy farms in Hungary, and he states that it is basically attained via a price mechanism with the help of marketing contracts. Recently, Fertő *et al.* (2005) dealt with the new governing structure of milk/dairy sector and its policy issues regarding small and medium entrepreneurs. Popovics and

Tóth (2005) analyze characteristics of the Hungarian dairy sector and its structural changes and examine price transmission mechanism of the Hungarian milk chain.

2.3 Definition, roles and types of contracts in agriculture

According to MacDonald *et al.* (2004: 3) *"Agricultural contract refers ... to contracts used to arrange for the transfer of agricultural products from farms to downstream users such as processors, elevators, integrators, retailers, or other farms"*. They argue that *"Contracts may be seen as a device to limit price and income risks (risk-sharing approach), or they may be regarded as a means to reduce the cost of using spot markets to arrange transactions (transaction-cost approach)"* (MacDonald *et al.*, 2004). In our study we use the latter, focusing mainly on transaction cost characteristics of vertical coordination and contracts.

Four methods of organizing transactions exist in (US) agriculture: spot (or cash) markets, production contracts, marketing contracts and vertical integration. The main elements of marketing contracts are delivered quantities, product specification and compensations and quality control. They also include a so-called basic price formula. Production contracts contain assignment of responsibilities and products, compensation, contract length and delivered quantities (MacDonald *et al.*, 2004).

Based on the contractual features mentioned above (Williamson, 1985), classical, neoclassical and relational contracts can be distinguished depending on the possibility to renegotiate the whole or at least parts of the contracts. Marketing contracts show the features of classical contracts, but – depending on contract period – neoclassical characteristics can also appear.

The structure of contractual agreements may vary with the objectives of the contracting parties, underlying production relations, and the nature and size of informational and strategic impediments to contract formation and enforcement. As a consequence, theory provides no unifying structure for the specification and testing of contract design hypotheses (Masten and Saussier, 2000).

Contracts can be categorized as informal (oral or handshaking) or formal (written) agreements. If the parties have a strong social capital and shared cultural heritage, as well as a stable legal environment, oral negotiations work fine.

One can distinguish contracts according to their terms. Basically short, medium and long term contracts can be found in agriculture. The length of the contract period depends on the volume of milk delivered, bargaining power of the producer and also the regulation by the state. A longer contract period and tighter collaboration are common in relational contracts. Joskow's study provided evidence that the benefits of contracting vary with the duration of contracts (Joskow, 1987).

3. Brief description of the EU and Hungarian dairy sector

Dairy production is the foremost agricultural activity of the EU, accounting for around 18% of the total value of the Community agricultural output. Since Hungary is a full member of the EU, its milk and dairy products are part of the common market organization. Regulation, which came into force in 1968, is unified and compulsory for each member state. The EU dairy sector can be characterised by close vertical coordination, 66% of sales occur within the framework of some kind of contracts. Sales via cooperatives exceed 70%, but this differs in the member states. Support systems presume a well-operating marketing channel without malfunctions, which should be primarily established by effective coordination among market actors, partly by the cooperation of those with the same interests (*e.g.* cooperatives) and partly by appropriate market operation. Instead of perfect competition, this market shows oligopolistic characteristics.

Import milk has an increasing market share. Most of the milk delivered to processing companies is high quality, but more and more small producers, failing to meet the high quality requirements sell their milk on the black or grey markets. Hungary's EU accession has fundamentally affected the national quota handling and its institutional background. Domestic reference quantity is 1,947,300 t. Nationwide; 90% of the total milk produced within the framework of the milk quota is sold under some kind of contractual arrangements, directly to the processor.

Hungarian milk farms can be classified into three main groups with respect to concentration, technology, number of employees and heads of stock. The basic elements of market organization of milk and milk products are intervention prices, quota system, supplementary state supports, quantitative regulation, producers' loss-reducing support, *etc.* This paper focuses only on the quota, as the main element of regulation as it affects the producers directly; intervention and buying affect them only indirectly through the processors.

More information and detailed analyzes on the Hungarian dairy sector and its market development can be found in Szabó and Tóth (1999), Szabó (1999), Szakály (2003), Mészáros and Popovics (2004), Fertő *et al.* (2005), Popovics and Tóth (2005).

Based on conditions fixed in a certain contract, we distinguish in Hungarian practice between marketing contracts (price, quantity, quality, term of delivery are fixed), production contracts (up to a certain degree they regulate the production process as well), pre-financing or input-supplying contracts (buyer provides seeds, *etc.*) and general or so-called framework contracts (containing only general data, so they can be seen as a declaration of intent).

Because of the circumstances in Hungary (lack of trust and bargaining power on behalf of the producers), most of the agreements are written contracts. Nevertheless, processors often arbitrarily change the terms of contracts using their power and cause hold-up problems for the producers, who have relation-specific investments. Establishing producers' organizations (producers' groups, cooperatives) can be a solution and it may allow them to increase their bargaining power in a number of cases, at least as a marketing tool.

4. Hypotheses

We grouped our 7 hypotheses into the three areas of governance structure, contract features and bargaining power-cooperation field. We tried to explain the complexity of different contracting practices, and the latent and/or apparent motivation behind contractual arrangements, in order to discover how restrictive factors affect governance structures. Hence we posed the question of whether the whole sample can be divided into two or more homogeneous subgroups from the point of governance structure. Based on Table 1, we employed three variables in order to form groups in the case of milk transactions.

Table 1. Predictions on milk transaction features pergovernance structure.

Governance structure	Asset specific investment	Specified selling price	Bargaining power
Spot market	No	No	No
Neoclassical contract	Yes	Yes	No
Relational contract	Yes	Yes	Yes

We assumed the sample could be divided into homogeneous subgroups based on governance structure. So our first hypothesis was:

Hypothesis I: *We can put farmers participating in homogeneous milk transactions into three, significantly different subgroups based on governance structure.*

To test our hypothesis we used a *non-hierarchical cluster procedure*. Confirmatory cluster analysis could determine whether variables related to governance structure are significant in forming groups.

Hypothesis II: *Variables related to governance structure (asset specificity, bargaining power and contract determination) are significant in grouping/categorizing the farmers participating in the survey.*

As a next step, we investigated the characteristics of individual clusters from the point of supply chain relationships.

Hypothesis III: *The reason for changing partners is the same in subgroups as in the total sample.*

The expected effect on contract period of variables affecting the formation of subgroups is summarized in Hypothesis IV:

Hypothesis IV: *One or all of the variables that are significant in the formation of the subgroups significantly influence the contract period.*

We applied a great number of variables in the survey in order to identify the main factors influencing governance structure and contractual features. In this phase we attempted to reduce the number of variables and set up Hypothesis V:

Hypothesis V: *Space composed by the whole set of variables of contracting features can be reduced to two or three dimensions and respondents can be separated on the basis of the reduced dimensions.*

We applied linear regression to test Hypothesis VI, which captures the correlation between contract features and bargaining power:

Hypothesis VI: *Changes in bargaining power can be explained by the variations of contract features.*

We tried to reveal any relationships between bargaining power and the volume of milk sold, hence to show a significance of producer-owned organizations.

Hypothesis VII: *The volume of sold milk has a positive, significant effect on bargaining power.*

5. Empirical research: the survey and the sample

To investigate producer-processor contracting characteristics and to test our hypotheses, a questionnaire was designed and data were collected from Hungarian milk producers from each county. The aim was to obtain a database so that proxy variables could be constructed. Survey data contain nominal, ordinal, interval scales matching the corresponding theoretical indicators best. 1900 members of the Hungarian Dairy Product Council (HDPC) served as a database for a postal survey. We cut the upper and lower 10% of the sample, considering the quantity of the quota. HDPC's members own 75% of the total milk quota. This is distributed among the 1900 members. All the members have some kind of contractual relationship(s) with the processors. The membership of HDPC is made up of joint companies (528) and producers delivering directly to processors (1368). We posted questionnaires to 300 of these members in the second quarter of 2005. After sending out a total of 300 standardized questionnaires, the size of the sample to be evaluated was 68. The questions in the questionnaires were classified into six groups: basic data on the firm examined, characteristics of contract(s) applied, bargaining power (of the producers), (changes of) relationships with trading partners, (specific) investments, and access to information. The preparation of the survey was supported by the Hungarian Dairy Product Council (HDPC). A total of 84 variables were applied. The key measures were statistically evaluated. Table 2 contains the names and descriptions of variables applied in the course of testing the hypotheses, as well as the means, scales and dispersions used.

6. Results of empirical research

6.1 Hypotheses on governance structure

We based our hypotheses related to governance structure on different features of clusters. First, to determine the number of clusters, we applied hierarchical clustering. With this method we tried to reveal hidden structures in the sample on the basis of theoretical considerations, *i.e.* how to organize observed data into meaningful structures. The ranking was done on observed units. The reduction process was stimulated by a dendrogram. On this basis, we could paraphrase our first hypothesis relating the number of sample groups from the point of view of governance structure.

To test our hypothesis we used a non-hierarchical cluster procedure, which mostly includes partitioning methods that create disjunctive clusters. Initial clusters were made by giving the number of the future groups, which was based on the hierarchical cluster method described above. We then tested our hypotheses concerning the number of clusters (Hypothesis I) and the role of employed variables (concerning specific investments, bargaining power and contract determination, Hypothesis II) .

As a first step we employed all three variables in clustering. An F-test revealed the significance of the individual variables in grouping. Table 3 illustrates the results of the F-test and the variables' significance ($p<0.05$). This data confirms our Hypotheses I and II from the subgroup

Table 2. Variables applied in analysis and their statistical characteristics.

Variable's name	Variables composed	Questions included	Unit	Mean	σ
INV_ASS	Specific investment = (INV_ASS + INV_REL) * 2	Have you invested assets specifically in milk production in the last five years?	y:1; n:2	1.29	0.459
INV_REL		Have you invested in market relations specifically in milk production in the last five years?	y:1; n:2	1.50	0.611
BARG_PRICE	Bargaining power = BARG_PRICE + BARG_BENEF + BARG_DELIV	Do you have any influence on selling price?	1-5	1.41	0.717
BARG_BENEF		Is there any possibility to obtain other economic benefits in addition to the price?	1-5	1.57	0.852
BARG_DELIV		Do you have any influence on determining place and time of delivery?	1-5	2.35	1.255
SIZE	SIZE	Size of animal stock of cows	head (cow)	129.6	-
MILK_SOLD	MILK_SOLD	Approx. amount of milk delivered in 2004.	kg	823,626	295,200
F/P JOB	F/P JOB	Owner is full or part-time worker at the farm.	1 or 2	1	1
EMPLOY	EMPLOY	Number of employees at the farm.	head	18.62	-
AGE	AGE	Age of farmer.	year	44.00	17.34
EDUCATION	EDUCATION	Final level of education.	1-7	4.1	2.167
REASONS_SELL	REASONS_SELL	Reasons for selling to purchaser.	see Table 6		
CONTR_VAL	CONTR_VAL	Since when is the contract valid?	1-5	3.57	1.65
DEF_TIM	DEF_TIM	Is the contract time defined?	y:1; n:2	1.03	0.243
CONC_PR	Contract definition = SUM (CONC_PR; CONTR_TIM; PRICE_CHANGE; CONCR_QU; CONCR_DEL)	Is the selling price specified in the contract?	y:1; n:2	1.43	0.498
CONTR_TIM		How long is the contract period?	1:one year; 2: more years	1.18	0.517
PRICE_CHANGE		Does the price change if the quantity is divergent?	y:1; n:2	1.51	0.611
CONCR_QU		Is the quantity to be sold specified?	y:1; n:2	1.07	0.315
CONCR_DEL		Are the days of delivery specified?	y:1; n:2	1.18	0.455

Table 3. ANOVA table on the role of variables in subgrouping.

	F-test	Significance
Contract specifications	24.42	0.095
Bargaining power	61.044	0.000
Specific investments	42.837	0.000

perspectives, *i.e.* three homogenous subgroups can be formed by means of the governance structure's variables.

Table 4 summarizes the main characteristics of the clusters on the basis of variables that describe the features of governance structures.

A relatively low value (yes: low values, no: high values) of the mean of contract specification shows that contracts signed by the members of this group (n=25) contain fixed selling price, fixed quantity of milk and fixed delivery date. Moreover, the highest level of asset- and relation-specific investments (low values) were implemented during the period in question. Those producers are labelled as "high investors". These types of contracts are the most flexible in the whole sample, however these arrangements result in low bargaining power, and this group has to suffer from market risks due to the lack of fixed contractual conditions. This fact can cause hold-up problems (Hendrikse and Veerman, 2001; Royer, 1999), which can be caused by low bargaining power as well. On the other hand, their asset- and relation specific investments might be secured by definitive marketing conditions.

The second cluster (n=23) can be described by a higher value (*i.e.* lower level) of specific investments (yes: low values, no: high values). In terms of contract specification, the agreements do not specify the selling conditions in the majority of the cases. Contract conditions can be formed jointly; their extremely strong bargaining power supports this fact. The group is labelled as "bargainers". Looking at Table 5, clusters show marked differences from a sociological point of view.

Table 4. Statistical characteristics of clusters on the basis of grouping variables.

Variables and statistical indices		Cluster 1 (n=25) High investors	Cluster 2 (n=23) Bargainers	Cluster 3 (n=20) Vulnerable producers
contract specification (price, selling and delivery conditions; min: 4 max: 8)	mean	6.16	6.65	6.9
	median	6	7	7
	dispersion σ	1.312	1.07	1.02
bargaining power (min: 3 max: 15)	mean	4.88	7.52	3.4
	median	5	7	3
	dispersion σ	1.39	1.344	0.885
specific investment (min: 2, max: 8)	mean	3.76	6.43	6.9
	median	4	6	6
	dispersion σ	0.880	1.7	1.02

Table 5. Clusters from a sociological point of view.

	Cluster 1 High investors	Cluster 2 Bargainers	Cluster 3 Vulnerable producers	Whole sample
Size (head of cattle in 2004)	120.44	205.83	53.40	129.6
Milk sold (kg)	811,931	1,295,433	295,082	823,626
Full or part time job? (1=full, 2=part time), median!	1	1	1	1
Number of employees	11.32	38.39	5.00	18.62
Age	41.48	43.17	47.95	44.00
Education (1-7)	4.00	4.91	3.45	4.1

The bargainers cluster sells the biggest quantity of milk, their enterprises have the highest number of employees, and they have the highest degree of education. These facts could contribute to their bargaining power.

High value (low level) of specific investments and extremely low bargaining power characterizes the last cluster (n=20), labelled as "vulnerable producers". Hardly any of the contractual conditions are specified. Since the level of specific investment is very low, their low bargaining power might have been derived from their vulnerability. According to Table 5, they have the lowest level of education, their ages are the highest and their enterprises show the most atomized structure.

The next phase of our investigation was the exploration of selling reasons, within the clusters and in the whole sample. We applied a scale containing five different grades (1: not true, 5: totally true) to measure the reasons for selling to chosen purchasers. In all clusters, a valid contract is the most important reason for selling (Table 6).

Table 6. Reasons for selling per cluster and in the whole sample. Most important selling reasons are italic.

Reasons for selling (1: not true, 5: true)	Cluster 1	Cluster 2	Cluster 3	Whole sample
Habit	3.84	2.70	4.00	3.25
Reliability	4.00	4.30	*5.00*	4.07
Geographical location	4.28	4.35	*5.00*	4.31
Personal contacts	4.12	3.78	*5.00*	3.78
Vulnerability	4.32	3.61	*5.00*	4.04
Valid contract	*4.56*	*4.65*	*5.00*	*4.66*
Favourable price	3.20	2.52	3.50	2.93
Delivery conditions	4.20	3.83	5.00	3.85

Our next Hypothesis (III) to be tested says that the reason for changing partners is the same in subgroups as in the total sample. (1: no partner change, 5: very frequent changes in the last 10 years). After comparing intra-class and interclass means (1st cluster: 1.25; 2nd cluster: 1.57; 3rd cluster: 1.33; whole sample: 1.34) we found that our hypothesis was confirmed, so there is no difference in the frequency of partner changes in the whole sample and in sub-samples.

The expected impact of variables having a role in the formation of subgroups is summarized in Hypothesis IV. First, with the help of linear regression, we examined the fit by the mean of coefficient of determination which shows the tightness of correlation: R=0.114 and R^2=0.013. Considering the value of coefficients of determination, only 1.3% of the variance of contract period is explained by the variables applied. None of the variables reflecting governance structure indicators proved to be significant. Test statistics and parameter estimations indicate (Table 7) that the hypothesis that one or more variables have a significant role in determining the contract period cannot be accepted at 0.05 percent significance level, since none of the variables was proven to be significant.

6.2 Hypotheses in connection with contract characteristics

As previously demonstrated, we applied 84 variables to measure governance structure characteristics and contract features. Since these variables measure the same two or three theoretical concepts, we can safely reduce their number. Through explorative multidimensional scaling, we gained information about the differences between the respondents on the basis of applied contracts. The degree of correspondence between the distances among points implied by MDS is measured (inversely) by a stress function[1]. Our aim is to create an objective scale in a reduced-dimensional space, *i.e.* the representation of the cases in a lower dimensional space than the original, six-dimensional space. In case of two or three dimensions the fit is good:

$$S_{3\ \text{dimensions}} = 0.033$$

$$S_{2\ \text{dimensions}} = 0.001$$

Table 7. Parameter estimation with linear regression.

Variables	Standardized coefficient ß	t-values	Significance
Constant		2.493	0.015
Contract specification	-0.28	-0.222	0.825
Bargaining power	0.107	0.857	0.394
Specific investments	-0.008	-0.066	0.948

Note: Contract period is the dependent variable.

[1] $S = \sum_{r<s} (d_{rs} - d_{rs})^2 / \sum_{r<s} d_{rs}^2$. S<0.05, fit is good, while S>0,20 fit is satisfactory.

The following key variables form the original, six-dimension space and they are involved in MDS: DEF_TIM, CONTR_TIM, CONTR_VAL, CONCR_PR, CONCR_QU, CONCR_DEL (See Table 2).

There are two things to look for in interpreting an MDS picture: clusters and dimensions. Clusters are groups of items that are closer to each other than to other items. Dimensions are item attributes that seem to order the items in the map along a continuum. The axes are, in themselves, meaningless and the orientation of the picture is arbitrary. Figure 1 shows that the groups can clearly be isolated in the new, three-dimensional space, confirming our Hypothesis V, which proposes that space composed of the whole set of variables of contracting features can be reduced to two or three dimensions and respondents can be separated on the basis of the reduced dimensions. (It is rather difficult to illustrate the natural structures in a three-dimensional space, so we illustrated them in two dimensions, see Figure 1.)

We applied linear regression to test our next hypothesis (Hypothesis VI) on the correlation between contract features and bargaining power. It proposes that change in bargaining power can be explained by the variations of contract features. Table 8 summarizes the results.

Since t-values and significance levels show that none of the variables are significant in explaining the variations in bargaining power, we reject Hypothesis VI.

6.3 Hypothesis in connection with bargaining power

On the basis of parameter estimation, one can see that the higher the volume of milk sold, the better the bargaining power is in determining contract conditions (Table 9), so Hypothesis VII has been confirmed.

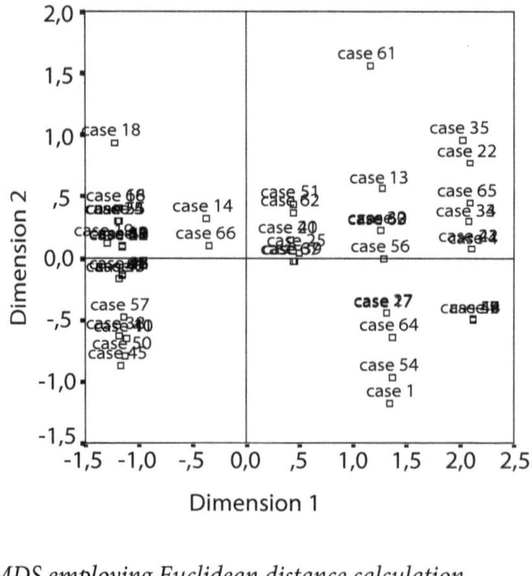

Figure 1. Result of MDS employing Euclidean distance calculation.

Table 8. Parameter estimation by linear regression for the correlation between contractual relationships and bargaining power.

Variable	Standardized coefficient ß	t-value	Significance
Constant		0.991	0.326
DEF_TIM	0.131	1.029	0.308
CONTR_TIM	-0.161	-1.175	0.244
CONTR_VAL	-0.146	-1.184	0.241
CONCR_PR	0.137	1.014	0.314
CONCR_QU	0.160	1.215	0.229
CONCR_DEL	0.046	0.376	0.708

Note: Bargaining power is the dependent variable

We have to point out the role of cooperatives and producer groups in integrating small, individually weak agricultural units (Van Dijk and Mackel, 1994; Ollila, 1994; Bijman, 1998; Hendrikse and Veerman, 2001; Van Bekkum and Van Dijk, 1997; Van Bekkum, 2001, Sykuta and Cook, 2001). The development of countervailing power – even regionally – through the disposal of the milk collected by cooperatives can result in strengthening market competition (*e.g.* 'radiation effect' on prices). This might raise the members' income.

7. Conclusions

Some advanced predictions of contract theory and transaction cost economics have been hypothesized and empirically tested by an overall survey in the Hungarian dairy sector. Different methodological tools in the framework of multivariate analysis provided a qualitative basis to explain the focal research questions. Cluster analysis provides interesting results since we learned that different contracting practices co-exist depending on transaction characteristics in spite of the strictly regulated legal environment.

We found that enhancing economic strength through building up countervailing power also raises the level of bargaining power regardless of the cluster type. Because of attributes, such as asset specificity and uncertainty, the hold-up problem is really significant in the dairy and fruit and vegetable sectors; therefore, the proportion of cooperatives is high in those industries. Producers' organizations, especially dairy cooperatives and producer groups, provide a lot of economic (*e.g.* strengthening bargaining position) and social advantages (*e.g.* securing independence) for

Table 9. Relationship between the volume of sold milk and bargaining power.

Variable	Standardized coefficient ß	t-value	Significance
Constant		12.548	0.000
QUANT	0.285	2.398	0.019

Note: Bargaining power is the dependent variable

producers in a great number of countries. Unfortunately only a few have been set up in Hungary so far and these are still in a very embryonic stage.

The findings show how contracting arrangements can be segmented according to asset-specific investments, bargaining power and price determination. Our results could help some market and state institutions integrate reasonable theoretical assumptions in policy design, with special respect to heterogeneous producers in terms of by transactional background and economic practice. Table 10 summarizes the main hypotheses, methods used and results of the empirical research.

Acknowledgements

Different parts of the research were supported by OKTK (Project No. A/0118/2004), Hungarian Scientific Research Fund, OTKA (Project No. No. F038082 and No. T048779) and Bolyai Janos Research Scholarship. The authors are grateful to the Hungarian Dairy Product Council (Tej Terméktanács), particularly to Erzsébet Bakos for her invaluable help in conducting the survey. An earlier version of the study was presented at the Conference on "Transition in Agriculture - Agricultural Economics in Transition II", at the Institute of Economics, Hungarian Academy of Sciences, October 28-29, 2005. The authors are grateful for the comments and suggestions of the participants, especially the presenter József Tóth, and for the advice provided by Imre Fertő. The authors are also grateful for the comments provided by the participants of the Scientific Seminar, Institute of Economics, Hungarian Academy of Sciences, December 8, 2005.

Table 10. Summary of results and methods employed.

		Hypotheses	Methodology	Result
Governance structure	I.	We can put farmers participating in homogeneous milk transactions into three, significantly different subgroups based on governance structure.	non-hierarchical clustering	✓
	II.	Variables regarding governance structure (asset specificity, bargaining power and contract determination) are significant in grouping/categorizing farmers participating in the survey.	K-means clusters, F-tests	✓
	III.	The reason for changing partners is the same in subgroups as in the total sample.	K-means clusters, comparing means	✓
	IV.	One or all of the variables that are significant in the formation of the subgroups significantly influence the contract period.	linear regression	X
Contract features	V.	Space composed by the whole set of variables of contracting features can be reduced to two or three dimensions and respondents can be separated on the basis of the reduced dimensions.	multidimensional scaling	✓
	VI.	Changes in bargaining power can be explained by the variations of contract features.	linear regression	X
Bargaining power	VII	The volume of sold milk has a positive, significant effect on bargaining power.	linear regression	✓

References

Aust, P., 1997. An institutional analysis of vertical coordination vs vertical integration: The case of the US broiler industry. Michigan State University Staff Paper, June 1997.

Banker, D. and J. Perry, 1999. More farmers contracting to manage risk. Agricultural Outlook, Economic Research Service, US Department of Agriculture, January-February.

Bash, K. and S.P. Davies, 1998. Vertical integration in agribusiness foreign investment. In: J. Royer and R. Rogers (eds.). The industrialization of agriculture: Vertical coordination of the U.S. food system. Aldershot, UK: Ashgate, pp. 295-317.

Bárdos, K., 2004. Vertikális koordináció és tranzakciós költségek: a magyar húsmarha-szektor esete (Vertical coordination and transaction costs: the case of Hungarian beef sector). PhD-értekezés. Budapesti Corvinus Egyetem, Budapest, Hungary, 170 pp.

Bijman, W.J.J., 1998. Internationalisation of European dairy companies: strategies and restrictions. In: G.W. Ziggers, J.H. Trienekens and P.J.P. Zuurbier (eds.), Proceedings of the third International Conference on Chain Management in Agribusiness and Food Industry (Ede, 28-29 May 1998). Wageningen: Management Studies Group, Wageningen Agricultural University, The Netherlands, pp. 769-779.

Boger, S., 2001a. Agricultural markets in transition. Shaker Verlag, Aachen, 277 pp.

Boger, S., 2001b. Quality and contractual choice: a transaction cost approach to the Polish hog market. Eur. Rev. of Agr. Econ. 28, 241-261.

Crocker K.J. and S.E. Masten, 1988. Mitigating contractual hazard: unilateral options and contract length. Rand J. of Econ. 19, 327-343.

Fertő, I. and G.G. Szabó, 2002. The choice of supply channels in hungarian fruit and vegetable sector. Selected paper presented to American Agricultural Economics Association 2002 Annual Meeting, Long Beach, California, July 26-31, 2002, 15 pp. ("AgEcon Search" - http://agecon.lib.umn.edu)

Fertő, I., G.G. Szabó and K. Bárdos, 2004. Transaction cost economics considerations regarding the Hungarian fruit and vegetable sector. Studies in Agricultural Economics 99, 101-114.

Fertő, I., Cs. Forgács, A. Juhász and Gy. Kürthy, 2005. Regoverning markets in the Hungarian dairy sector. IAMO – Forum 2005: "How effective is the invisible hand? Agricultural and Food Markets in Central and Eastern Europe." Halle, 16-18 June 2005, CD-Rom: 40-56

Frank, S.D. and D.R. Henderson, 1992. Transaction costs as determinants of vertical coordination in the US food industries. Amer. J. of Agr. Econ. 74, 941-950.

Fraser, I., 2004. An analysis of wine grape supply contracts in Australia. Paper Presented to the 78th Annual Conference of the Agricultural Society, 2nd – 4th April 2004, U.K.

Hendrikse, G.W.J. and C.P. Veerman, 2001. Marketing co-operatives and financial structure: a transaction costs economic analysis. Agricultural Economics 26, 205-216.

Hobbs, J.E., 1997. Measuring the importance of transaction costs in cattle marketing. Amer. J. of Agr. Econ. 79, 1083-1095.

Joskow P.L., 1987. Contract duration and relationship-specific investment: empirical evidence from coal markets. Amer. Econ. Rev. 77, 168-185.

Loader, R., 1997. Assessing transaction costs to describe supply chain relationship in agri-food systems. Supply Chain Manegement 2, 23-35.

Lyons, B.R., 1996. Empirical relevance of efficient contract theory: inter-firm contracts. Oxford Rev. of Econ. Policy 12, 27-52.

MacDonald, J., J. Perry, M. Ahearn, D. Banker, W. Chambers, C. Dimitri, N. Key, K. Nelson and L. Southard, 2004. Contracts, markets, and prices. organizing the production and use of agricultural commodities. USDA Economic Research Service. Agricultural Economic Report Number 837, 81 pp.

Masten, S.E. and S. Saussier, 2000. Econometrics of contracts: an assessment of developments in the empirical literature on contracting. Revue D'Économie Industrielle 92, 215-236.

Mészáros, S. and P.A. Popovics, 2004. Price transmission and its analysis in the milk and dairy sector. AKII-MTA, Studies of Agricultural Economics 101, 5-21.

Montaigne, E., D. Sidlovits and G.G. Szabó, 2005. Examination of contracting relationships in the Hungarian Wine Industry. 2nd International Conference on Economics and Management of Networks, September 15-17, 2005, CORVINUS University of Budapest, Hungary, 21 pp. (http://www.univie.ac.at/EMNET/download/MontaigneSidlovitsSzabo.doc)

Ollila, P., 1994. Farmers' cooperatives as market coordinating institutions. Annals of Public and Coop. Econ. 65, 81-102.

Peterson, H.C. and A. Wysocki, 1997. The vertical coordination continuum and the determinants of firm-level coordination strategy. Michigan State University, USA, Staff Paper, No. 97-64, 18 pp.

Popovics, P.A. and J. Tóth, 2005. Analysis of price transmission and the asymmetric effect of prices in the hungarian dairy sector. IAMO – Forum 2005: "How effective is the invisible hand? Agricultural and Food Markets in Central and Eastern Europe." Halle, Germany, 16-18 June 2005, CD-Rom: 1-16.

Royer, J.S., 1999. Co-operative organisational strategies: a neo-institutional digest. J. of Coop. 14, 44-67.

Shelanski, H. and P.G. Klein, 1995. Empirical research in transaction cost economics: a review and assessment. J. of Law, Econ., and Org. 11, 335-61.

Sykuta, M.E. and M:L. Cook, 2001. A new institutional economics approach to contracts and cooperatives. Amer. J. of Agr. Econ. 83, 1273-1279.

Szabó, G.G., 2002. New institutional economics and agricultural co-operatives: a Hungarian case study. In: Local society & global economy: the role of co-operatives. S. Karafolas, R. Spear and Y. Stryjan (eds.). Editions Hellin ICA International Research Conference, Naoussa, Greece, pp. 357-378.

Szabó, M., 1999. Vertikális koordináció és integráció az Európai Unió és Magyarország tejgazdaságában. (Vertical coordination and integration in the milk sector in Hungary and the EU) AKII, Budapest, Hungary, Agrárgazdasági Tanulmányok 9, 92 pp.

Szabó, M. and J. Tóth, 1999. Agricultural market development and government policy in Hungary – the case of the milk/dairy sector. Budapest, Hungary, 59 pp.

Szakály, Z., 2003. A magyar tejtermékek választéka, minősége és marketingje az EU csatlakozás küszöbén. (Assortment, quality and marketing of Hungarian dairy products on the eve of EU accession) Tejgazdaság 63, 55-76.

Van Bekkum, O.F., 2001. Cooperative models and farm policy reform. Assen, Van Gorcum, The Netherlands, 231 pp.

Van Bekkum, O.F. and G. van Dijk (eds.), 1997. Agricultural cooperatives in the European Union. Assen, Van Gorcum, The Netherlands, 195 pp.

Van Dijk, G. and C.J., Mackel, 1994. A new era for co-operatives in the European agro-food industries. In: The world of co-operative enterprise 1994. Celebratory Edition. Rochdale Pioneers - 150 Years. The Plunkett Foundation, Oxford, pp.75-85.

Williamson, O.E., 1985. The economic institutions of capitalism. Free Press, New York, USA, 445 pp.

Young, M.L. and J.E. Hobbs, 2000. Public policy responses to increased vertical linkages in agri-food supply chains. Research Discussion Paper No.43. Montana State University-Bozeman; University of Saskatchewan.

Part 3. Horizontal collaboration

Producer groups as a form of entrepreneurship and a tool for strengthening the competitive position of private farms in Poland

M. Adamowicz and M. Lemanowicz

Abstract

The farmer's role in the contemporary market has been changing in Poland as the production of food has increased more quickly than its consumption. Development of computer networks, the transport infrastructure as well as food preservation techniques have caused borders to disappear between the markets of individual regions, countries and continents that make up the global food market. Agriculture in Poland is subjected to a large extent to the rules of the market. Using all marketing possibilities becomes the only way to beat the competition. Opportunities to improve market position should be sought in horizontal integration, which consists of creating producer groups and organizations. Through group activity it is possible to gain a lot of advantages that are difficult to attain individually. The paper presents conditions that contributed to development of agri-producer groups in Poland. The major part of the paper describes results of empirical research conducted in 62 agri-producer groups in Poland. In each group both the leader and individual members were interviewed, leading to 279 interviews in total. The research results confirm that many advantages are achieved by farmers who integrate their activities.

Keywords: producer organizations, agri-producer groups, horizontal integration, competitiveness

1. Introduction

Since 1990 economic conditions in Poland have undergone many changes. Before 1990 food and agricultural policies were subject to a command economy. A system of contracting secured markets for the producers and provided them with limited access to goods.

When the economic system began to change, the free-market economy became the driving force for economic processes and it necessitated efficient management. Individual farms were too weak to meet the requirements of the market economy, leading to a chain reaction: first both the quantity and quality of agricultural production had to be adjusted to fit market demand, which caused a "market shock" among producers, who – until then – had not been used to acting in open markets and operating without fixed prices. Moreover, opportunities to sell surplus products decreased because the state had an insufficient budget to intervene in the market. Responsibility for the results of production decisions was transferred entirely to managers of farms.

The need to strengthen the competitive position of farmers led to the establishment of new forms of organizational cooperation. Horizontal integration of agricultural producers is the way to overcome both cost and sales barriers. The poor negotiating position of individual farms compared to firms in the next level of the distribution channel can be improved through horizontal integration activities. Horizontal functional integration means creating agri-producer groups. By acting in a group, farmers can gain a lot of benefits that would be either very difficult or even impossible to achieve alone, such as a stronger competitive position in the market, improved economic efficiency of agricultural production, better opportunities to create a base for storage, trade and processing. Producers could also benefit if they adapted and developed a more market-oriented approach.

2. Crisis of agri-cooperatives in Poland after 1990

The downfall of agri-cooperatives strongly influenced the situation of agri-producers during the transition period. In the former command economy system, cooperatives held a very strong position, as the system defined for them the sources and conditions of purchasing goods and commodities. It guaranteed low cost credits and defined limits of basic costs such as: salaries, rents and transportation fees. On the other hand, the system also defined the level of sale prices and profit margins. In the past, cooperatives (Dyka, 1998):

- held a dominant position in the sector of trade and services for individual farms (purchasing, supply of means of production, production services);
- held a dominant position (or even a monopoly) in the sector of consumption goods supply for households in villages and small towns;
- had a prominent role in developing food processing and food industry;
- took actions aimed at organising collective agri-production (until 1956 there were even different forms of compulsion);
- granted credits for individual agri cultural enterprises and farms.

The cooperatives' the range of activities was very wide, but they were not capable of taking risk and using marketing tools. This is why they found it difficult to continue their activities in the free-market economy, and all social and economic changes resulted in a considerable decrease in the range of their activities.

The crisis of agri-cooperatives started at the beginning of the 1990s with a drastic decrease in cooperatives' share of business in the retail and catering sectors, in the purchasing of agricultural products, and in the agri- and food processing sector.

The reason for these changes was that agri-cooperatives could not adapt to free-market rules. Employees and members of cooperatives lacked knowledge about free-economy rules and abilities to compete with newly established private enterprises. Moreover, at the beginning of the 1990s new companies enjoyed preferential terms in the market, which made the cooperative crisis even worse. These developments weakened the competitive position of Polish farmers, but the following factors within the farms themselves also affected and continue to limit their position:

- Many small farms produce and sell small batches on the market.
- Seasonal character of products and the lack of storage facilities. Constant demand for agri-products causes either a scarcity or surplus on the market, which leads to fluctuations in prices of agri-products and increases the economic risk taken by an individual farmer.
- Varying quantity and quality in batches.
- Anonymity of suppliers, making it difficult to create groups of regular purchasers.
- Lack of advertisements for agricultural products.
- Weak negotiating position of Polish farmers in relation to wholesale purchasers of their products.
- Lack of differentiation among products sold to the market, e.g. no brands.
- Concentration of sales in one distribution channel, making the producer dependent on one customer only.

3. Concept and functions of agri-producer groups

The present condition of Polish agriculture proves that a return to the methods of group farming is necessary. Individual producers are economically too weak to act in the market as strong actors who can meet their customers' demands in the fields of quality, quantity and marketing. For this reason, farmers have to lose the image they still have of cooperatives, which is based on how cooperatives operated under the previous system. In the past the rules of cooperative democracy that protected the members' right to make decisions affecting the cooperative were not always respected.

The process of creating producer groups started again in Poland at the beginning of 1990s, but did not become more common until the end of the decade when more people understood the necessity of such activities to improve the competitive position of producers. In Polish literature, the terms "producer group" and "marketing group" were not used, nor was any definition of these concepts given, until the "Law of on agri-producer groups and their unions" was passed on September 15, 2000 (Journal of Laws, 2000).

Producer groups and marketing groups were not considered to be legal entities, but were seen as organizations whose main aim is to introduce new commodities produced on members' farms onto the market. They were seen as a way for farmers to expand activities within their own farms.

According to Małysz (1996), a producer group is created in a bottom-up way, voluntarily in order to collectively sell products. Selling jointly allows farmers' groups to improve their bargaining power and consequently obtain higher prices and pay lower prices for purchased agricultural production resources. Farmers operating together have easier access to market and scientific information, external financial sources and also better possibilities to invest. Change can only take place when profits gained can occur be invested in preliminary processing, storing and transport.

In summary, the main aim of a producer group is to sell raw food to a trading or processing company. The most important advantage for a member in such an organization is the possibility to reach agreements for a larger volume of commodities with specified quality requirements. Raw food produced by groups can be collectively purchased, transported and distributed.

A marketing group is interpreted in literature slightly differently. According to Małysz (1996), it is created when a producer group starts its activity outside a "primary market". The first change is a division of one distribution channel into several: it can be a local wholesale market, a regional wholesale market or a commodity exchange. Another basic change that has to take place during this transition is replacement of distribution management with marketing management..

One must also restore the confidence of farmers and other entities in cooperative forms of management. Over the last years we can observe changes in the mentality of farmers, with respect to mutual management. At present Polish farmers are convinced that because of their income and the Polish agrarian situation it is necessary to organise producer groups to the meet market demands and to become active actors. That is why all ideas and forms of strengthening the competitive position of farmers are of such great importance. Improving distribution systems by integrating distribution channels so that a prominent role is played by agri-producers has many economic benefits, most importantly (Karasiewicz, 2001):

- less disparity in negotiating power between the agri-producer and primary wholesaler;
- new opportunities for farmers to enter next stages of market system;
- increased share for agri-producers in the final price of agri-products;
- development of other primary wholesale entities (wholesale markets, commodity markets).

All aforementioned benefits can be gained thanks to functions fulfilled in the agri-market by producer groups. These functions include:

- planning and adjusting production to demand, especially regarding quality and quantity of supplied agri-products;
- concentrating supply of agri-products, which enables farmers to strengthen their position during negotiations with representatives of the next stages of the distribution channel;
- making bigger batches of agri-products;
- implementing quality standards, resulting in higher quality of agri-products;
- creating and promoting proper brands;
- actively selling, i.e. searching for new ways to sell both in the country and abroad;
- participation of capital in establishing new primary market entities, e.g.: wholesale markets and commodity markets;
- a bigger range of provided logistic services connected with storage and transportation of agri-products;
- improving quality and value of agri-products (packaging);
- promoting new production technologies and methods of cultivation that increase management efficiency and are environmentally friendly;
- implementing preliminary processing of agri-products (e.g. mills, abattoirs, processing plants owned by agri-producers);
- concentrating demand for the means of production, such as farming machines, fodders, seeds and fertilisers, thus enabling agri-producers to negotiate better purchasing terms;
- credit and insurance services for members.

4. Legal regulations concerning agri-producer groups in Poland

The law governing the establishment and conduct of producer groups in Poland is titled "Law of the 15th of September, 2000, on agri-producer groups and their unions" (Journal of Laws, 2000).

The law did not create any new organizational or legal form (agri-producer group or producer group). It only defined conditions that must be fulfilled if cooperating farmers want to apply for subsidies from public finances. The phrase "agri-producer group" used in the law relates to any organization whose target is to bring to the market all products made by their members. However, such an organization must be a legal entity. Polish law allows several forms of legal entities, *e.g.* cooperatives, associations and companies.

According to the law, groups can be set up by producers – physical persons running their own farms (in accordance with regulations on agricultural tax) and physical persons running farms in the field of special kinds of agri-production (art. 2). Additionally members must be producers of one specified agri-product or of a group of agri-products. The list of such products is enclosed with an amendment to the "Decree of the Minister of Agriculture and Rural Development of the 3rd of February, 2005 on the list of products and groups of products which can be made by producer groups, minimal annual volume of commodity production and minimal number of members in producer groups" (Journal of Laws, 2005). This is a domestic law, so groups established by its force may not be considered producer groups on the European Union level.

On July 31st, 2005, there were 102 agri-producer groups registered in accordance with the law, of which 24 produced fruit and vegetables, 15 cereals, 22 pigs, 11 tobacco, 7 poultry, 3 milk and 20 other products. There are in fact more groups operating on the market in Poland, but not all of them are registered. According to statistics of the Rural Cooperative Foundation, there are about 400 groups in Poland. Registration requires fulfilling many conditions defined by the aforementioned law and by the Decree of the Minister of Agriculture and Rural Development.

If registered, the groups can get financial support. Since the 1st of May, 2004, subsidies are given to applicants within the PROW Programme (Plan of Rural Areas Development).

Fruit and vegetable producers can establish groups in accordance with either the abovementioned law or the "Law of the 19th of December, 2003, on the organization of the fruit and vegetable market, hop market, tobacco and dried fodder market" (Journal of Laws, 2003).

Tentative or official recognition of producer groups according to the law means that they are also tentatively or officially recognised as producer groups by the law of the EU. These groups can receive the same financial support as producer groups in other member countries of the EU. Until 31 May 2005, there were 30 tentatively recognized fruit and vegetable groups and 9 recognised producer organizations in Poland. The process of registering groups in accordance with domestic law and the law on common organization of fruit and vegetable markets is very dynamic. New producer groups and producer organizations are registered all the time.

5. Research methodology

Empirical research was carried out in 62 agri-producer groups. Five voivodships (administered regions) were selected: Lublin, Świętokrzyskie, Mazovia, Kujawy-Pomerania and Podlasie were selected. Producer groups were established in all of these regions in the early 1990s. Selection was based on fair representation of groups that operate within different commodity markets: fruits, vegetables, cereals, pigs and milk. On producer group was dominant in each voivodship. In each group of the 62 groups, the leaders and generally 4-5 individual members were interviewed, resulting in a total of 279 interviews. At the time of the interviews, the groups were not registered by the provincial government, and thus were not entitled to receive financial support form public resources.

The general aim of the research was to investigate how group actions improve competitiveness of farmers in the fresh produce market. Within this aim, we also identified the following specific objectives:

1. to define benefits and advantages resulting from group actions taken by farmers;
2. to indicate changes that took place on farms of group members, regarding improvement of their competitiveness;
3. to define the competitive position of organised groups of agri-producers in the market and factors that contribute to it.

The research included 29 groups operating on the fruit and vegetable markets, which represented 46.8% of the total number of groups interviewed. Groups of pig producers represented 29% (18 groups), milk producers 12.9% (8 groups), and cereal producers 6.5% (4 groups).

At the time, 400 groups were operating in Poland. Many of them are not registered as agri-producer groups because they find it difficult to fulfil the legal requirements concerning, for example, the volume of commercial production. These groups therefore do not receive any financial support. The average farm area in the surveyed population was 21.3 ha, while in the case of fruit and vegetable farms it was 13 ha (at the same time the average farm area in Poland was about 7 ha).

6. Results of empirical research

6.1 Changes in farm area that took place after farmers joined the group

Counteracting the dispersion of farms and concentration of supply are two conditions that determine the strength of the competitive position of Polish agri-producers. Concentration of supply is the main aim of setting up producer groups, which are an example of functional integration, *i.e.* joined efforts of many producers who would not be able to produce a sufficient volume of goods of the same quality if they worked individually.

The area of farms increased after farmers joined the group in 56% of surveyed households (156 households). The area of arable land did not decrease among any of them, but the other 44% of farmers did not take any action aimed at enlarging their area. On average, farm area increased by 15% after farmers joined the group. In 2001 the average area of a surveyed farm was 21.3 ha, while before joining the group it was 18.53 ha. The area of farms producing different kinds of products changed in different ways (Figure 1).

During the same period, no increase in average farm area was observed among farmers not operating in groups.

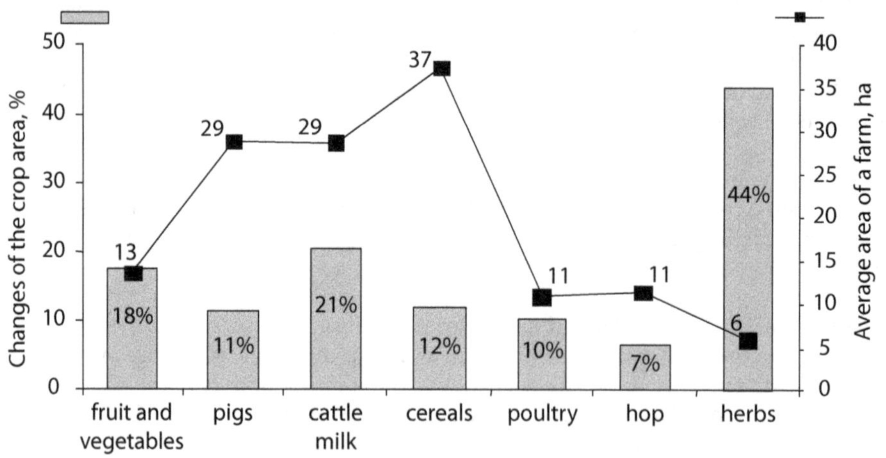

Figure 1. Changes in the area of surveyed farms per sector after joining producer groups (Source: Author's research).

6.2 Changes in the distribution of output after joining

Distribution is mainly aimed at supplying the final purchaser with requested goods at a time and place convenient to them as well as at the best possible price. Thus, distribution means transferring the product from the producer to the consumer through distribution channels. The research allowed us to identify the main changes that took place in distribution channels after farmers joined producer groups. To identify the changes, we calculated the percentage of farmers using individual channels before and after joining. These data enabled us to outline two general trends: farmers are sell their products less in bazaars (retail) and less to agents. These are both positive developments. Selling at bazaars is highly inefficient due to the high costs compared to the profits gained, and eliminating private agents, who concentrated the production by making bigger batches out of the small supplies sold to them by individual farmers, effectively shortens the distribution channel. At the same time, the number of farmers supplying goods directly to processing plants increased by 13.9% and those supplying goods to retail shops by 4.3%.

Figure 2 shows changes in the percentage of farmers using different distribution channels.

Apart from the percentage of farmers using each distribution channel, we calculated the average share of each channel in distribution of the output (Table 1) in order to better illustrate changes in the system of sales. Analysis of data enclosed in the table allowed us to elaborate the hierarchy of distribution channels, based on the share they hold in the sale of output. Direct sale to processing plants dominates as the share of this distribution channel is 55%. Private agents take second place, but their share is only 12.7%, representing a decrease of 50% after farmers joined the group. This demonstrates the strong tendency to shorten distribution channels. Direct sales at bazaars also declined. Before the farmers joined producer groups, bazaars held a share of 9.0% of total sales, which fell to 4.1% after the farmers took collective action. At the same time, sales

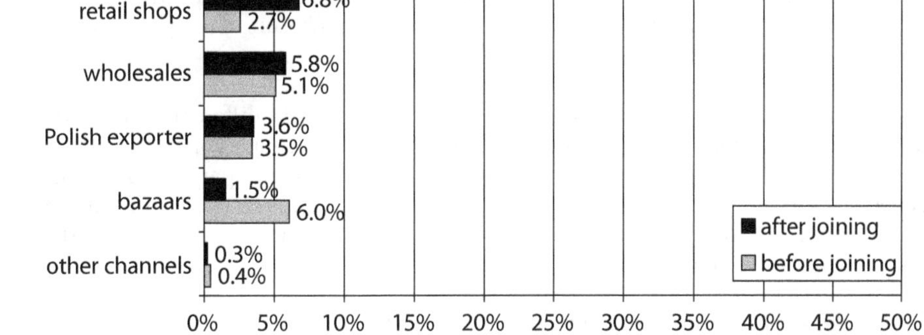

Figure 2. Changes in the percentage of farmers using different distribution channels after joining a producer group (Source: Author's research).

Table 1. Average share of various distribution channels in sale of output before and after farmers joined the group (%).

Specification	Before joining	After joining	Change
Wholesale markets	7.3	10.1	2.8
Wholesalers	4.1	6.0	1.9
Retail shops	3.1	5.7	2.6
Bazaars	9.0	4.1	-4.9
Processing plants	44.0	55.0	11.0
Exporters	5.0	6.4	1.4
Private agents	27.5	12.7	-14.8
Total	100	100	

Source: Author's research.

to wholesale markets, retail shops, wholesalers and exporting companies (excluding processing plants) increased noticeably.

6.3 Contracts between groups and purchasers

Contracts between agri-producer groups and purchasers are a form of vertical integration in agri-business. Contract integration is the most common form of vertical integration in agri-business, because this form does not change ownership relations and is widely accepted by farmers. Growing production risks, price fluctuations as well as increasing competition from other producers encourage the development of contract integration. In Poland contracting is known as a common system of contract agreements that provide the food industry and trade with raw material. It is also a universal way to influence a particular sector of economy. The contracts established in Poland's previously closed economic system did not differ much from those accepted in a market economy, but the contents varied. Depending on their form and range, contracts affect vertical relations to a small or great degree. Casual contracts on the supply of products are not very significant in this respect. Production contracts that include elements of agri-production control play a more important role, as they influence the organising and planning of production, access to investment resources, training for agri-producers and access to credit. Most surveyed groups (72.6%) had contracts with purchasers of their products. Among these groups 69% had yearly contracts and 40% long-term contracts.

Table 2 illustrates the percentage of groups that have contracts with different purchasers, based on the kind of production. Groups producing fruit and vegetables have contracts with different purchasers. Most of such groups have contracts with processing companies (71% of groups), retail chains (33.3%) and wholesalers (29%). Agri-producer groups in this branch are the only ones that operate in foreign markets (19% have contracts with entities in eastern markets and 14% with entities in EU markets).

6.4 Factors determining success of members of the groups

Farmers decided to join producer groups because they wanted to gain benefits they could not gain by acting individually. The surveyed agri-producers listed difficulties in finding markets, the

Table 2. Percentage of groups in various branches that have contracts to supply products (total respondents = 62 groups).

Branch	Whole-sellers	Processing plants	Retail	Exporters	Eastern markets	EU markets	Other
fruit and vegetable	28.6	71.4	33.3	9.5	19.0	14.3	-
pigs	-	92.9	-	-	-	-	7.1
cattle - milk	-	100	-	-	-	-	-
cereals	-	100	-	50.0	-	-	-
poultry	100	-	-	-	-	-	-
hop	-	-	-	-	-	-	-
herbs	-	-	-	100	-	-	-
Total	15.6	80.0	15.6	8.9	8.9	6.7	2.2

Replies do not equal 100% for groups that have contracts with several purchasers at the same time.
Source: Author's research

need to eliminate agents from the distribution chain, high costs of production, *etc.*, as the main reasons for cooperating with other farmers. On the other hand, the main target of their actions was to improve their financial situation.

Results of the research helped to define the main benefits gained as a result of group work and so they are called "success factors". Figure 3 shows how farmers responded to the question of

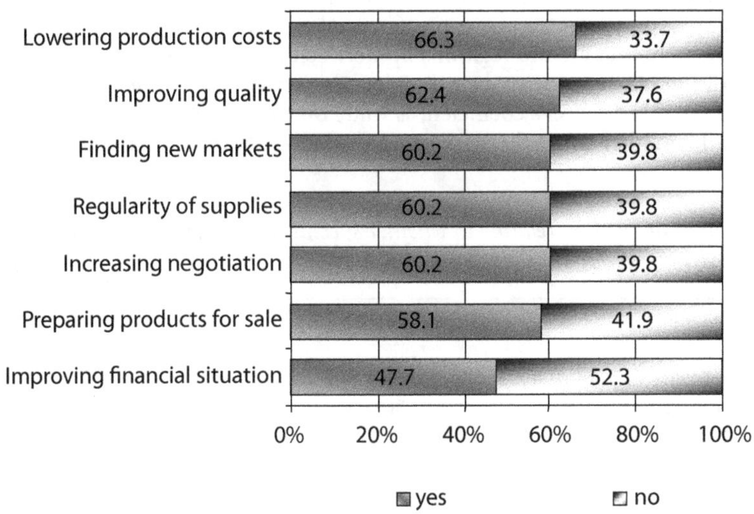

Figure 3. Distribution of answers to questions on whether joining a producers group helped the farmers obtain individual success factors (total respondents = 279 members of groups) (Source: Author's research).

whether joining a producers group helped them gain specific benefits that led to an improvement of their situation.

The factors listed in the figure are not the reasons for, but rather the benefits gained from, joining a producer group. About 48% of farmers interviewed reported that their financial situation had improved, which is clearly one of the most important success factors. The financial situation of farmers who operated alone on the market were more likely to deteriorate rather than improve.

7. Factors determining improvement of competitive position of agri-producer groups

The competitive position of individual producer groups is shaped by their resources and their abilities to use them for building a competitive advantage in the market. Their position in the market depends to a large degree on the situation on the members' farms, as the quantity and quality of the output originate from farm resources. When agri-producers combine output from multiple farms, they also enrich it with other resources or actions (*e.g.* marketing actions), which cannot be undertaken by each member individually. Groups try to use their resource advantage to build up their market advantage.

The research results revealed factors that in the opinion of the groups' leaders determine the strength of the groups' market position.

The majority of the leaders surveyed (62.9%) claimed that the competitive position of their groups became stronger, while the remaining 37.1% claimed that there was no improvement resulting from actions taken by the group in the market. Table 3 shows factors (and percentage of answers) that in the opinion of leaders contributed to the improved market position of their groups.

Based on the leaders' answers, there is thus sufficient reason to believe that joint action does indeed create opportunities to improve the farmers' market position. We used V Cramer's coefficient (Table 4) to analyse the relationship between factors determining improvement of competitive position (independent variables) and improvement of the position (dependent variable). A base for a V Cramer's coefficient is value of a X^2 statistic. It reaches values from an interval [0.1] and is higher when there is stronger relation between examined features. For all

Table 3. Distribution of answers relating to whether specific factors helped improve the competitive position of surveyed groups (total respondents = 62 leaders of groups).

Specification	yes (%)	no (%)
Meeting product quality requirements	69.4	30.6
Taking advice from specialists	50.0	50.0
Purchasing agri-products from other farmers	56.5	43.5
Contracts for goods supply	72.6	27.4
Taking marketing actions	66.1	33.9
Regularity of supplies	69.4	30.6
Meeting product quantity requirements	710	29.0

Source: Author's research

analysed variables probability accompanying X^2 was lower than 0.05, which indicates that these relations are essential from a statistical point of view. Thanks to the use of this coefficient it can be shown that there are statistical relations between a particular operand variable and a dependent variable (improvement in groups' market position).

V Cramer's coefficient allows us to conclude that strong market position was determined mostly by the factor of meeting quantity requirements. Hence concentration of supply, which is one of the fundamental functions of producer and marketing groups, contributes more than other factors to strengthening the market position of the group. Furthermore, it is proof that farmers acting individually will not achieve such effects. That is why individual farms find it difficult to strengthen their market position. Quite a strong relation exists between the factor of regular supplies and contracts with customers and the position of the group in the market.

7.1 Model of logistic regression for strengthening market position of groups

The influence of particular and combined factors on a dependent variable can be determined through logistic regression analysis. The probability of occurrence of phenomenon (for example improvement in competitive position of groups) is a dependent variable in a proposed regression model. This probability depends on the value of a dependent variable, which is measured on a nominal scale (0-no, 1-yes). This coding system makes it possible to estimate probability of phenomenon with different combinations of levels of operand variables.

In the model we identified factors that make the market position of groups stronger.

$$P = \frac{1}{1 + e^{-z}},$$

$$z = \alpha_0 + \alpha_1 X_1 + \alpha_2 X_2 + \ldots + \alpha_k X_k$$

Table 4. V Cramer's coefficient for analysed variables determining stronger market position of the group.

Specification	V Cramer's coefficient	chi-square	Degreesof freedom	Importance
Meeting quality requirements of supplies	0.501	11.5	1	0
Taking advice from specialists	0.431	15.6	1	0
Purchase of agri-products from other farmers	0.605	22.7	1	0
Contract for supply	0.576	20.6	1	0
Marketing actions	0.538	17.9	1	0
Regularity of supplies	0.576	20.6	1	0
Meeting quantity requirements	0.612	23.2	1	0

Source: Author's research

When $\alpha_j>0$, it means that a variable X_j stimulates probability of occurrence of a particular phenomenon. When $\alpha_j<0$, it means that a variable X_j limits probability of occurrence of a particular phenomenon. When $\alpha_j=0$, it means that a variable X_j does not have any influence on probability of occurrence of a particular phenomenon.

Values of coefficients, except information about stimulating or imitating influence on a particular phenomenon, do not have any interpretation. There is a possibility of interpretation when an estimated equation is transformed into a quotient: called a chance coefficient.

The expression e^{α_j} (chance coefficient) has a similar interpretation to a coefficient α_j, except that the threshold value is 1. It also has an additional interpretation. It refers to the relative change in possibility of occurrence of phenomenon as a result of an activity described by a variable X_j, with an assumption of stability of other variables from an equation. The difference (expressed in percentage) between the value of expression e^{α_j} and the number 1 determines percentage increase (decrease) in occurrence of phenomenon, when the particular variable is 1.

The following equation is a product of our calculations:

$$P = \frac{1}{1 + e^{-(-6.8+2.5\,X_1+3.2\,X_2+2.6\,X_3+2.5\,X_4)}}$$

Table 5 shows estimated parameters of the regression model.

The positive value of all $\hat{\alpha}_j$ parameters of the model proves that they have a stimulating impact on the probability of strengthening market position of groups. Individual elements of the above equation have the following meaning for strengthening market position of the group:

$e^{2.5} = 12.1$ indicates that producer groups that are able to supply their products throughout the year increase the probability of strengthening their position in relation to customers by a factor of 11, provided that the impact of other factors remains under control.
$e^{3.2} = 23.8$ indicates that groups that meet quantity requirements with respect to their products increase the possibility of strengthening their position in relation to customers by a factor of 12, provided that the impact of other factors remains under control.
$e^{2.6} = 14.0$ means that groups offering products of the quality required by the customer increase the probability of strengthening their position in the market by a factor of almost 13, provided that the impact of other factors remains under control.

Table 5. Parameters of the regression model for strengthening competitive position of groups.

Independent variable	$\hat{a}_j$	Wald's statistics	Degrees of freedom	Importance	$e^{\hat{a}_j}$
Regularity of supplies	2.5	5.6	1	0.02	12.1
Meeting quantity requirements	3.2	7.5	1	0.01	23.8
Meeting quality requirements	2.6	5.7	1	0.02	14.0
Contracts for supplies	2.5	5.4	1	0.02	12.7
Const.	-6.8	12.8	1	0.00	

Source: Author's research

$e^{2.5} = 12.7$ indicates that groups that have contracts to supply products increase the probability of improving their market position by a factor of almost 12.

8. Summary and conclusions

Potential conflicts between agri-producers and the food industry (related mostly to quality, quantity, time, space and information) can be solved by establishing producer groups. Furthermore, setting up producer and marketing groups is a starting-point for vertical integration activities, giving the groups a possibility to take over the functions of primary wholesalers or even preliminary processing firms. Group support thus improves the balance of power in negotiations between agri-producer and other stages of distribution.

Results of the research allow us to conclude that:

1. Functional horizontal integration activities initiated changes in farms run by group members: these changes increased the competitiveness of the farms.
2. Group activities are mostly undertaken by farmers who own big farms and who are better educated than an average farmer in Poland.
3. Distribution channels changed through horizontal integration: they became shorter, private agents were eliminated, volume of direct sales to processing plants increased, farmers' share in wholesale markets and retail chains increased.
4. Overwhelming majority of groups sell their products based on supply contracts, however, they were mostly one-year contracts.
5. Joint efforts enabled groups to use advertising instruments to create demand for their products.

Results of the analysis also indicate that it is necessary to promote the programme of cooperation within a group and to promote usage of instruments supporting creation and development of horizontal integration activities.

References

Dyka, S., 1998, Co-operativeness. Pub. Warsaw School of Economics.

Karasiewicz, G., 2001, Systems of agri-food commodities distribution on the Polish market. Pub. Warsaw University.

Journal of Laws, 2000. Law of the 15th of September, 2000, on agri-producer groups and their unions. Journal of Laws of October 20, 2000, No 88, item 983.

Journal of Laws, 2003. Law of the 19th of December, 2003, on the organisation of fruit and vegetable market, hop market, tobacco and dried fodder market. Journal of Laws No 223, item 2221.

Journal of Laws, 2005. Decree of the Minister of Agriculture and Rural Development of the 3rd of February, 2005 on the list of products and groups of products which can be made by producer groups, minimal annual volume of commodity production and minimal number of members in producer groups. Journal of Laws No 30, item 258.

Małysz, J., 1996. Integration processes in agrobusiness. Pub. CDiEwR, Poznań.

The role of producer organizations in the process of developing an integrated supply chain; experiences from Quinoa chain development in Bolivia

G. Ton and J. Bijman

Abstract

Producers in developing countries are increasingly becoming integrated in international supply chains. As individual producers are too small to make the connection with (international) consumer markets, collecting and marketing of farm products as well as bargaining with domestic and international customers is done by collective producer organizations (POs). By exploiting economies of scale and reducing transaction costs, POs improve the efficiency and efficacy of agri-food supply chains. In developing countries, the process of supply chain development is often facilitated by state agencies and international development NGOs. These facilitators, however, do not always appreciate the role of the PO in the supply chain, as they do not reckon with the special characteristics of a PO. Being an organization that bases its efficiency on social relations, and having both an economic and an interest representation function, a PO should not be treated as just another commercial chain participant. This paper presents a case study of a chain development process for quinoa in Bolivia, showing how chain facilitators struggle with the special features of POs. The paper suggests that externally facilitated chain development processes will be more successful when the special features of a PO are taken into account.

Keywords: producer organization, co-operative, rural development, collective action

1. Introduction

Producer organizations (POs) are an important instrument for rural development, in their ability to support both economic development and social cohesion. They can also play a major role in national and international agri-food supply chains, as they are an intermediary between often a large number of individual farming households and a relative small number of customers (or suppliers). POs have several economic functions, such as collecting, processing and marketing agricultural products, implementing quality assurance programs, and giving advise and training to their members. By exploiting economies of scale and scope as well as by reducing transaction costs, POs can improve the efficiency and efficacy of agri-food supply chains (Rondot and Collion, 2001). Collective action of producers is often necessary to improve farm gate prices by reducing logistic inefficiencies, reducing transaction costs along the chain, and strengthening producer bargaining power. Initial access to premium price markets, like export markets or national supermarkets, requires a threshold number of producers to pool production and create sufficient volume to meet logistic and transport costs. In addition, producers have to meet the quality requirements of these premium markets. The horizontal and vertical coordination required can be obtained by various forms of chain organization, such as pre-financed contract agriculture, the establishment of a network of buying agents, or self-organization of producers in a producer organization. The necessity to organize producers in supply chains has lead to increasing attention to formal and informal organizational formats of POs in chains, and to analyses of the performance of existing POs in new and emerging supply chain dynamics (Reardon *et al.*, 2004; Hendrikse and Bijman, 2002; Chirwa *et al.*, 2005; Daviron and Gibbon, 2002).

In developing countries, the process of developing an integrated supply chain is often facilitated by actors not directly participating in the chain. Governmental agencies, domestic and foreign development NGOs and large international (donor) organizations often develop and execute programs to assist producers and POs in setting up vertical coordination systems with traders, processors and retailers. However, these externally supported multi-stakeholder programs may not always account for the autonomous goals and ambitions of the PO. External facilitators often look at the development of an integrated supply chain from a 'technical' point of view. The large body of literature on vertical coordination (*e.g.*, Royer and Rogers, 1998), international agri-food chains, and Supply Chain Management (*e.g.*, Lambert and Cooper, 2000) that has been built up over the last two decades, has strongly influenced public and private views on how to develop an efficient supply chain. Most of this literature has a normative character, prescribing the integration of formerly independent processes in order to gain efficiencies in logistics, information exchange and quality control.

Most producer organizations are not only economic organizations; they also have social and political functions. The efficient functioning of a PO is typically obtained by social mechanisms like trust, reputation and collective sanctions. This social network character of a PO, together with its voluntary membership and relative democratic control, makes it into an organization where the management cannot afford to take autocratic decision and just enforce these on the membership.

This paper will first describe the essence of a PO, and the global changes in the political and economic environment that POs have encountered in the last decades. Than, we will turn to our case study on quinoa chain development in Bolivia, to describe and analyse the participation of POs in an externally supported chain development process. The case is about quinoa, a crop grown in the highlands of Bolivia. Coordinated government supports to improve the quinoa supply chain has started in 2001. The case describes the way in which the insufficient account for the special social and political characteristics of the PO affected the multi-stakeholder process of quinoa chain development and lead to contradictions between the chain participants.

2. Producer organizations and its multiple functions

Different types of rural producer organizations have been distinguished based on different criteria like the type of farm that the members hold and the function that the organization has for its members. Another useful distinction can be made between community-based organizations on the one hand and commodity based organizations on the other hand. Community based associations, sometimes called self-help groups, are built around customary principles and ideas of promoting and protecting individual as well as collective well-being. These POs use combine economic, social and political functions, and represent the wider interests of the community in which they are based. Commodity based organizations, like marketing co-operatives, primarily concentrate more on their economic function. This function usually is the marketing (and processing) of products generated by the members' farms. Sometime a PO is established to support the production and marketing of one single cash crop (like the so called single-purpose co-operatives). Often, however, the PO mirrors the diversified farming systems of its members, taking care of marketing (and processing) of the different products coming from the members' farms and provides a number of different services to their members, from the supply of inputs to the marketing and processing of farm products, (like the so called multi-purpose co-operatives).

These PO structure a specific business model within a supply chain, and they have roles and interests that are qualitatively different than the sum of individual interests of their members. It has other interests than individual smallholders. Not the farm gate price is the prime concern, but the margin on commercialization of production. Profit is not their prime objective; instead they look foremost to ways to maximize the turnover of products from their members. Generating profit (or surplus) on economic activities is however a necessity as they struggle to meet ends as a intermediary enterprise: pay staff; pay organizational expenses of member consultations; pay taxes; *etc.* Their commercial activities are mostly realized in straight competition with private actors in the chain (middlemen).

In most developing countries even the single-purpose marketing co-operatives combine social, economic and political functions. The political function contains interest representation vis-à-vis local, region and central government. Where in most developed countries there is a clear distinction between the role of the farmer union - interest representation in the political arena - and the role of the co-operative - strengthening the economic position of the farmers by joint bargaining and marketing - these roles are usually combined in POs in developing countries.

The social character of a PO actually manifests itself in two ways. On the one hand, the PO has a social function within the area from which it draws its members. On the other hand, the PO itself is a social community or social network, where social processes regulate the relationships among members and between members and PO management. Formal institutions, like democratic decision-making and member control, can only be effective if complemented by informal institutions such as norms, trust and collective sanctions. For democratic decision-making and for effective and efficient member control, member commitment is indispensable. Commitment, particularly over a longer period, cannot be based only on the short term economic performance of the PO (as economic performance may vary over time). Commitment is a social process, determined by such factors as mutual solidarity, communication, and ideology.

This second element of any PO is clearly illustrated by the character of one particular type of PO, the co-operative. The International Co-operative Alliance (ICA) defines a co-operative as *"an autonomous association of persons united voluntarily to meet their common economic, social, and cultural needs and aspirations through a jointly-owned and democratically-controlled enterprise"* (Birchall, 1997: 65). In other words, a co-operative is both an association and an enterprise. It is in the association where the social processes of commitment and solidarity take place, and it is in the enterprise where the economic benefits for the members are gained. This dual character of the co-operative has long been acknowledged (see Draheim, 1955; Van Dooren, 1982), however, the recent emphasis on the economic function of the PO/co-operative may lead to a disregard of the social processes needed for good functioning.

One of the ICA principles (see www.coop.org) refers to the autonomy and independence of the co-operative: *"Co-operatives are autonomous, self-help organizations controlled by their members. If they enter into agreements with other organizations, including governments, or raise capital form external sources, they do so on terms that ensure democratic control by the members and maintain their co-operative autonomy"* (Birchall, 1997: 68). This principle is particularly relevant for POs receiving support from outside parties, such as the state and/or (foreign) NGOs. Any type of external aid to the PO carries the possibility of having to compromise its autonomy.

3. Changing policies and changing markets

The potential of POs to provide economic benefits to its members is partly determined by the economic and political environment in which the organization operates. Two major elements of this external environment are market conditions and public policies. In recent years, the scales of importance between these two has tilted towards market conditions. Global trends in liberalization of policies and markets has lead to increasing competition for all producers, whether in developed or developing countries. As farmers in developing countries are increasingly linked to world markets or at least no longer protected from global competition, POs have a function in supporting their members in developing niche markets or in strengthening their bargaining position vis-à-vis large, often foreign-owned, processors and retailers. POs face the challenge of reconciling the demands of (international) agri-food supply chains with the interests of their members. Many scholars indicate that smallholders, to remain part of the global food economy, will need to link with new and more complex ways of marketing their produce: traditional spot markets are increasingly internationalised or replaced by more complex retail mechanisms (Dolan and Humphreys, 2000; Reardon *et al.*, 2004).

Developing countries have a long history of strong government role in the development of POs (Van Dooren, 1982). POs have been used as an instrument for state development policies in many countries. As such, they have been established top-down, even in situations where potential 'members' were not interested in establishing a PO. Sometimes, POs were more like governmental agencies than autonomous membership organizations. POs have also been used to politically control particular groups or communities. In these POs, membership was not voluntary and control was not really in the hands of the members.

However, with increasing political and economic liberalization, this role of the state is being reduced or at least changed. Structural adjustment policies, often imposed by international lending organizations such as the World Bank and IMF, have induced a profound change in the way governments deal with POs. These changes posed both threats and opportunities of the PO. Formerly, because agricultural POs in many developing countries were strongly patronized by the state and used as a vehicle for attaining development goals, they were functioning in a protected environment. This meant on the one hand that they were granted monopolies, guaranteed prices, tax exemptions, financial assistance and support services, and on the other that they were subject to overbearing state influence and control so that they never became autonomous organizations (ILO, 1995). Structural adjustment affected a lot of these co-operatives, as they had become dependant for their survival on these support policies. As such the deregulation of trade, fiscal and monetary reforms, and currency devaluation led in many countries to a profound crisis in the co-operative sector. However, new opportunities appeared for co-operatives to transform themselves in 'real co-operatives', according to the ICA principles, that is voluntary, democratic member controlled organizations. The opportunities also lie in the new co-operative legislation that many countries have enacted, emphasising the co-operative as a private self-help organization belonging to their members, rather than as state-controlled distribution channels (Birchall, 2003).

The changing role of the state in most national economies, from intervention to regulation, implies that POs that were used to operate in a protected environment are now being exposed to the competitive forces of the world market. Over the last decades, agricultural producers and their organizations in developing countries have experienced a number of changes in their economic environment, posing both threats and opportunities for the sustainability of individual farms and collective organizations. These changes in agri-food markets are related to the more general trend of globalization (Reardon and Barrett, 2000). Growth in downstream components of the agri-food channel and multinationalization have set off significant changes in the agri-food system organization and institutions in developing countries. The most prominent changes include the development of more vertical coordination in supply chains, the rise of contractual exchange in place of spot markets, and increased attention paid to product quality and safety. Markets for agri-food products are becoming less commodity- or producer-driven and more customer-oriented or even buyer-driven (Gereffi *et al.*, 2005). Producers are thus required to deliver a specific product, at a specific quality at a specific place and time. In other words, marketing channels are turning into supply chains. In order to obtain the necessary vertical coordination among (potential) chain participants, many public and private development agencies have established support programs to facilitate the integration of small producers into (international) supply chains.

4. The Bolivia Quinoa supply chain development process

4.1 Bolivia and agricultural competitiveness

From 1999 on supply chain development appeared as the new paradigm for agricultural development in Bolivia, substituting former agricultural policy frameworks. It has been introduced in the Rural and Agricultural Policy under the Banzer Government (MAGDER, 1999) and further developed in the National Rural and Agricultural Strategy of the Mesa Government (MACA, 2004). Earlier policies focussed on production and productive infrastructure, while the new policy has a more integral orientation on developing markets for Bolivian products. This reorientation was motivated by the decline in competitiveness of Bolivian exports in national and international markets. After the 1985 structural adjustment package and the strengthening of regional integration (MERCOSUR), from 1995 onwards, Bolivia's traditional crops increasingly compete with cheap imports, especially in the urban markets. Domestic agricultural production has become less important in meeting urban food demand . Especially the hilly Andean region is loosing its former importance as breadbasket for cities and mining industry (Perez, 2003). However, in the last 30 years the rural population in Bolivia did only decrease in proportional terms, but not in absolute numbers (INE, 2002), and agricultural production still remains an important household strategy for securing livelihood needs in rural areas in a context of household pluri-activity (Zoomers, 1999).

In the globalizing interlinked agricultural markets Bolivian producers face growing competition. Instead of relying on spot markets and national agro-industry as the 'natural' market for their produce, they have to meet quality requirements and efficiencies 'intrinsic to' products imported from countries with more efficient production systems. A study on price formation in import supplied agri-chains and domestic production supplied agri-chains revealed important logistic advantages for import-supplied consumer markets (Prudencio and Ton, 2004). Urban spot market prices are roughly equal for imported or domestically produced food. The higher inefficiencies in domestic intermediary commerce increased transaction costs in the chain and resulted in relative low farm gate prices for Bolivian producers. The producers look for a variety of ways to survive this increasing competition. An important instrument for them is forming economic organizations (Baas and Rouse, 1997; Bosc *et al.*, 2001). These PO are increasingly visible and influential in developing agricultural policies in Bolivia.

4.2 Bolivian producer organizations

Bolivia counts a large number of economic producer organizations, called Organizaciones Economicas Campesinas (OECAs). Ramos (2002) found a total number of 635 organizations spread over entire Bolivia. This broad network of organizations constitute part of the "social capital" supportive for economic development and poverty alleviation (Grootaert and Narayan, 2004; De Janvry and Sadoulet, 2005). The existing network of POs and the embodied organizational experience is considered an important resource for rural development in the Andes region (Bebbington, 1999; Bebbington and Carroll, 2000; Muñoz *et al.*, 2004).

Some of these POs collectively market unprocessed farm output, but many try to realize value added by some form of processing, local branding or targeting premium-price export markets. In 1992, several POs constructed a national platform of organizations having the same 'identity': the National Coordination for Economic Organizations of Peasants – CIOEC. In 2000 CIOEC presented a policy agenda (CIOEC, 2000) during the PRSP National Dialogue, National Dialogue 2000, a nation wide consultation on the use of resources that had come available through debt relief for highly indebted poor countries (HIPC). In that document CIOEC presented the following self-definition of the newly emerging sector:

> *"The OECAs are a cluster of organizations with differing legal modalities: co-operatives, associations and communal enterprises. We have the characteristic to be a peasant organization, as we are indigenous and locally rooted small holders. We have the characteristic of being economic, as we get value added through taking position in the market and increasing production and income in rural areas. And finally, we have the characteristic of being organized through our associated activities." (CIOEC, 2000)*

Many of these PO constitute second level federated structures, like a regional *'central de cooperativas'* or a regional or national association of producers. While some of its economic functions are centred on the grass-root structure, where individual producer pool production and organize services like R&D and credit, the processing and exporting stage is usually prime concern for the federation.

4.3 Bolivian system of productivity and competitiveness

In order to develop ideas and instruments to strengthen agricultural competitiveness, the Bolivian government set up the Bolivian System of Productivity and Competitiveness (SBPC). The SBPC, which started operations in 2000, uses a product oriented chain development approach largely derived from Colombian experiences, concentrating government resources on those products where Bolivia has competitive advantage and market potential in international markets (MIDE, 2002; CIOEC, 2003). During 2001, SBPC selected an initial number of 14 promising supply chains. Research was motivated by the Andean Competitiveness Program, a joint effort of the *Corporación Andina de Fomento* and Harvard University to identify and develop competitive clusters, using the cluster theory developed by Porter (1985, 1998). Two of these clusters were selected for pilot projects: the oleaginous sector (soy beans, sunflower) in the Amazon Basin around Santa Cruz de la Sierra (in the east of Bolivia; see Figure 1) and the quinoa sector in the Highlands area (in the West).

Figure 1. Map of Bolivia (Source: www.sitesatlas.com).

The SBPC development process of a supply chain consists of four stages (MIDE, 2002):

1. Identification and mapping of the supply chain.
2. Organizational development and institutionalization.
3. Elaboration and enactment of a public-private multi-stakeholder agreement called the 'Bolivian Agreement on Competitiveness' (ABC), and
4. Implementation and monitoring the ABC.

Once the process has reached the fourth stage, SBPC considers that the supply chain actors will continue the project cycle of adjusting strategies and actions by monitoring and evaluating ongoing activities, and will have established a collective organization that will negotiate directly with governmental and non-governmental agencies for appropriate chain support measures.

4.4 Quinoa production and trade

Quinoa *(Chenopodium quinoa W.)* is a cereal grown on high altitude in the South American Andes almost exclusively produced by thousands of small producers. An estimated 16,900 households are engaged in quinoa production, grouping an estimated population of 84,500 persons (CIOEC-Quinua, 2000). Bolivia accounts for 46% of total quinoa production, followed by Peru with 42% (Brenes *et al.*, 2001a). As it can only grow in specific geographic and climatic conditions, quinoa production is relatively sheltered from international competition. Most Bolivian exports are informal and non-registered with destination Peru, while registered formal exports go to the USA and Europe. Table 1 present some figures on production and destination of quinoa produced in Bolivia. From 1985 demand for quinoa in developed country markets increased sharply, as did

Table 1. Bolivian quinoa production and destination in 2000 (Source: Brenes et al.*, 2001a).*

Farm output	metric tons	%
Total production	21,851	100
On farm consumption	14,233	65
Marketed surplus	7,618	35
Market destinations		
Marketed surplus	7,618	100
Registered exports	1,534	20
Estimated Non-registered exports	2,800	37
Domestic market	3,284	43

the export from the Andean countries. Bolivia is the country with the largest export to the North, while Peru is the largest consumer market (Jacobsen, 2003).

Detailed data on quinoa exports compiled by Laguna (2005) indicate that in 2004 an estimated 35% of Bolivian registered quinoa exports are realized by two producer organizations, ANAPQUI and CECAOT. The remaining exports are handled by a dozen privately owned exporting companies. Both producer organizations group a total of three thousand quinoa producing households (CIOEC-Quinua, 2000). Market shares for the producer organizations have fallen, but export volumes continue to rise after a serious set back in the 2000-2002 period. ANAPQUI is the National Association of Quinoa Producers, established in 1983 with its area of influence predominantly in the departments Oruro and La Paz. ANAPQUI is a federation of 7 local quinoa associations along Bolivia's immense Highland salt lake *Salar de Uyuni*. CECAOT is the Central of Agricultural Co-operatives Operation Earth, established in 1974 with activities concentrated in Potosí. It groups 13 local co-operatives in the Nor Lipez Province. Both organizations have received support from external donors and have developed their commercial relations to a large extent supported on Fair Trade niche market price premiums, and the development of certified organic quinoa supply chain. Both organizations participate in the national coordination of economic peasant organizations (CIOEC-Bolivia) and in the national platform on organic agriculture (AOPEB).

Figure 2 shows the development in export volume by Bolivia's leading export firms. Between 1995 and 1999 ANAPQUI and CECAOT increased their share in the Bolivian supply of quinoa for exports from 47% to 52, with an estimated export value of US$ 1.6 millions. Despite a serious set-back in exported volume in 2000 and 2001, presently ANAPQUI is still the export leader in Bolivia, though JATARY in only a few years has become their principal competitor in the market. (Brenes *et al.*, 2001a; Laguna, 2005).

4.5 Identification and mapping of the supply chain

During the process of diagnosing the current quinoa chain, a stakeholder mapping has been done (Brenes *et al.*, 2001a). The diagnosis mentions producer organizations as important actors in the quinoa chain. When it refers to their actual performance the study takes a rather critical

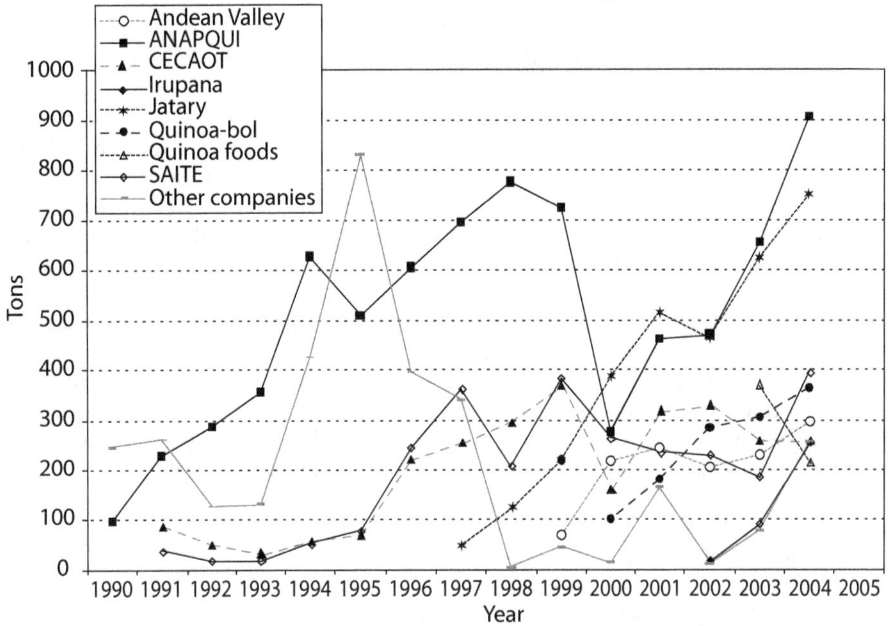

Figure 2. Evolution of volume of Bolivian quinoa exports (metric tons), per producer organization (Source: Laguna, 2005, annex 3).

perspective, echoing the findings of a case study on ANAPQUI (Laguna, 2003) about inefficiencies in commercial and organizational management. Being broad membership organizations with a combination of social and economic objectives, producer organizations are considered to be limited in their ability to react timely to changing market demands and to properly translate customer quality requirements to the primary producers. They openly questioned the viability of POs as processors and exporters in the quinoa chain (Brenes *et al.*, 2001a: 14 and 33). The increased international price for ecological quinoa from 1997 onwards supported the formation of new strategic alliances between recently created national export firms with international capital." *These firms have more management capacity and flexibility to respond to changes in the markets than the producer organizations we mentioned, and are competing them out.*" (Brenes *et al.*, 2001a: 23)

The quinoa cluster diagnosis provided the input for multi-stakeholder workshop discussions to analyse bottle-necks in chain performance. From the outset SBPC tried to position the newly created private export firms as the dynamic actors in chain development and developed pre-proposals in line with this preference for private entrepreneurship. The producer organizations did not agree with this diagnosis. They had developed their own quinoa development strategy during the 2000 National Dialogue (CIOEC-Quinua, 2000) and stressed the importance of POs for farm gate price regulation and for reliable quality control and certification.

"ANAPQUI and CECAOT are the only exporters in Bolivia that have an internal quality control and certification policy. The private firms buy in the Challapata spot market and export them as organic. (..) Other export firms get their quinoa at lower price and use fraudulent certification practices

(mixing organic quinoa with conventional quinoa). Although the producer organizations are not decisive in regulation of farm gate prices, their position in the international market for organic quinoa gives them the possibility to contribute to improving the living conditions of the producers ." (CIOEC-Quinua, 2000)

The quinoa cluster diagnosis has been discussed in several meetings, where it proved difficult to 'bridge' the gap between the two opposing views on the dynamics of the quinoa cluster. The contradictions were difficult to manage, as one of the major private firms was set-up by a former ANAPQUI marketing staff employee that had taken over its commercial contacts with some of its major clients in the United States. During the workshops, the producer organizations kept stressing their historic role in the development of the quinoa chain, and the SBPC kept trying to position the new intermediary firms as prime actors in future quinoa chain. Their leaders of the POs felt increasingly uncomfortable with the chain development dynamics and had difficulties to legitimize the emerging results to their membership (Ton, 1995). It proved difficult for them to make commitments to the other chain actors and support organizations, feeling trapped in a dynamic they could not agree with. The *ex ante* prejudiced image of dynamic and pro-active private firms versus slow and defensive producer organizations became reality.

4.6 Institutional development

SBPC looked for the rapid institutionalization of a chain governance organization, as an inter-professional committee that included all important actors of the quinoa supply chain (MIDE, 2002). The first Committee of Quinoa Competitiveness was elected in 2002 after the first chain analysis workshops in order to negotiate quinoa development support with the government. President of the committee became a representative of the producer organization CEDEINKU, and vice-president became a representative of ANDEAN VALLEY, an important exporting firm. The composition of the committee was quite ideal from a SBPC point of view, as CEDEINKU represented the producers segment and ANDEAN VALLEY represented the (private) trade sector, while governmental support agencies obtained other board functions.

However, from the point of view of the other POs the composition of the board reaffirmed model of division of labour they did not agree with. Governmental agencies and private companies saw them as being producers not traders and exporters. However, both ANAPQUI and CECAOT derived their identity and legitimacy to a large extent from being collective efforts to gain control over more remunerative links in the quinoa chain, adding value to the raw material produced in their local branches, engaging in processing and exporting. Moreover, other POs did not consider CEDEINKU as a good representative. CEDEINKU had been one of the local branches of ANAPQUI but during the chain development process it entered in a direct trading relationship with Andean Valley for the exports of quinoa. The prominence of CEDEINKU in the chain governance organization posed a threat to ANAPQUI, as their example challenged the internal legitimacy of ANAPQUI as specialised marketing channel for its constituency.

During the diagnostic workshops, POs were assigned to subgroups dealing only with issues in production and public infrastructure, leaving processing and exports to subgroups dominated by specialised private players. Instead of supporting the efforts of POs in strengthening vertical coordination between quinoa producers and its (foreign) customers (especially through the Fair Trade system), the SBPC supply chain development approach was biased towards functional segmentation in the chain, thus increasing vertical competition. The existing vertical coordination in the quinoa chain under Fair Trade quality conventions was considered - by the SBPC - a problem instead of a successful business strategy of POs, as shown by the following quotation from Brenes *et al.* (2001b: 15):

> *"The quinoa consultative committee did function well, although some of its participants did not compromise themselves strong enough, because of existing difficulties in the relationship of producers with some of the industrialist and exporters, as a result of the existence of a 'solidarity market' that clearly distorts the free market for quinoa".*

After the first workshops in 2000 and 2001, ANAPQUI and CECAOT gradually became more pro-active actors in the chain development process. They were elected into the presidency of the quinoa chain committee in 2004, when this organization was renamed the National Quinoa Competitiveness Committee. This national committee became composed of two elected regional Committees of Quinoa Competitiveness. The quinoa private firms lost prominence in these committees and established a separate entity formalized in 2005 as the Chamber of Quinoa Exporters.

4.7 The Bolivian Agreement on Competitiveness (ABC) for quinoa

The third step in the chain development process is the elaboration and enactment of a public-private multi-stakeholder agreement called the 'Bolivian Agreement on Competitiveness' (Acuerdo Boliviano de Competitividad, ABC). The quinoa ABC provides the legal framework for further activities in quinoa chain development, as a public-private partnership. It consists of an ' shared vision' by all chain actors and complemented with an activity matrix where responsibilities of public and private actors were defined. Some tangible agreements were the designation of a special day to promote quinoa productions and consumption in Bolivia, and the target of the specific market share of the POs in quinoa exports. The 'shared vision' consisted in joining efforts to get a doubling of yields per hectare in 2012, reaching a production target of 40,000 tons were 50% would be managed through POs. This explicit mentioning of POs in the ABC has been the fruit of their more active participation in the chain development workshops from 2002 onwards.

The discussions during the implementation of the ABC in 2003-2004 were strongly influenced by two proposals of supporting development agencies. The first one consisted in the creation of a special revolving fund that would provide co-investment for a huge centralized processing unit. The proposal suggested new ownership structures, prescribing collective investments (in joint ventures) with only minority shares for the POs. This was interpreted by the POs as an intent to create unequal competition; a project that could compete their plants out and threatened their survival as dominant processor.

A second contentious proposal was a credit support proposal that consisted in pre-financing quinoa production on a village basis, with the requirement of a formal trade agreement between a groups of producers and an exporting firm. The POs were afraid that this credit would erode their membership and loyal supply base. CECAOT and ANAPQUI had invested a lot in creating the social capital of decentralized collection units. To be able to serve a large membership without incurring high pre-financing costs, they relied on trust. They had developed a supply chain in which producers accepted to wait some time before final payment of quinoa deliveries. However, this trust relationship could be eroded by the proposed system of pre-financing through local middlemen. These middlemen generally use advance payments and other patronage mechanisms to bind producers before harvest, while the POs work on the basis of loyal members that wait for (premium) price payment until ANAPQUI has finished transactions with its clients. This increased competition in a context of relative cheap working capital would urge POs to pay their producers in cash at the moment of quinoa delivery to the plant, and sharply increase their financing costs. The proposal would benefit individual producers in the short run, but result in structurally increase transaction costs for the POs active in collection and transformation. These credit proposals were severely questioned by peasant leaders, not in the least because the special features of the social and political reality in Bolivian rural areas from 2000 onwards, where the rural – urban, producer - intermediary dichotomy increasingly structures political discourse.

Interestingly, many constraints in the quinoa chain were not resolved by concerted public private cooperation through the National Quinoa Competitiveness Committee. While implementation of the ABC stranded in lengthy consultancies and bureaucratic obstruction, the POs used another political arena to get results for lowering transaction costs in the quinoa supply chain. ANAPQUI had been an important supporter of a national peasant march in July 2003, when the government responded to demands from the POs in a national pledge of peasant and indigenous organizations. This response included several measures that resolved important constraints in the domestic quinoa market. The most important one had to do with government procurement policies to make POs eligible for the provision of food for nutritional programs, which had until then been the preferential market for firms inscribed in the Bolivian Chamber of Commerce (Mendoza and Ton, 2002). The POs also managed to change the policy for nutritional programs, gaining a niche market by the explicit preference for Andean cereals like quinoa as a component in school breakfasts.

These results of political action 'outside' the formal quinoa chain governance committee reaffirmed their identity of being, next to an economic instrument also an instrument for interest representation. ANAPQUI and CECAOT use their political profile to increase member loyalty in quinoa marketing vis-à-vis local middlemen. These political activities thus strengthened both their position as an economic actor and their recognition as a social and political actor when dealing with support agencies. Late 2005 they also effectively changed the design of the two financing and support interventions described above. Following the changes in the Bolivian political power balance and the rise of MAS, the political party intimately related with the national peasant movements, several support agencies adapted the conditions and procedures for the support schemes to the interests of the POs.

5. Conclusions

The quinoa case is a clear example of the complex and sometimes conflictive interaction of government and (international) development NGO's on the one hand and producer organizations on the other hand. During the process of developing a quinoa chain several issues appeared as controversial, rooted in different perceptions of the role of producer organizations in economic development.

The chain development efforts initially produced the result the facilitators aimed at. A professional organization was established to govern the whole quinoa chain, consisting of private firms, governmental institutes and producer organizations all sitting down to develop a shared perspective on development. However, when specific support activities has to be decided on, conflicting economic interests between the various chain actors came to the surface, and decision making in the inter-professional organization intended for chain coordination became increasingly difficult.

Perhaps many contradictions in the process of developing a quinoa chain could have been prevented if a less mono-functional interpretation of the role of POs would have been used. The initial actor mapping in the quinoa supply chain defined the position of POs merely as representing members in their functional role as producers. However, most POs are vertically integrated into marketing and/or processing, thus substituting for (and thus competing with) private firms providing the same services. If this ambition and development strategy of POs would have been made explicit in the diagnosis phase of the quinoa chain development process, the economic interests of different actors would have been more transparent.

The quinoa case study illustrates two issues that are essential for a correct understanding of the identity and business model of producer organizations in developing countries, and thereby for developing approaches to help small producers strengthen their position in domestic and international markets (Roduner and Gerrits, 2006). First, the function assigned to a producer organization by the chain development facilitators – whether they are government agencies or foreign development NGOs – may be opposed to the function that the PO self-defines and considers as its core identity and ambition. The PO may not agree on a proposed division of labour in the chain that may be 'optimal' for others but threaten their survival and organizational mandate. Many POs strive to vertically integrate into marketing and/or processing, to gain a larger share of the chain value added and to reduce the risks inherent to dealing with private trading and processing companies. Second, POs, in defending the interests of their members, not only bargain with other actors in the supply chain, but also negotiate with governmental bodies to obtain favourable conditions for the production and marketing of their members' products. Some development issues will be discussed directly with chain partners, others will be treated in other arenas and through national political alliances outside the chain development process.

Given the wider responsibility and activity portfolio of a PO, its organizational strength is largely derived from its ability to deliver results to their constituency in different domains, and from the ability of PO leaders to skilfully manage the sometimes conflicting legitimacy demands in these domains. Historically, in Northern Europe and North American the political representation of producer interests and the economic business development has been separated into different organizational formats. Interest representation has been the domain of farmer unions while economic business development has been the task of co-operatives. In most developing countries, however, these two objectives are present within one and the same organization. The loyalty of members towards the PO as an economic organization is influenced by the role and success of the PO in political interest representation. PO leaders do not only respond to the member demands related with chain performance, but also to their social and political demands. Coordination of different domains (Van der Ploeg, 1990)is an inherent feature of a PO, and the organizational strength of a PO is related to the ability of board members to manage sometimes conflicting or competing legitimacy needs.

Multi-stakeholder processes around pro-poor value chain development will have to grasp with the special features of POs as well as with the special position of a PO in the value chain. A PO is a member-controlled organization, where the leadership is democratically chosen and where membership is voluntary. These structural features imply an important role for internal social and political processes (of elections, commitment and accountability). The strategic (re)orientation of the PO is therefore always a compromise between management ambitions and member demands. External facilitators of chain development that only focus on (economic) management ambitions will not get a good picture of feasible solutions. In addition, strengthening the role of a PO in the supply chain will most likely challenge the existing division of labour, as the PO is a instrument for producers to appropriate a larger share of the value added in the total chain.

References

Baas, S. and J. Rouse, 1997. Poverty alleviation: the role of rural institutions and participation. Rome: FAO.

Bebbington, A.J., 1999. Capitals and capabilities: a framework for analyzing peasant viability, rural livelihoods and poverty. World Development 27, 2021-2044.

Bebbington, A.J. and T.F. Carroll, 2000. Induced social capital and federations of the rural poor. Social Capital Initiative Working Paper No.19. Washington DC, World Bank.

Birchall, J., 1997. The international co-operative movement. Manchester, etc., Manchester University Press.

Birchall, J., 2003. Rediscovering the co-operative advantage: poverty reduction through self-help. Co-operative Branch International Labour Office, Geneva.

Bosc, P.M., D. Eychenne, K. Hussein, B. Losch, M-R. Mercoiret, P. Rondot and S. Mackintosh-Walker, 2001. Reaching the rural poor: the role of rural producer organisations in the World Bank rural development strategy, CIRAD-ODI.

Brenes, E.R., K. Madrigal and F. Crespo, 2001a. El cluster de la soya en Bolivia: diagnóstico competitivo y recomendaciones estratégicas. Documentos de Trabajo. Programa de Competitividad, CLACDS-INCAE.

Brenes, E.R., K. Madrigal and A. Condo, 2001b. Reporte final del proyecto: agrotecnología en Bolivia, Documentos de Trabajo. Programa de Competitividad, CLACDS-INCAE.

Chirwa, E., A. Dorward, R. Kachule, I. Kumwenda, J. Kydd, N. Poole, C. Poulton and M. Stockbridge, 2005. Walking tightropes: supporting farmer organisations for market access. ODI Natural Resource Perspectives. No. 99, November 2005.

CIOEC, 2000. Agenda para el Desarrollo Estratégico de las Organizaciones Económicas Campesinas: propuestas para la lucha contra la pobreza y la reactivación económica, La Paz, CIOEC-Bolivia,

CIOEC, 2003. Cadenas de Valor y Agricultura Campesina: interrogaciones al Sistema Boliviano de Productividad y Competitividad. La Paz, CIOEC-Bolivia.

CIOEC-Quinua, 2000. Propuesta del sector quinuero para la cumbre del CIOEC dentro del proceso de Dialogo Nacional, mimeo, La Paz.

Daviron, B. and P. Gibbon, 2002. Global commodity chains and the african export agriculture. Journal of Agrarian Change 2, 137-161.

De Janvry, A. and E. Sadoulet, 2005. Achieving success in rural development: toward implementation of an integral approach. Agricultural Economics 32 (s1), 75-89.

Dolan, C. and J. Humphrey, 2000. Governance and trade in fresh vegetables: the impact of UK supermarkets on African horticulture industry. Journal of Development Studies 37, 147-176.

Draheim, G., 1955. Die Genossenschaft als Unternehmungstyp. Göttingen.

Gereffi, G., J. Humphrey and T. Sturgeon, 2005. The governance of global value chains. Review of International Political Economics 12, 78-104.

Grootaert, C. and D. Narayan, 2004. Local institutions, poverty and household welfare in Bolivia. World Development 32, 1179-1198.

Hendrikse G. and J. Bijman, 2002.On the emergence of new growers' associations: self-selection versus market power. European Review of Agricultural Economics 29, 255-269.

ILO, 1995. Co-operative law and the regulatory role of the state. International Labour Organisation. Meeting Of Experts On Co-operative Law, Geneva, Switzerland)

INE, 2002. Censo Nacional de Población y Vivienda 2001, La Paz.

Jacobsen, S.S., 2003. The Worldwide Potential for Quinoa (Chenopodium quinoa Willd.). Food Review International 19, 167.

Laguna, P., 2003. ¿Capital social o caja de pandora? Contestación y deformación de la acción colectiva en comunidades y organizaciones económicas campesinas de cara a la mercantilización de la quinoa. Ganadores Fondo Minká de Chorlaví 2001, Santiago de Chile.

Laguna, P., 2005. Identificación de necesidades de apoyo al fortalecimiento de la participación de las organizaciones económicas campesinas en la cadena de la quinua boliviana. Agriterra, Arnhem.

Lambert, D.M. and M.C. Cooper, 2000. Issues in supply chain management. Industrial Marketing Management 29, 65-83.

MACA, 2004. Estrategia Nacional de Desarrollo Agropecuario y Rural (ENDAR), Ministerio de Asuntos Campesinos y Agropecuarios, República de Bolivia, La Paz.

MAGDER, 1999. Plan Nacional de Desarrollo Agropecuario y Rural, Ministerio de Agricultural Ganadería y Desarrollo Rural, La Paz.

Mendoza, M. and G. Ton, 2002. Manual de Procedimientos Jurídicos y Tributarios para Organizaciones Económicas Campesinas, Sucre, CIOEC-SNV

MIDE, 2002. Sistema Boliviano de Productividad y Competitividad: perfil del proyecto. La Paz. Ministerio de Desarrollo Económico.

Muñoz, D., B. Cruz and M. Canedo, 2004. Small farmers economic organisation and public policies: a comparative study. La Paz. DFID-IIED.

Pérez, M., 2003. Apertura Comercial y Sector Agrícola Campesino: la otra cara de la pobreza del campesinado andino, La Paz: Centro de Estudios para el Desarrollo Laboral y Agrario (CEDLA).

Porter, M.E., 1985. Competitive advantage, Free Press, New York.

Porter, M.E., 1998. Clusters and the new economics of competition. Harvard Business Review 76 (6).

Prudencio, J. and G. Ton, 2004. Integración Regional y Producción Campesina: La Urgencia de Políticas de Soberanía Alimentaria, La Paz. CIOEC-Bolivia.

Ramos, A., 2002. Guía de OECAs de Bolivia por Municipio: Directorio Nacional Munipalizado de las Organizacioens Económicas Campesinas, PADEER-COSUDE-CIOEC, La Paz.

Reardon, T. and C.B. Barrett, 2000, Agroindustrialization, globalization, and international development. An overview of issues, patterns, and determinants. Agricultural Economics 23, 195-205.

Reardon, T., P. Timmer and J. Berdegué, 2004. The rapid rise of supermarkets in developing countries: induced organizational, institutional, and technological change. In: Agrifood Systems 1, 168-183.

Roduner, D. and A. Gerrits, 2006. The role of donors in value chain interventions: compilation of insights. Online debate value chains in rural development, http://www.sdc-valuechains.ch, AGRIDEA.

Rondot, P. and M.H. Collion, 2001. Agricultural producer organizations; their contribution to rural capacity building and poverty reduction, World Bank.

Royer, J.S. and R.T. Rogers (eds.), 1998. The industrialization of agriculture. Vertical coordination in the U.S. food system. Aldershot, Ashgate.

Ton, G., 1995. Legitimising peasant organisations. Proceedings Agrarian Questions Congress Wageningen, pp. 1721-1735.

Van der Ploeg, J.D., 1990. Labor, markets and agricultural production, Westview Press, Boulder Colorado.

Van Dooren, J.P., 1982. Co-operatives for developing countries. Amsterdam: Royal Tropical Institute.

Zoomers, A., 1999. Linking livelihood strategies to development: experiences from the Bolivian Andes. Amsterdam: Royal Tropical Institute (KIT) and Centre for Latin American Research and Documentation - Centro de Estudios y Documentación Latinoamericanos (CEDLA).

Horizontal cooperation among small and medium-sized supermarkets as a tool for strengthening competitivenes

F.A. Ghisi, D.P. Martinelli and T. Kristensen

Abstract

The lack of research and documentation related to the benefits of a horizontal network as a decisive mechanism for competitive sustainability of small and medium-size supermarkets is notable. This phenomenon has received increasing attention in the last five years and studies have been predominantly of an exploratory nature. This paper explores an important discussion about horizontal cooperation of small and medium-sized supermarkets. It critically analyzes the formation of cooperation, identifying the potential benefits and challenges that influence sustainability. The paper further presents the results of empirical research that can give valuable insights to actors of the grocery industry who are trying to maintain their businesses. The methodology of the study is based on five stages that complement each other, including both quantitative and qualitative research. Findings can also be used by researchers to better understand the essence and advantages of cooperation between small and medium-sized supermarkets and the possible complications inherent in this type of partnering effort.

Keywords: grocery retail, business network, competitiveness

1. Introduction

The Brazilian supermarket sector in the 1990s was characterized by the presence of large national and international chains. The large chains have intensified their domination of the markets throughout the decade of 2000s through a line of acquisitions as a strategy of growing, local and small supermarket companies; trying to enlarge their business at the neighbourhood level; and opening new, smaller stores. The Brazilian supermarket sector also became much more attractive to international chains in the 1990s due to the stability of the Brazilian economy. The sector attracted particular interest from the American, Dutch, Portuguese and French capital chains that gained market share in recent years in Brazil.

With the advent of the new international chains, independent small and medium-sized grocery retailers were disappearing from the Brazilian market. It became difficult for these companies to compete with the large chains that would easily offer lower prices. The independent grocery retailers who decided to maintain their businesses began to build competitive sustainability by unifying their selling efforts with other small companies in a horizontal cooperation agreement. The cooperative arrangement allowed these stores to generate efficiency gains, which led to substantial economic benefits and cost savings. This system seems to be a viable alternative to increasing market power for small and medium stores, which then reverts the market domination of the large chains. The economic arrangement also leads to higher performance - bringing to the firms new ideas and solutions to problems. It is a collaborative way to accomplish shared goals, reduce costs, and exchange experience and results that positively impacts competition.

The horizontal cooperation that was initiated in the 1990s allowed small and medium supermarket stores to secure large purchases to achieve higher volumes, discounts, and improved purchasing power. With the success of this cooperative purchasing effort, companies began to extend their joint actions and diversify other activities as well. For example, the companies started joint training of employees, studying the market and consumer habits, determining commercial aspects related to the sale of products, sharing transportation expenses, doing marketing campaigns and jointly producing private labels.

The new focus of the purchasing framework led to the concept of 'buying groups', which eventually became a 'business network' that exceeded the mere buying actions of companies. The horizontal – or business network- is known in Brazil as one of the most important routes for the survival of the small and medium-sized supermarkets. However, its importance is not exclusive to the supermarkets in Brazil. This model is also found in countries such as Italy, France, Spain, USA and Denmark, as well as in other business sectors such as drugstores, bakeries, and furniture manufactures.

2. Networks: principal concepts

The concept of organizational network presented in Table 1 is inclusive. Among the existent typologies, it can be defined as an organizational network, inter-organizational network, interpersonal network, intra-organizational network, social network, governance network, hybrid arrangement, strategic block, partnership, strategic alliance, joint venture and industrial district. The common characteristic among these concepts is that two companies or partnerships work together in search of a competitive advantage (Bridgewater, 2002). However, Bridgewater stands out with regards to several specific points. For example, the partners (1) know the value of the contribution of each partner; (2) choose the actors involved carefully; (3) know the other partners well, develop their trust and confidence; and (4) understand and respect the individuality of each integrant company of the network.

Ribaut *et al.* (1995) make a reference similar to the network concept, but clearly relate it to the agglomerations of companies that share the objective of strengthening participants' activities without financial bows to each other. Table 1 summarizes some concepts of 'network', considering the selected authors' vision, in chronological order.

Santos *et al.* (1994) present two types of network organizations: the vertical network, in which the relationships happen between a company and its commercial partners (*e.g.* suppliers, distributors, wholesalers and service support), and the horizontal network, in which the relationships among companies that operate in the same industry or market have similar skills, and decide to adopt a network approach to exploit economies of scale. Participants act in the same level of the supply chain. The combination involves purchasing, training, technology sharing, marketing or export development and international trade.

Complementing the literature, O'Donnell *et al.* (2001) point out that most research focuses on vertical network, emphasizing the relationship among supplier-retailer (channels). With regard to the horizontal network, Bengtsson and Kock (1999) point out that horizontal networks do not receive the same kind of attention as vertical network studies; however, the authors recognize that both are equally important in the current global market – a fact also observed by O'Donnell *et al.* (2001). One argument for the focus on vertical networks is related to the fact that economical benefits along the entire supply chain are more perceptible in vertical networks. In the horizontal network, earnings are based more on information and social benefits as well as the constructed relationships among informal players that are not easy to measure (Bengtsson and Kock, 1999; O'Donnell *et al.*, 2001).

Table 1. Concepts of 'network', considering the different selected authors' vision in chronological order.

Authors (Year)	Origin	Term	Network definition
Miles and Snow (1986, 1992)	USA	Network Organizations	Integrate entities which rely on other companies to perform basic design, supply, manufacturing, marketing and/or distribution functions on a contractual and relationship basis
Jarillo (1988)	Spain	Network	Long-term purposeful arrangements amongst distinct but related for-profit organizations that allow the firms in them to gain or sustain competitive advantage vis-à-vis their competitors outside the network
Borys and Jemison (1989)	USA	Hybrid Arrangement	Arrangements that offer a wide range of solutions to such problems as operational inefficiency, resource scarcity, lack of facilities to take advantage of economies of scale
Parkhe (1991, 1993)	USA	Strategic Alliances	Voluntary interfirm cooperative agreements, often characterized by inherent instability arising from uncertainty regarding a partner's future behaviour and the absence of a higher authority to ensure compliance
Lipnack and Stamps (1994)	USA	Network Organization	Exchange relationship between firms that work together independently, based on trust and reciprocity, to achieve common goals
Smith (1996)	Australia	Business Network	Group of companies or individuals connected by collaborative arrangements focus on a defined purpose
Yoshino and Rangan (1996)	USA	Strategic Alliances	Cooperation and competitive alliance between two or more firms involving shared control and continuing contributions to all partners, to gain from one another's market or industry knowledge
Holmlund and Tornroos (1997)	Finland	Business Network	Firm's interaction with other firms that leads to an integration of its resources. They might become interdependent over time as the firms continue to interact. Business networks are evolving social and economic systems that are intrinsically linked to time
Jones et al. (1997)	USA	Network Governance	A core of autonomy firms are densely tied to each other and a periphery whose members have more ties to core members than each other. They create service and products based in implicit contracts that can be adopted conforming the situations
Buono (1997)	USA	Network	Groups of organizations whose boundaries blur together in day-to-day operations - is increasingly portrayed in strategic terms as a powerful tool for firms to strengthen their position in the global economy
Gulati (1998)	USA	Strategic Alliances	Voluntary arrangements between firms involving exchange, sharing, or co-development of products, technologies, or services. They can occur as a result of a wide range of motives and goals, take a variety of forms, and occur across vertical and horizontal boundaries

Table 1. Continued.

Authors (Year)	Origin	Term	Network definition
Porter (1998)	USA	Strategic Alliances	Alliance between firms in response to competitive pressure in order to achieve competitive advantages through low-cost leadership, differentiation or focus strategies
Shaw (1999)	England	Network	Important tool which can improve the marketing effectiveness of entrepreneurial organizations, such as by developing new products, introducing firms to new clients, widening their resource base and improving their pricing structure
Rank and Wald (2000)	Germany	Network	Well-defined, persistent, and structured set of semi-autonomous corporate actors engaged in numerous mutual exchange relationships in order to jointly reach the common network objectives
Gulati et al. (2000)	USA	Strategic Network	Firm's set of relationships, both horizontal and vertical, with other organizations -- suppliers, customers, competitors, or other entities -- including relationships across industries and countries
Cândido and Abreu (2000)	Brazil	Network Organization	Structure, in which can participate companies, who due to the limitations of dimensional, structural and financial order, cannot assure the conditions of survival and development. They are formed by a non rigorous cellular structure and it includes aggregate value activities
Hakansson and Ford (2002)	Norway, United Kingdom	Network	Structure where a number of nodes are related to each other by specific threads. A complex business market can be seen as a network where the nodes are business units and the relationship between them are the threads. Both (the threads and the nodes) have their own particular context

The focus of this study on horizontal networks makes it important to further discuss the contribution of Bengtsson and Kock (1999). The authors subdivide horizontal networks according to the type of prevalent relationship - focusing on the degree of competition/cooperation existent within the network. In each relationship, it becomes possible to identify differences in exchanges, power and dependence relationships among partners, trust, norms, objectives and proximity or distance between them.

The cooperation among actors in the network makes it possible to attend to several needs and objectives that would be difficult to reach individually. It also reconciles mutual interests and propagates the rise of resources and information (Powell, 1990; Amaro, 2000; Premaratne, 2001). A horizontal network has the potential to generate positive synergies and build competitiveness for small and medium-sized companies. According to Premaratne (2001), the formation of the network can help small companies compensate for the lack of resources - money and information, and other non material (social) benefits. It also allows the organizations involved to acquire - through the network - abilities and complementary resources that are essential for their competitiveness and survival in the current global market.

3. Horizontal cooperation of small and medium-sized supermarkets

In accordance with Gonçalves (2000), when the process of internationalization of the retail sector intensified in Brazil, many specialists foresaw the end of small retail companies. They pointed to the difficulty of surviving the new focus of the market and to resisting domination of large supermarket chains, which were the result of joint ventures and acquisitions. Now, small firms face enormous challenges in trying to survive long enough to bring innovations to the market and have recognized that the network is a critical avenue to accessing the resources and competencies needed to survive and succeed.

Along the same line of argument, Marqui *et al.* (2002) point out that this perspective became even more critical in the 1990s with the intensification of the acquisitions of small and medium grocery stores by the large chains. This threatened the stability and longevity of several companies that acted in the retail field and pressured them to find alternatives to deal with this reality and challenge. Among the possible options to change this situation, Gonçalves (2000) emphasizes that small and medium-sized supermarkets had few alternatives that would combine professionalization of management; specialization and knowledge for better understanding the consumer's needs; and associativism. In the absence of a solid system of management, those organizations found in horizontal networking a great alternative, which combined new survival means and managerial practices (Marqui *et al.*, 2002).

The potentials benefits to be gained through this 'joint working force' led to the development of several horizontal networks, which were created by the small and medium-sized supermarkets. Retail stores were one of the pioneers of this partnership form. The success of the first experiences in the retail sector led to the appearance of similar initiatives in areas such as drugstores, construction materials, butcher shops, bakeries and others.

The collaborative effort has now existed for some time. The case of the association of Key Foods supermarkets serves as an important example. Founded 60 years ago in New York, Key Foods gathered 130 medium-sized companies. In France, Inter Marché, an association of small supermarkets consolidated 2,400 stores of 2,100 companies and has more than 60 years of history.

Allevato Junior (2001) argues that in the United States today, associations in the retail field consist of hundreds of companies, and mentions the case of the Iberian Peninsula.

In Brazil, the horizontal networks in the supermarket sector appear as 'buying groups'. The main focus of the horizontal networks was to buy shipper products. The first buying group was formed 24 years ago in the city Vitória in Espírito Santo State in Brazil, in the North of the country. According to De Pinedo Kasper (Termômetro do Varejo, 1998), there were, in 1998, approximately 30 buying groups in Brazil. The first one in São Paulo city appeared in 1994, and its success stimulated similar initiatives in other sectors as drugstores, construction materials, butcher shops, bakeries and others.

The horizontal network allows for the accomplishment of larger-scale purchases and also makes the reduction of costs possible - on average up to 6% (Drumond, 2003). In one of the interviews conducted by a consulting company (ActAsOne Ltd) with the horizontal network called ARCOS (one of the most well-known horizontal networks in Brazil), it is important to consider some aspects before initiating a collaborative joint action with competitors (Allevato Júnior, 2001). Recommendations are listed in the Table 2.

Table 3 characterizes the horizontal networks in the supermarket sector, summarizing the principal benefits and challenges faced in the process of their creation. The table also reflects fundamental actions required from the actors for the sustainability of the network.

4. Methodology

The purpose of research is to provide answers to questions through the application of scientific methods, which are applied to increase the validity of the information to be used in the dissemination of results, ensuring the research is thorough and unbiased. Although research does not always result in trustworthy and unbiased information, the use of scientific methods increases the probability of success (Selltiz *et al.*, 1965). Research is also a meticulous investigation of and search for facts and principles. The research methodology used for this study can be divided into five complementary stages (Figure 1):

Table 2. Recommendations to supermarkets that have interest in initiating a horizontal network.

Select the business partners appropriately because the partnership requires joint work that not many companies are prepared to deal with.
Join the other supermarkets with a clear sense of partnership and be willing to compromise.
Define the politics of performance and the objectives of the horizontal network being created.
Become aware that it is often necessary to change business management styles to something that is more conducive to the larger group.
Be aware that the horizontal network will be an extension of the stores and it will need investments for its operationalization and maintenance.
Select specific professionals to purchases the products and deal with the marketing of the horizontal network. As many of the small supermarkets are family businesses, they have to establish a limit for performance and influence.

Source: Ghisi (2005)

Table 3. Characteristics, barriers, benefits and fundamental actions.

Main Characteristics	Barriers	Benefits	Fundamental Actions
Formed by small and medium-sized companies	To enlarge the focus of the horizontal network. The idea of its formation has to be more than just 'joint buying'	To share some savings in costs, exchange experiences and market knowledge, and get competitive advantages	To enlarge the concept of 'buying group' to 'business group', once they consider other business actions in this process
Companies that joined essentially to buy products. Later they enlarge the main focus for other collaborative actions, such as marketing, training, *etc.*	To adapt the mix of products to each associate's needs and to standardize stores	To reduce costs of distribution and of the prices of the products acquired, and get advantages in the negotiation process (large-scale)	To hire specific professionals to manage the horizontal network, and to establish an internal regiment to facilitate the administration of the association formed
Slow growth, but continuous ripening	Double taxation and high rates paid by the small companies to the government	More competitive price, better mix of products and increase of sales	To adopt technological tools to facilitate process and requests
It is one way to survive for small and medium-sized retail	Operational problems and distrust on the part of the actors involved in the horizontal network	Increase of the credit line to buy products with the suppliers and financiers	To enlarge the communication among the actors, to develop the team spirit
Main objectives are to gain competitiveness and survive in a dynamic market	Conflicts between the actors' interests, the president's laws that often focus on his/her own benefit, and lack of trust between the actors involved	Possibility to improve the training and growth of the professionals' of the company, with reduced costs	To stimulate the actors involved to have a similar managerial culture and convergent objectives
The horizontal network is a national and international trend in several market segments	Difficulty of direct access to the big industries	Larger inclusion of the marketing actions	To defend the associates' interests
Effective strategy to balance the relationship with suppliers and to distribute costs in united operations	Loss of the identity of the own company in favour of the characteristics of the group	Larger visibility in the market	To create a consensus among the actors about what they intend to obtain – focus on objectives
Strategy that allows the invigoration of the small and medium-sized retailers	To create a culture focused on information sharing	Improvement in the actors' administrative management, greater access to new technologies	To stimulate the consensus of the decisions; sharing of information that results in the actors' cohesion

Source: Ghisi (2005)

Stage 1 – Bibliographic and secondary data research

To improve theoretical knowledge of horizontal cooperation, a bibliographic and secondary data research was conducted. The researchers used journals, books, and key retail institutions for secondary data research. It is important to note that even considering the importance of the theme studied, we noticed that there are few references in the international academic literature about supermarket buying groups, or rather, the business networks. We found a large number of material that focuses on networks between firms, especially vertical cooperation, as well as in the retail field, such as concentration of the sector. However, we found little material that focuses on the interrelation between both themes.

Stage 2 – Qualitative Research, exploratory approach

After a review of the literature, the team conducted an exploratory research with three 'business networks' formed by small and medium-sized grocery retailers in Brazil, seeking a practical understanding of the theme; the definition of the methodological stages; and clarification of the problem, propositions, questions and variables of the study. The data was collected through personal interviews. The sample was chosen in consideration of some business networks that act regionally in the state of São Paulo. Additionally, the networks had been formed at least two years earlier, so that they would have the required knowledge for this research. The personal interviews were based on a 20-question semi-structured questionnaire, with some flexibility, as it was an exploratory study. This stage was realized between January and March 2002. A summary of this stage is noted in Table 4.

Stage 3 – Qualitative research

In stage 3, a qualitative study was conducted with six business networks of small and medium-sized grocery retailers. The cases selected were based on the following requisite: business networks formed at least two years earlier that could supply information on and share their experience with the formation process. The data was collected through personal interviews with managers of the six business networks. The six selected business networks were some of the 10 biggest businesses in invoicing according to the Brazilian business network ranking. This ranking was published in

Table 4. Summary of stage 2 methodology.

Unit of analysis	Three (3) business networks of small and medium-sized supermarkets located in the state of São Paulo
Research approach	Exploratory research
Data collected	Data collected: personal interviews Instrument: semi-structured questionnaire
Main objectives of stage 2 research	Empirical understanding of the theme (practical)
	Definition of methodology steps
	Definition of the main problem , propositions, research questions and variables
Realization period	Beginning: January 2002 End: March 2002

the most popular retail magazine in the country – 'SuperHiper'. The idea was to research a few business networks that had aggregate knowledge and know-how in the formation of business networks. The resources/funds to conduct this research were limited, so the team was not able to visit all of the business networks investigated. Data from four networks located more than 300 km away from the city where the study was being conducted were therefore collected by e-mail and telephone. The questionnaire had 26 open-ended questions that focused on the formation of the business network as well as the gains and difficulties the business faced. The data was collected between the first and the third trimester of the year 2003. Table 5 shows a summary of stage 3.

Stage 4 – Quantitative research

A quantitative study with 21 business networks in several Brazilian States was also conducted to analyze the propositions, investigate the correlation among several variables and to describe the characteristics of the business networks. The researchers contacted the Brazilian Retail Committee to have access to all business networks in Brazil. All 150 business networks identified were contacted, but just 21 business networks provided feedback to our numerous tentative contacts. The respondents were the managers and directors of the business networks, the ones who were more involved with the main questions required in this research. The questionnaires were composed of 25 multiple-choice questions and 4 open-ended questions. They were sent via e-mail, and in some cases, the interview team conducted personal interviews with the respondents. Some contacts were made by telephone to gain a better understanding of the respondents' answers. A letter signed by the Brazilian Retail Committee was also sent to all research participants in an attempt to ensure participation and commitment from them. This stage was conducted between the December 2003 and September 2004. The data was tabulated and analyzed using the SPSS statistics program. A summary of stage 4 is presented in Table 6.

Stage 5 – Qualitative research

Finally, at stage 5, qualitative research was conducted to identify within the Danish experiences potential actions that could be adapted to the Brazilian national context and serve as examples of success in this field. The interviews conducted at this stage provided insights and knowledge about the 'business network' in an international context and improved understanding of the

Table 5. Summary of stage 3 methodology.

Unit of analysis	Six business networks of small and medium-sized supermarkets with knowledge and experience in horizontal formation partnerships
Research approach	Qualitative research
Data collected	Data collected through personal interviews, e-mail and telephone Structured questionnaire
Method of research	Case studies
Main objectives of stage 3 research	Identify new elements and dimensions for a better understanding and analysis of correlations for all variables Construct the research instrument to be used in the quantitative research Investigate the operation of specific successful cases
Realization Period	Beginning: January 2003 End: September 2003

Table 6. Summary of stage 4 of the methodology.

Unit of analysis	21 business networks of small and medium-sized supermarkets
Research approach	Quantitative research
Data collected	Data collected through e-mail and telephone Structured questionnaire
Method of research	Survey
Main objectives of stage 4 research	Description of the Brazilian horizontal networks' characteristics, in the supermarket sector Verification of the correlation between all variables
Realization period	Beginning: December 2003 End: September 2004

impacts on the supermarket sector. The contrast between the two countries analyzed – Brazil and Denmark – was relevant to understand the beginning of business networks in a developing country (Brazil) and the issues faced by business networks in a country characterized for a long time by cooperatives (Denmark). The team intends to study other Europeans countries in the future, as there are others business networks in the supermarket sector with know-how and experience in horizontal cooperation in Europe. The University of São Paulo, Brazil, supported travel to Denmark to conduct personal interviews. During the first month in Denmark, the focus was on understanding the Danish retail field. The team made several contacts with professionals, professors and managers to find the companies to be investigated. The interviews were scheduled and conducted in January 2005. Three business networks with the required profile were investigated (personal interviews). A semi-structured questionnaire was elaborated to collect the data during this stage. The questions were similar to the ones applied in stage 3, with minimum modifications. Moreover, a personal interview was conducted with one of the managers of the Retail Institute Scandinavia, which is a retail organization that promotes knowledge of the retail field in Denmark, Sweden and Norway. A specific questionnaire was made to conduct this last personal interview that focused on the trends within business networks in the retail field. A summary of stage 5 is shown in Table 7.

Figure 1 shows the stages of the methodology developed for this study and the proposal of each step.

Table 7. Summary of stage 5 methodology.

Unit of analysis	Three business networks and one retail association
Research approach	Qualitative research
Data collected	Data collected through personal interviews Semi-structured questionnaire
Method of research	Case studies
Main objectives of stage 5 research	Identify successful international activities that can be adapted for the Brazilian context
Realization period	Beginning: December 2004 End: February 2005

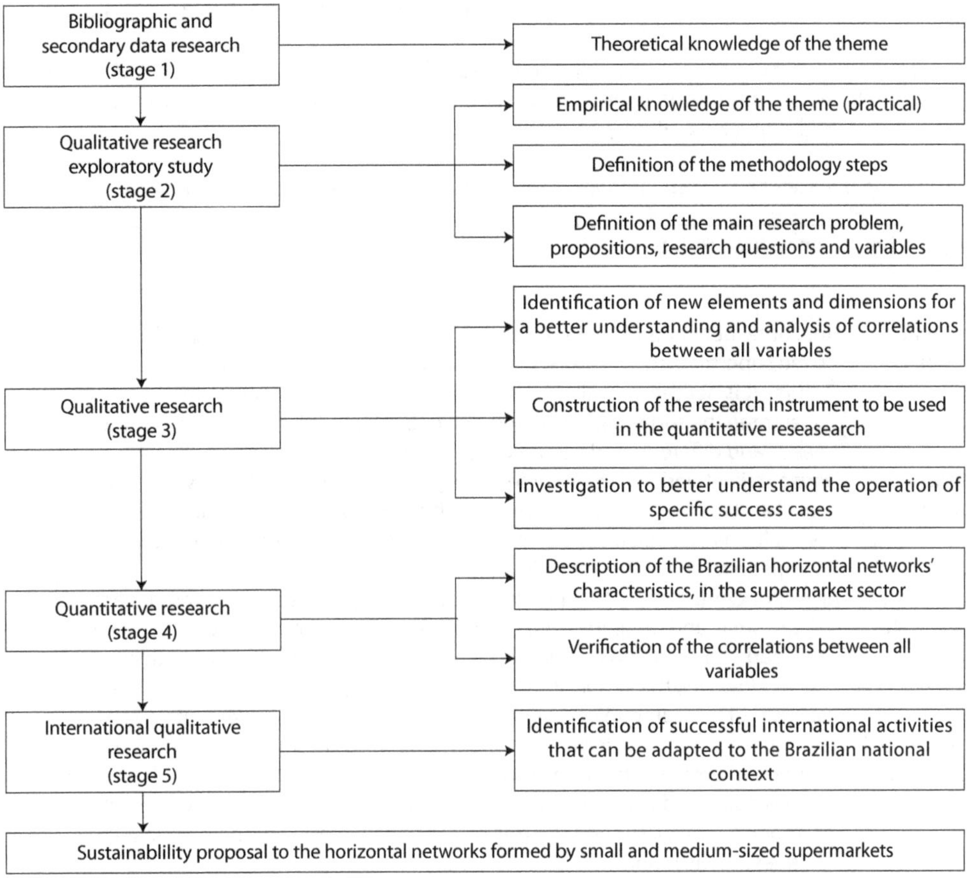

Figure 1. The stages of the methodology of the study.

5. Results of empirical research

This article is a summary of a PhD dissertation developed between the years 2002 and 2005. Since the data collection involved 5 stages, each stage was analyzed separately, and then a consolidated analysis was conducted to have a better idea of the total results. Due to the vast amount of material collected during the interviews and applied questionnaires, the research team opted to use, in this article, an integrated description of the results. A complete discussion of each stage can be found in the dissertation. The idea here is to summarize the data obtained, focusing on providing answers to the main questions of this study – based on the most common answers provided by the managers who participated in the interviews and responded to the questionnaires in each stage. All the respondents of this research received a summary of the study. The proposition was confirmed based on the results of the four research stages conducted (stages 2 to 5). Note that the letter "Q" is used to represent *question* and "Pr" is used to illustrate the *proposition* at the beginning of the research.

Q1: What is the importance of the horizontal network for small and medium-sized supermarkets?
Pr1: Horizontal networks represent an efficient alternative for small and medium-sized supermarkets to maintain their businesses and face competition in the sector, which is characterized by the presence of large companies.

Table 8. Importance of horizontal networks for small and medium-sized supermarkets.

Stage of the research	Summarized results
Exploratory Research	Horizontal networks can be seen as a strategic alternative used by small and medium-sized supermarkets to ensure their sustainability in the supermarket sector. But of course it is not the only alternative they have. Other aspects should be considered to guarantee the supermarkets' competitiveness, such as increased focus on consumer needs and differentiated services.
Qualitative Research	Horizontal networks were discussed as a great alternative to the survival of the small and medium-sized supermarkets, but it should not be considered the only alternative, because as an isolated strategy, it does not guarantee the competitiveness of the actors involved.
Quantitative Research	All respondents believe that horizontal networking is an efficient alternative that allows small and medium-sized supermarket to gain competitiveness and deal with the market dominated by a limited group of big chains that hold a high market share. However, the simple integration of other supermarkets does not guarantee the survival of any company, because it has to be seen as a complementary strategy.
International Qualitative Research	The horizontal network formed satisfies all the needs of small and medium-sized supermarkets to maintain their sustainability. The support offered by the network involves simple discounts of products and also management support. The network is important for all supermarkets involved to gain efficiency and maintain their business in the market.
Proposition	Confirmed. The horizontal network represents an efficient alternative for the small and medium-sized supermarkets to keep their businesses and deal with the concentration of big chains. But this alternative has to be seen as one complementary action among other necessary activities that should be developed by the supermarket owners, focused on the professionalization and differentiation of their services.

Q2: What do small and medium-sized supermarkets gain by being members of a horizontal network?
Pr2: As members of a horizontal network, small and medium-sized supermarkets can buy products at a lower price from suppliers than if they purchased individually.
Pr3: As members of a horizontal network, small and medium-sized supermarkets can have a better mix of products in the stores.
Pr4: As members of a horizontal network, small and medium-sized supermarkets can have access to some suppliers that they were prevented from working with before as a result of the low volume of products acquired.
Pr5: As members of a horizontal network, small and medium-sized supermarkets can increase sales.

Pr6: As members of a horizontal network, small and medium-sized supermarkets can increase invoicing.
Pr7: As members of a horizontal network, small and medium-sized supermarkets can reduce the costs involved in the acquisition of products.

Table 9. Gains obtained by small and medium-sized supermarkets.

Stage of the research	Summarized results
Exploratory Research	The retailers started to buy cheaper products, the mix of products was maximized, the retailers had the same access as the large chains, which was previously not available. The retailers had sales and invoicing increases (sales force) and could reduce their costs (in buying products).
Qualitative Research	The retailers started to buy cheaper products and had the same access to markets as the larger chains. They also benefited from sales and invoicing increases, lower costs and stocks reduction in the stores. They gained market visibility, could expand promotional activities together, standardized the layout in the stores belonging to the group, developed joint training and marketing research, professionalized stores and exchanged market information between the members of the network.
Quantitative Research	The retailers started to buy cheaper products because of the increased volume acquired (in most studied cases, the price reduction was between 5.1% and 10.%). Considering the mix of the products, however, the percentage varied between 5.0% and 25.0%. Most of the researched companies started to buy products from suppliers that they didn't have access to before. They also reported a sales increase (ten networks researched has a sales increase greater than 25.0%). In all cases investigated, they confirmed an invoicing increase (twelve had an increase greater than 15.%). Most of the networks investigated (twelve cases) reported a cost reduction of 5.1% to 15.%.
International Qualitative Research	Based on the gains reported by the members of the network, it can significantly improve conditions negotiated with suppliers, increase sales (even considering the variations in the researched cases, it was always greater than 10%); decrease costs of the acquired products (between 8.0% and 25.%); increase invoicing (between 5.0% and 15.%); increase the mix of products (5.1% to 10.0%); allow joint training of staff, marketing and promotion of products; and promote development of label products, professionalization of the business, sharing of costs, standardization of the stores of the group, integrated logistics and creation of a fidelity card.
Proposition 2	Confirmed. In most cases, the retailers started to buy cheaper products after they became a part of the horizontal network.
Proposition 3	Confirmed. When the retailers extend the number of suppliers that they negotiate with, the consequence is a greater variety of products and commercialized brands in the stores.
Proposition 4	Confirmed. When the retailers start to buy products jointly, the negotiated volume is bigger, and that allows them access to suppliers that they didn't have before (some suppliers just sell products if the retailer buys a minimum volume).
Proposition 5	Confirmed. The sales increase can be related to the cheaper products offered in the stores (the retailer buys cheaper products as part of the network and sells its cheaper too).

Proposition 6	Confirmed. The invoicing increase of the stores is a consequence of the retailer's improved competitiveness. After becoming a member of the horizontal network the retailers' needs are supported by the network.
Proposition 7	Confirmed. The cost reduction is related to the possibility of sharing costs between the members of the horizontal network.

Q3: Which factors influence the success of horizontal networks in a negative way?
Pr3: A lack of cooperation and favouritism among actors in the horizontal network negatively influence its success.

Table 10. Factors that negatively influence the success of the horizontal network.

Stage of the research	Summarized results
Exploratory Research	Lack of involvement and commitment of the actors within the horizontal network, and the existence of individualistic actions that limit joint benefits.
	Difficulty of developing integrated logistics.
	Lack of trust and credibility among some suppliers about the horizontal network.
Qualitative Research	Union of very heterogeneous companies.
	Lack of understanding among the actors that they should be allied and not competitive.
	Lack of standardization among the involved stores of the horizontal network.
	Problems in the structure of the horizontal network that limit their ability to attend to all the retailers' needs.
	Difficulties in professionalizing the horizontal network.
Quantitative Research	Lack of commitment of some retailers to the network.
	Cultural differences among some actors.
	Double taxation.
	Difficult relationship with big industries.
International Qualitative Research	Cooperation and commitment of the actors involved.
	Investments and changes. Some associates resist investing and changing when certain necessary improvements are needed.
	Lack of common purposes and likeness.
Proposition 8	Confirmed. The commitment is, without a doubt, one of the most critical aspects of success in the horizontal networks. Issues with commitment appear in the initial creation of the network, and often stay in the following phases, though to a smaller degree. The problem also persists in the mature networks, as was noticed in the international cases examined by the research team.

The following joint activities are currently carried out by many of the international horizontal networks and could be adopted and adapted to the Brazilian national context.

- Provide managerial service support to the associates, or other consulting services. Recommendations and support services should focus on invigoration of competition among the actors involved.

- Invest in institutional marketing to reinforce the image of the network and assist in integrating it into the market.
- Develop a clear definition of the objectives, responsibilities and planned decisions. This step will assist in developing a common mentality, as differences among the players are not related to purchases, but rather to the relationship channel and integrated marketing.
- Invest in greater volumes of production of private labels that result in improved marginal gains.
- Amplify the number of associates. Some networks should focus on expansion and inclusion of new retailers, which will allow the network to increase bargaining power and increase earnings.
- Establish a contract among the actors in the horizontal network to guarantee greater involvement and commitment of its members.

Q4: Which factors influence the sustainability of the horizontal network?
Pr: - (no propositions)

Table 11. Factors that influence the sustainability of the horizontal network.

Stage of the Research	Results – resume view
Exploratory Research	- (Not approached in this stage of the research)
Qualitative Research	Network efficiency.
	Active participation of network partners.
	Active participation of and commitment from the actors involved in the horizontal network
	The characterization of the horizontal network (it should not act like a 'wholesaler').
	Credibility of the horizontal network in the market.
Quantitative Research	The partner's vision (the retailers have to be entrepreneurs).
	Commitment of the actors to the horizontal network.
	Competitiveness of businesses considering the domination of the larger chains within the market.
	Credibility of the industries in relation to the horizontal network.
	Profitability of the operations accomplished by the horizontal network (still limited).
	Economic situation of the country and stability of the supermarket sector (not very stable).
	High competition in the supermarket sector and price wars practiced by the big chains.
	High fees paid by the horizontal networks.
	Insolvency of some retailers involved in the horizontal network.
	Benefits to society of the horizontal networks .
	Professionalism of the horizontal network.
	Fiscal responsibilities of the members
	Pro-active posture of the members (some just react to the problems and believe that the horizontal network is going to resolve all their problems).
	Standardization of the logistics is fundamental. Lack of standardization makes the delivery of goods difficult.
	The partner's vision about the need of technological innovation and modernization (not always desired but in many situations it is necessary).

	Volumes of purchases (a low number of members in a horizontal network often limits the earnings).
	Migration of the consumers to big chains, to find better offers and prices.
International Qualitative Research	Common objectives, responsibilities and planned decisions. The members of the horizontal networks need to have common objectives, to 'speak the same language' and to follow the same direction.
	Competitive prices. The horizontal network should always negotiate competitive prices with suppliers and maintain the low prices to their customers.
	Keep offering to the consumers what they need. It is necessary to develop the culture among the actors of the horizontal network that the customer has to be always satisfied.
	Acquisition and continuous investment in innovation technology focused on the improvement of the actor's processes.
	Investment in marketing, reinforcing the image of the horizontal networks.
	Number of members. Only with a high number of members will the horizontal network be able to offer and sustain services; otherwise, the earnings are limited.

6. Conclusions and recommendations

The dynamics of the global economic, social and political changes stimulated diverse small and medium-sized companies - from different sectors of the industry - to reconsider their performance and led many companies to reflect on the creation of alternatives to preserve their businesses. Beginning in the 1990s, there was a noticeable intensification of joint actions. Consolidated inter-organizational partnerships began and have since stimulated the strengthening and the sustainability of small and medium-sized enterprises. This scenario has imposed the necessity of seeking innovations, new managerial methods, tools and strategic programs in an attempt to find answers to the specific problems facing small and medium-sized companies. As a result, horizontal networks have an important role to play in the sustainability of these companies. The networks help to harmonize competition and enhance cooperation. This form of associativism has led to a collaborating working force that mobilizes individual firms around common projects – focusing on competitive advantage and stimulating the development and the professionalization of their members. The network model aims for a common solution to similar problems faced by small and medium-sized enterprises and has been implemented in several segments of the market, such as construction materials, drugstores, and, on a larger scale, in the supermarket sector.

National and international experiences also reinforce the potential for increased earnings obtained with the creation of horizontal networks, which often lead to consolidation and growth among these associations. Research has revealed the rise of many horizontal networks in the supermarket sector, which confirms further evidence that this model is being used in several countries. This model of joint action can be considered as an efficient alternative for competitiveness that is still to be adopted by other supermarkets around the world.

The evidence provided by this research study reinforces the idea that the creation of these partnerships or network unions is motivated by the potential to purchase cheaper products from the suppliers. A supermarket becomes a member of a horizontal network to start purchasing cheaper products. Cheaper products are available as a result of the higher volumes that can be acquired jointly, leading to lower costs (economy of scale). Horizontal networks essentially become 'buying groups', as the joint effort is restricted to acquiring improved negotiating power and the potential of profits that this partnership can provide to its members.

Actually, most networks in the retail field in Brazil are characterized by small independent retailers. In most European countries there are also other kinds of business networks formed not only by small retailers but also big chains, such as Carrefour. Most of the buying groups in Brazil have around 40 or at most 60 members. In many Europeans countries it is common to see a network with more than 1000 members.

As more members get involved in the business network, more potential gains are possible, because the power of the group is maximized and they can share costs and the need of investments between the actors (scale). In the same way, the bigger the business network is, the more complex the union will be, and it will become difficult to manage the group, especially because of the diverse cultures of the participants.

It is important to note that the network eventually matures as the organizational culture becomes more set and the actors become more committed. As this maturing process occurs, new value-adding activities are incorporated into the network. For example, the partners begin to exchange key information; develop joint training programs and marketing activities; develop private labels together; share technology; professionalize the business management; make some improvements in innovation; and modernize their stores. This research study revealed that some of these activities occur in many of the horizontal networks, however, others have more limited coordination.

While the research demonstrates that horizontal networks in the supermarket sector are an interesting idea and could serve as an important alternative for the sustainability of small and medium-sized supermarkets, the study also showed that it is not simple to transform this joint action into reality. This type of partnership requires a high level of maturity among the key partners to surpass the many barriers of working with competitors.

Commitment of the partners in the network is, without a doubt, one of the most limiting factors that impact the success of the horizontal network. The challenges of generating true commitment among the partners tends to appear in the initial creation of the network and will often remain (to some degree) in the later stages. The research also showed the importance of having mature networks – to assimilate the idea of cooperation and cost-sharing, individual actions have to be substituted by joint actions, which require arduous work as well as a different way of working together.

It was noted that some retailers have difficulties working together with other retailers, even seeing partners as a rival and not a partner. Many actors in the horizontal network simply expect to receive benefits. On the other hand, other cases demonstrate the assimilation of partners into the network and an adoption of the values and principles that permeate the horizontal network. In those situations, the retailers recognize the importance of understanding the purpose of the network and the importance of its active role in the construction of the inter-organizational relationships, as well as the need to mobilize in the direction of common projects and ideas.

Horizontal networks in the international supermarket sector are quite developed and advanced. The dissemination of the horizontal networks in Denmark happened in the 1930s and today they are in a mature apprenticeship phase. Other European countries have also reached the point of mature partnerships with a high degree of sophistication. Their processes are professionalized, there is clear communication and commitment among the partners and they develop all business activities together. Moreover, companies engage in technological, logistic and information systems integration; negotiation and payment schedules are centralized in the horizontal network centre, under the specialized professionals' care. The success of the existing networks in Brazil contributes to the addition of new partners to the network and expands their businesses in other areas.

In Brazil, the horizontal network is often formed by independent retailers. The companies become a part of a horizontal network as a 'last choice' in an effort to maintain their businesses and survive within the sector. In these cases, the companies are usually experiencing financial and managerial problems. In Denmark, most cases of horizontal networking in the supermarket sector happened among independent retailers and the action strengthened and consolidated partners within the Danish market. This was also the case for other European countries. The supermarkets that decided to be a part of a horizontal network needed financial and managerial stability. In this context the future partner sees the horizontal network as a potential for becoming more efficient. It is important to note that many horizontal networks also analyze the potential partners and their business practices before allowing them to enter into the network. If a supermarket has a strong financial deficiency, it will be difficult for this company to become a member of the network, as investments will be required to gain some important changes to competitiveness.

In Brazil, the horizontal networks in the supermarket sector strengthened around 1990, and although some networks already reflect a certain evolution, most are still in the initial phase – and the companies have probably not begun to address the limitations discussed in this paper. Maturity and growth of these businesses will likely happen on a larger scale in the medium run.

Empirical findings suggest that this kind of network plays an important role in the survival of the small and medium-sized supermarkets and it can be viewed as a powerful way to achieve effectiveness. Surpassing the limitations, especially the cultural aspects, this network can promote positive effects, technical and economic progress and knowledge exchange between the parties involved. As a result, for this concerted action to succeed, it needs to include a proposal with economic, social and cultural perspectives, but at the same time to focus on developing clear objectives, to improve the qualification of the processes and the professional training of its members.

Bearing in mind the cases studied, it is possible to list the main managerial implications of the findings. Based on the results of the research, it can be concluded that the following steps are necessary for a business network to succeed:

- *Preparation and planning the business network.* The first step in creating a business network is to understand why it is beneficial to form this association, what is its focus, who will be the members involved and what are their expectations. Statutes (articles of association) are necessary that define the members' duties and rules and that formalize the relationship between the partners. It is important to clarify what the advantages of being a member of the network are and what efforts and costs are required. Then, the members have to understand the role of the network and define who will be on the management board of the business network to conduct all the business. The members then have to write an action plan for the network. They have to define the scope of the association's actions and establish communication channels.

- *Starting the business network*: After defining the scope of the business network, it is essential to check the actual performance of the organizations involved and the standards required for the business network to succeed. In most cases, it is necessary to adapt the internal processes and infrastructure, such as technology and human resources. It is important to create synergies between all partners to start to develop the actions together. Note that these networks are carefully built over time and going too fast makes the whole construction fragile. So, they can start, for example, by buying products together. At this stage it is fundamental to reduce the instability of the relationship and create trust.

- *Developing joint actions*: In this stage it is necessary to establish and provide the consummate effort associated with a high commitment of the members. They must therefore first understand how the members perceive their relationship with the business network and identify the factors that are most valued by them, the fulfilment of which will have an impact on their commitment. When trust is obtained, the members become more inclined to cooperate. Activities focused on maximizing trust and establishment of long-run projects can facilitate the members' integration and their commitment. A focus on reducing the cultural distance between the members in a business network is also required. When trust and commitment are built, it will be possible to maximize and consolidate joint activities, such as training and marketing. At this stage it is also relevant to define the stores' standards and professionalize all the services involved.

- *Establishing a long-run relationship*: Not many business networks last in a long-term perspective. But if they do, the members share a high level of experiences, based on a trusting relationship. The uncertainty of the relationship is minimal and they already share information, reanalyze responsibilities and make the necessary adaptations to maximize gains and effectiveness. The joint actions are intense at this stage. Some suggestions for making the business network stronger that were identified in the field research are to invest in institutional marketing to strengthen the business network image at the market; to offer managerial support services to the members of the network (for example, consulting services); to extend the volume of private label products commercialized by the business network for a better margin profit; to extend the number of members of the business network to obtain better economies of scale and negotiating power. It is also important to consolidate the establishment of partnerships with suppliers and strengthen the relationship with the consumers. Finally it is important to continuously improve technology, integrate and modernize the businesses.

A number of issues touched on in this paper deserve further research. The authors will continue to investigate different factors that influence the competitiveness of the business networks, studying some successful and unsuccessful cases by applying the content analysis methodology and conducting research in other countries. The final objective is to create a model to better understand the evolution of the business network.

References

Allevato Jr, J.L., 2001. Acting as one: associativismmo no ramo supermercadista - práticas, desafios e tendências. Estudo realizado pela área de pesquisa da ActAsOne Ltda. em janeiro.

Amaro, J.N., 2000. Redes de cooperação produtiva e clusters regionais: oportunidades para as pequenas e médias empresas. Fundação Vanzolini. São Paulo, Atlas, 168 pp.

Bengtsson, M. and S. Kock, 1999. Cooperation and competition in relationships between competitors in business networks. Journal of Business & Industrial Marketing 14, 178-193.

Borys, B. and D.B. Jemison, 1989. Hybrid arrangements as strategic alliances: theoretical issues in organizational combinations. Academy of Management Review 14, 234-249.

Bridgewater, S. and C. Egan, 2002. International marketing relationships. New York, Palgrave.

Buono, A.F., 1997. Enhancing strategic partnerships. Journal of Organizational Change Management 10, 251-266.

Cândido, G.A. and A.F. Abreu, 2000. Os conceitos de redes e as relações interorganizacionais: um estudo exploratório. Anais do 24° Encontro da ENAMPAD. Rio de Janeiro: ANPAD.

Drumond, A., 2003. A união faz a força: pequenos e médios supermercados criam centrais de compra para enfrentar a concorrência das grandes redes internacionais. http://www.superavit.com.br/materias.as?cad=Mercado.

Ghisi, F.A., 2005. Fatores críticos na sustentabilidade das centrais de negócios do setor supermercadista. Tese (Doutoramento) - Faculdade de Economia, Administração e Contabilidade, Universidade de São Paulo, São Paulo.

Gonçalves, J., 2000. Super Censo 2000: retrato fiel do setor. Revista SuperHiper 26, n.301, 10-24.

Gulati, R., 1998. Alliances and networks. Strategic Management Journal 19, 293-317.

Gulati, R., N. Nohria and A. Zaheer,2002. A strategic networks. Strategic Management Journal 21, 203-215.

Hakansson, H. and D. Ford, 2002. How should companies interact in business networks? Journal of Business Research 55, 133-139.

Holmlund, M. and J.A. Tornroos, 1997. What are relationship in business networks? Management Decision 35, 304-309.

Jarillo, J.C., 1998. On strategic networks. Strategic Management Journal 9, 34-41.

Jones, C., W.S. Hesterly and S.P. Borgatti, 1997. A general theory of network governance: exchange conditions and social mechanisms. Academy of Management Journal Review 22, 911-945.

Lipnack, J. and J. Stamps, 1994. Rede de informações. São Paulo: Makron Books.

Marqui, A.C., A.B. Guirro and E.M. Merlo, 2002. Vantagens decorrentes da formação de associações de compras: um estudo de caso. Anais. XXXVII Cladea. Porto Alegre.

Miles, R.E. and C.C. Snow, 1986. Network organizations: new concepts for new forms. California Management Review 28, 62-73.

Miles, R.E. and C.C. Snow, 1992. Causes of failure in network organizations. California Management Review 34, 53-72.

O'Donnell, A. A. Gilmore, D. Cummins and D. Carson, 2001 The network construct in entrepreneurship research: a review and critique. Management Decision 39, 749-760.

Premaratne, S.P., 2001. Networks, resources, and small business growth: the experience in Sri Lanka. Journal of Small Business Management 39, 363-371.

Parkhe, A., 1991. Interfirm diversity, organizational learning, and longevity in global strategic alliances. Journal of International Business Studies 22, 579-601.

Parkhe, A., 1993. Strategic alliances structuring: a game theoretic and transaction cost examination of interfirm cooperation. Academy of Management Journal 36, 794-829.

Porter, M., 1998. Clusters and the new economics competition. Harvard Business Review 76, 77-90.

Powell, W.W., 1990. Neither market nor hierarchy: network forms of organization. Research in Organizational Behavior 12, 295-336.

Ribault, J.M., B. Martinet and D. Lebidois, 1995. A gestão das tecnologias. Coleção gestão & inovação. Lisboa, Publicações Dom Quixote.

Rank, K.O.N. and A.M.A. Wald, 2000. A methodological framework for the analysis of network organizations. In: J. Hagedoorn and S. Lundan (eds.), 26th Annual Conference of the European International.

Santos, S.A., H.J. Pereira and S.E. Abrahão França, 1994. Cooperação entre as micro e pequenas empresas. São Paulo, Sebrae.

Selltiz, C., M. Jahoda, M. Deutsch and J.M. Cook, 1965. Métodos de pesquisa das relações sociais. São Paulo, Herder.

Shaw, E., 1999. Networks and their relevance to entrepreneurial/marketing interface: a review of the evidence. Journal of Research in Marketing & Entrepreneurship1, 24-40.

Smith, S., 1996. The role of trust in SME business network relationships. Honours thesis, Queensland University of Technology.

Termômetro do Varejo, 1998. Associativismo: pequeno varejo cria um novo modelo de crescimento. Termômetro do Varejo, n.3, jul.

Yoshino, M.Y. and U.S. Rangan, 1996. Alianças estratégicas. São Paulo, Makron books.

Part 4. Supply Chain Management

Logistics outsourcing in food supply chain networks: Theory and practices

H.I. Hsiao, J.G.A.J. van der Vorst and S.W.F. Omta

Abstract

The food industry is becoming an interconnected system with a large variety of complex relationships, reflected in the market place by the formation of (virtual) Food Supply Chain Networks (FSCN) via alliances, horizontal and vertical co-operation, forward and backward integration in the supply chain. This has resulted in the rise of new types of Logistics Service Providers offering their advanced services to food manufacturers. The question is what logistics activities should be sourced out to which type of logistics service providers under what specific conditions? This paper investigates the outsourcing decision from three main streams of approaches, Transaction Cost perspective, Resource-Based View and Supply Chain Management theory. Case studies help building up a conceptual framework for three levels of logistics outsourcing in FSCN, namely execution, value-adding and planning level. Our preliminary results indicate that the degree of logistics outsourcing in a particular firm depends on the logistics strategy, supply chain complexity, asset specificity, transaction uncertainties and core activity analysis.

Keywords: outsourcing, Supply Chain Management, logistics service provider, chain orchestration, food industry

1. Introduction

There are a number of developments that put dynamic requirements on the performance of the food system initiating a re-orientation of companies in the Agriculture and Food industry regarding their roles, activities and strategies (Van der Vorst *et al.*, 2005). Demand and supply are no longer restricted to nations or regions but have become international processes. Consumers in Western-European markets have placed new demands on attributes of food such as quality (guarantees), integrity, safety, diversity and associated information (services). There is an increasing concentration in agribusiness sectors, an enormous increase in cross-border flows of livestock and food products, and the creation of international forms of co-operation. FSCN are particularly characterized by increased complexity and decision-making uncertainty related to demand and supply. Global sourcing of perishable fresh produce combined with a request for frequent, fast deliveries of increasing product assortments create shelf life and delivery time constrains, which increase the complexity of inventory planning, transportation planning and asset utilizations. In order to satisfy the increasing demands of consumers, government, business partners and NGO's and to obtain the "license to produce and deliver", companies continuously have to work on innovations in products, processes and forms of co-operation. The food industry is becoming an interconnected system with a large variety of complex relationships, reflected in the market place by the formation of (virtual) Food Supply Chain Networks (FSCN) via alliances, horizontal and vertical co-operation, forward and backward integration in the supply chain and continuous innovation (Van der Vorst *et al.*, 2005). The latter encompasses the development and implementation of enhanced quality, logistics and information systems. One of the consequences of these developments is the rise of new types of Logistics Service Providers (LSPs) that offer complete supply chain solutions. A LSP is defined as an external company who performs logistics activities between two or more facilities in the chain or performs logistics activities across different supply chains.

The increasing interest in the role of LSP in FSCN is not only due to the increased interest in Supply Chain Management (SCM), but also due to the increased growth in logistics outsourcing of the last few years (Capgemini, 2003). SCM asserts that organizations along the supply chain need to reconfigure their operations by internal integration and external integration in order to accommodate changes to their logistics strategy. In order to achieve a SCM approach, organizations are looking for solutions from LSPs to achieve seamless supply chain operations.

Outsourcing is an abbreviation for "outside resource using." Logistics outsourcing means that an organization uses a LSP to carry out an activity which is originally performed in-house (Lieb *et al.*, 1993). It represents a specifically defined contractual relationship, which requires the LSP to meet specified performance criteria set by the buying organization (Bolumole, 2001). Compared to traditional transportation providers, new types of LSP with complete supply chain solutions are encouraged to play a part in the strategic direction and performance of the entire supply chain. In a way, they are becoming true supply chain members (Bolumole, 2001).

Hertz and Afredsson (2003) mentioned that future development of LSPs can be classified by the abilities of general problem solving and customer adaptation. Based on their research, we distinguish three main types of LSPs:

- Standard LSPs: the companies who provide standard and traditional services, such as carried-based, warehouse-based, forward-based.
- Integrated LSPs: these companies provide value-added services and also provide at least two standard services.
- Supply chain solution providers: this is the new type of LSP, called supply chain solution providers or "4PL (fourth-party logistics)". They have higher problem solving ability and higher ability of customer adaptation by involving a high integration with customers, often in the form of taking over customers' whole logistics operations.

Most of the research on logistics outsourcing focus on advantages and risks of outsourcing in traditional logistics relationships. The development of supply chain partnerships, increased focuses on core business processes, reduction on distribution cost, and time saving have been cited as the main reasons why organizations choose to outsource (Razzaque and Sheng, 1998; Norek and Pohlen, 2001; Sohail and Sohal, 2003). However, very little attention is devoted to the new type of LSP and its potential effects on Supply Chain Networks; also less research is done on the implications for its role in FSCN (Bourlakis and Weightman, 2004). This research attempts to bridge these research gaps by developing a research framework for levels of logistics outsourcing in FSCN.

2. Research design

This research aims at developing a logistics outsourcing decision framework for food manufacturers that indicates what logistics services should be sourced out to which type of logistics service provider under what specific conditions. The main research question is: What decision-making criteria are considered by food manufacturers when outsourcing a certain logistics activity to a LSP?

First, a broad literature review was undertaken in order to gain a better understanding of important variables and the relationships between them and to construct a preliminary research framework.

The most important issue in this activity is to discuss general outsourcing approaches and identify important theories related to this subject. Result of this first activity is an outsourcing framework.

Second, we carried out an explanatory multiple-case study. The primary purpose of this case study is to explain why the decision of outsourcing a certain activity is taken in FSCN. We will do this by applying the outsourcing framework established from the literatures review into food industry and determining the relationship between level of outsourcing and the decisive factors (Voss *et al.*, 2002). Companies selected in the Netherlands are from the dairy industry, meat industry, and three logistics services providers who provide services to a variety of food companies. The food industry is a diverse industry. The logic for the use of multiple-case studies is that we wish to obtain similar results or produce predictable contrasting results from these diverse food companies to generalize a framework for whole food industry. The prime sources of data are semi-structured in-depth interviews. The unit of analysis is the decision making process in a particular company. Logistics managers or general managers were interviewed using a set of well-designed questions. In some cases multiple informants were used to improve the reliability of data. The interviews were taped and transcribed, and the transcripts were scanned to identify similarities and differences, thereby paving the way for identifying consistency patterns and developing plausible explanations in helping to form the hypotheses and to build internal validity. The interview with each person lasted about 60 minutes and covered the following issues:

- Details of the interviewee.
- The business context within which the company is operating.
- The factors to be considered in the outsourcing decision of logistics activities.
- Future plans for outsourcing.

The next sections present complexities of FSCN (section 3), and the classification of logistics activities (section 4). Section 5 introduces several outsourcing approaches. Case study results are presented in the section 6. In section 7 and 8, results are analyzed and theoretical and empirical insights are discussed. The paper concludes with a summary of our main findings and suggestions for future research in section 9.

3. Increased complexity of FSCN and the involvement of LSP

In line with Mentzer *et al.* (2001) we define a food supply chain as a set of three or more entities (organizations or individuals) directly involved in the upstream and downstream flows of fresh produces, industrial food products, services, and/or information from a source to a customer. From operational point of view, a supply chain network is used here to refer to a network of facilities that produces raw materials, transforms the raw materials into finished goods at processing centers and delivers the finished goods to customers through distribution warehouses. SCM aims at managing the supply chain network as a single entity to plan and control the total goods and information flows from suppliers through end users effectively and efficiently.

The structure of FSCN has been subject to change in the last decades, which has increased the complexity of managing the FSCN. FSCN are complex networks because one chain actor is usually involved in different types of supply chains, collaborating with different chain partners. Perishability of food resulting in quality losses places pressure on time and temperature control for each food actor at each node. Specific process characteristics, long production time and large food product varieties places specific demands on inventory planning and capacity utilization. Further, the trend of internationalization of FSCN creates difficulties on the identification of food products due to the lack of uniformity of standards. Hence, a comprehensive tracking system creating visibility of pipeline combined with a standard identification technology of food products are certainly needed in international business. Finally, concerns of food safety, environment protection, and energy use have made the EU set a number of targets on tracking and tracing (Walther and Spengler, 2005).

Overall, the complexity of FSCN has placed many pressures on the food actors. It is vital for FSCN to work as a single entity to plan and control the logistics of food products in a more effective and efficient way and supply consumers with high quality and secure food. Outsourcing some of logistics functions and involving of LSP into to the networks might help to solve the complexity problems and achieve superior efficiency and effectiveness.

4. Classification of logistics activities

To study the level of outsourcing in FSCN, we should classify the activities first that can be outsourced by food manufacturers. In this paper we explicitly classify logistics activities into three levels: execution level, value-adding level and planning/control level (see Table 1). A series of physical logistics activities in a food manufacturing plant starts from receiving materials, storing materials, processing the food, and distributing products to customers. Activities at execution level include transportation (inbound and outbound transportation) and warehousing. Labelling, packaging and some selected manufacturing activities (*e.g.*, materials mixing) are in the categories of value-adding activities. The purpose of planning and control is to ensure operations run effectively and produce products and services as expected (Slack *et al.* 2001), and react to unexpected events in the pipeline (such as materials shortage, stock out, shipment delay, food disease outbreaks, food product recalls) to maintain performance. Such activities are inventory planning/control, purchasing and production planning/control, transportation planning/control and network design.

5. Review of outsourcing approaches

Outsourcing has been an important issue for decades. It has been investigated from different perspectives due to its multidiscipline nature. Three main streams of approaches are often discussed: logistics strategy and SCM theory; transaction cost theory; and the Resource Based View. Table 2 summarizes the results of a literature survey on reasons/considerations for logistics outsourcing from these theoretical approaches.

5.1 Logistics strategy and Supply Chain Management theory

Logistics competitive strategy

We hypothesize that the selection of activities to be outsourced is influenced by the company's logistics competitive strategy, because the logistics strategy influences the control of logistical activities and thus the design of logistics systems (Van Goor *et al.*, 2003). A business strategy forms a comprehensive master plan stating how the corporation will achieve its mission and objectives (Wheelen and Hunger, 2002). Adopting Michael Porter's (1980) framework, a typology of competitive strategies can be used to classify type of food companies. At a generic level, logistics competitive strategy can be distinguished into lean and agile strategies. *Lean* strategy seeks to achieve a lower price than competitors whilst trying to maintain similar value of product or

Table 1. Logistics activities that can be outsourced by food manufacturers.

Classification	Logistics activities	Concept
Execution level	Warehousing	Storage; receiving; order-picking
	Transportation	Delivery
Value-adding level	Product labelling/packaging/ selected product manufacturing activities	
Planning/control level	Inventory management	Sales forecasting; stock control; tracking and event control
	Transportation network management	Route planning and scheduling; carriers selection; mode selection; rate negotiation; tracking and event control
	Production management	Material-requirement processing; production planning; tracking and event control
	Network design	Location and site analysis

service to that offered by competitors (Johnson and Scholes, 2002), in other word, having low cost as main logistics objective. *Agile* strategy seeks to provide products or service unique or different from those of competitors in terms of dimensions widely valued by buyers (Johnson and Scholes, 2002). Its logistics aims are to achieve higher market share than competitors (which in turn could yield cost benefits) by offering better services (flexibility) at the same price. *Leagility* seeks both differentiation and a price lower than of competitors (flexibility and low cost). In brief, we argue that the degree of outsourcing by food companies will depend on their orientation of the logistics competitive strategy.

Supply chain complexity and outsourcing

Rao and Young (1994) mentioned several outsourcing considerations: product complexity, process complexity, and network complexity. They refer to complexity as the degree of volume or size, such as product complexity (perishability of finished products, size, density), demand fluctuation, process complexity (sensibility of time, manufacturing cycle), network complexity (number of trading companies countries, continents). For example, it is assumed that the higher the number of trading parties or countries involved in the supply chain, the more likely the decision to outsource. More suppliers and customers involved and more geographic dispersion implies that many more logistics transactions must be managed. Derived from these statements, other factors which influence logistics operations might also relate with decision making of outsourcing, such as delivery frequency, distribution volume or demand fluctuation. For example, low distribution volume might result in low resource utilization within the firm. Thus, distribution cost will be increased due to low assets utilization. Based on these arguments, we could hypothesize that outsourcing decisions might be favoured for certain supply chain settings, and complexity in the settings would make firms likely to outsource a certain activity.

Table 2. Summary of reasons/considerations for outsourcing.

Sources	Anderson (1985)	Rao and Young (1994)	Damme and Amstel (1996)	Robertson and Gatignon (1998)	Leiblein and Miller (2003)	Wilding and Juriado (2004)	Capgemini (2005)
Reasons and considerations for outsourcing	From different perspectives: - Environmental unpredictability - Transaction specificity - Difficulty of evaluating performance (from Transaction Cost perspective)	- Centrality of logistics activities to core competence of the firm - Logistics complexity: Product complexity (e.g. size or density) Process complexity (e.g. time sensitivity or manufacturing cycle time) Network complexity (e.g. trading partners, countries, continents)	- Increased assortment of articles - Rapid growth of company - Activity did not have strategic value - Activity have an irregular pattern - Transport and storage of certain products requires special skills (e.g. dangerous goods)	- Technological uncertainties - Behavioural uncertainties (from Transaction Cost perspective)	- Cost reduction - Service improvement - Strategic flexibility - Focus on core - Change implementation	- Competencies of 3PL - Operational flexibility - Cost reduction - Focus on core business - Avoiding investment - Expansion to new markets	User respondent: - Pressure cost reduction - Pressure to service enhance - Focus on corporate core competencies - Industry experience - Unique service available from 3PL providers - Globalization Non-user respondent: - Logistics is core competency - Logistics is too important to outsource - We have more expertise - Time/effort would not decrease - Service levels would not be realized

Table 2. Continued.

Sources	Anderson (1985)	Rao and Young (1994)	Damme and Amstel (1996)	Robertson and Gatignon (1998)	Leiblein and Miller (2003)	Wilding and Juriado (2004)	Capgemini (2005)
Research types industry	Survey research Electronic manufacturer	Case study research Wide range of industry	Conceptual Wide range of industry	Survey research Manufacturer	Survey research Wide range of industry	Survey research Wide range of industry	Market survey Wide range of industry
Activities to be outsourced	Internal salesman vs. outside Wide range of logistics functions Most commonly outsourced activities with more than 50% uses rate:	Terminal activities Internal R&D vs. outsourcing Handling activities Administration activities Warehousing activities Transportation activities	Production in-housed vs. outsourcing	Primary transport		Secondary transport Additional storage during peak periods Fleet management Relabelling, repacking	Outbound transportation Warehousing Customs clearance Inbound transportation Freight forwarding Transportation management Freight bill payment Cross-docking shipment consolidation

5.2 Transaction cost perspective

The principle question "why is an activity organized within firms and not purchased?" is often discussed based on Williamson's transaction cost theory (Williamson, 1985). Transaction Cost Economic (TCE) at its core focuses on the costs of completing transactions by one institutional mode rather than another (Williamson, 1975). The transaction, a transfer of a good or service, is the unit of analysis. The theory claims that a transaction is likely to take place within the hierarchy of the firm (is carried out in-house) if subject to uncertainty and transaction-specific investments. Alternatively, transactions with less uncertainty and transaction-specific assets are likely to take place based on market contracting (outsourcing). Two main principle attributes of transactions are often applied in the outsourcing decision: asset specificity and uncertainty (technological uncertainty and behaviour uncertainty). For example, Robertson and Gatignon (1998) examined the decision to use internal R&D versus alliances in the development of technology by using asset specificity, technological uncertainty, behavioural uncertainty. In conclusion, TCE claims that the decision to organize an activity within the firm or in a market can be answered by the characteristics of the transaction, notably uncertainty and asset specificity.

5.3 Core activity analysis from resource-based view (RBV)

There is an often-heard justification for outsourcing decisions, which is "keep core competence and outsource non-core activities?" The "what is core and what is non-core" question can be discussed through the Resource-Based View (RBV). RBV and TCE have a different domain of interest on answering why firms exist and why they differ. They are interested in different aspects of an economic activity; RBV focuses on the analysis of a firm's resources/capabilities, while TCE focuses on the transaction costs (Madhok, 2002). Insinga and Werle (2000) asserted that the core activity should be kept internally because this activity is central to the company successfully serving the needs of potential customers in each market. They further identify that an activity with potential to yield competitive advantage is a core activity. Thus, the core activities should not be outsourced.

6. Case study results

The logistics outsourcing decision is difficult to evaluate, especially for FSCN. In this report, we present several cases to uncover the main decision criteria that decide on the level of logistics outsourcing. The general company information and interview results are displayed case by case at Table 3a (food companies in AI to AIII) and 3b (logistics service providers in BI to BIII). Case AI and Case AII are two different product groups of one large food company. Case AI produces over three hundreds of fresh milk products with annual production volume of 300 million liter. Geographic network is not complex because the factory, suppliers and customers are mainly located in The Netherlands. However, operating complexity is quite high when the factory has nearly 900 customers, and delivers once a day within 24 hours of lead-time. The activities being outsourced so far include warehousing and transportation network management. The main considerations for the outsourcing decision are: (1) seeking a reduction of logistics cost because of price war between supermarkets; and (2) not keeping activities that do not belong to the core business. Company AI considers the outsourcing decision to be correct because costs have been reduced and customers are highly satisfied with on-time delivery services. The other cases can be seen from the Tables, and the cases are ordered according to the variables of interest, so that the differences among high, medium, and low cases can be easily seen.

Table 3a. Summary of the results: food companies.

Cases	AI	AII	A III
Food sector	Dairy manufacturer	Dairy manufacturer	Veal slaughterhouse
turnover	600 million euros		25 millions euros
Employees	5,000		160
Activities been outsourced ■ Planning activity ▲ Value added activity ● Execution activity	(■ Transportation network management used to be outsourced but now taken back in-house) ● Transportation ● Warehousing	● Transportation network management ● Transportation ● Warehousing	● Transportation
Products	Milk products (milk, yogurt)	Juice, coffee milk	Veal products
Product varieties	High (300-400SKU)	High	High (few hundred)
Demand fluctuation	Low	Medium (juice has high peak in summer)	Low
Number of suppliers	Low (2)	Low (6)	Low (1)
Number of customers (yearly)	High (900)	Medium (250)	High(800)
Geographic diversion (suppliers, factory, customers)	Low (NL, NL, NL)	Medium (NL,NL, other Western European countries)	High (NL, NL, other Western European countries, and Eastern European)
Average shipment-size	High (Full-truck load)	High (Full-truck load)	Low (Less-than truck load)
Delivery frequency	High (Once per day)	Medium (Once per week)	High (once per day)
Total production volume	High (300 million liter/year)	High (500,000 pallets/year)	High (50,000 calves/week)
Average lead-time required	Short (24 hour)	Medium (48 hour)	Medium (48 hours)
Considerations of outsourcing decision	Cost saving due to the price war between supermarkets	Cost reduction	Do not own transport vehicles.
	Most of planning activities are the core activities	Competition in the international market	Warehousing won't be outsourced because it is very important
		The activities not outsourced are the core activities	
Current logistics performances	Cost were reduced and customers are highly satisfied with on-time delivery	Cost were reduced and customers are highly satisfied with on-time delivery	Customers are satisfied with on-time delivery.
Challenges	Price competition between retailers remains	Price competition between retailers remains	Global market expansion Retailers are more demanding Customers place orders in the last minutes
Future plan of outsourcing	Remain the current situation	Remain the current situation	Remain the current situation

Table 3b. Summary of the results: logistics service providers.

Cases	BI	BII	BIII
Sector	Logistics service provider	Logistics service provider	Logistics service provider
Turnover (2004)	100million euros	5 billion euros	100 million
Employees	500	80,000	-
Activities been outsourced /services offered ■ Planning activities ▲ Value-added activities ● Execution activities	● Transportation ● Warehousing ● Quick express ▲ Value-adding services	■ Inventory management ■ Global freight management and forwarding ● Warehousing and distribution ▲ Kitting, pre-packaging Others: returns management; aftermarket parts distribution	■ European forwarding and world-wide forwarding ▲ Value-added services; ▲ Contract manufacturing ● Warehousing ● Transport Others: Return logistics
Considerations of outsourcing decision from customers	Reduction of distribution cost and materials cost Logistic service providers have professional knowledge	Activities to be outsourced are not important to the clients	Cost reduction Core businesses are not outsourced.
Challenges received from customers		Global source of supply Increased product varieties	
The future of outsourcing	More outsourcing are expected, and more companies are likely to outsource the management of information	More outsourcing	More outsourcing, but most planning activities still are remained in-housed

7. Cross-cases analysis and discussions

The paper aims to identify hypotheses of decision making factors for a certain logistics activity in FSCN. The purposes of case analysis are to examine the relationship between the choice of the logistics activities and the key drivers. Through cross-cases analysis, the advantages of generalizability can be achieved (Miles and Huberman, 1994). The strategies used here are case-oriented analysis and variable-oriented strategies. With cross-case analysis, the replications can be examined whether the pattern found in one case matches in another cases. With variable-oriented strategies, the variables built from the outsourcing framework can be examined in the cases one by one (Miles and Huberman, 1994). Based on the case study results, we will now discuss the five main hypotheses.

7.1 Logistics competitive strategy and choice of logistics activities

Cases AI, AII, and AIII show that cost reduction is one of the main considerations when food companies outsource execution activities. These results fit well with the findings of Bolumole, who argues that outsourcing for operational and cost based reasons will tend to restrict LSPs' involvement to the basic logistics functions (Bolumole, 2001). More evidence was found on the relationship between execution activities and low cost competitive strategy, but less was found on higher levels of outsourcing.

As for high levels of outsourcing, Bolumole (2001) states that organizations that are motivated to outsource for the purpose of facilitating the integration of supply chain members tend to perceive the LSPs' functions as an integral part of that supply chain strategy. Thus, from our empirical results and literatures, we could propose the following statements. Companies with lean strategy tend to outsource execution level activities, while companies with agile objectives appear to be differentiating themselves in terms of service (flexibility). Differentiation can be achieved by enhancing their operational capabilities to respond quickly and effectively to future changes in the marketplace. This strategy is similar to postponement; it is an organizational concept whereby some of the activities in the supply chain are not performed until customer orders are received (Van Hoek, 2001). For example, to manufacture food products until the last minute and to achieve delivery reliability, companies tend to outsource some value-added activities. Leagile is the third type of strategy; leagile companies emphasized similar objectives as agile companies with the exception that they put flexibility and maximization of product value as their top objectives as well. Thus, companies with leagile objective are likely to outsource the planning/control actives. Based on these statements, we propose the first hypothesis:

H0 : When a firm outsources its planning activities, the corresponding execution activities will also be outsourced.
H1a: A food manufacturer with a lean strategy is likely to outsource execution activities.
H1b: A food manufacturer with an agile strategy is likely to outsource value-added activities.
H1c: A food manufacturer with leagile strategy is likely to outsource planning and control activities.

7.2 Supply chain complexity and outsourcing decisions

The literature suggests that product characteristics (size or density), network characteristics (trading partners, countries or continents), processing characteristics (*e.g.* time sensitivity or manufacturing cycle time) and market characteristics influence outsourcing decisions (Rao and Young, 1994). The observations from Case BII show that food companies face, amongst others, the following future challenges: increased global market expansion; global source of supply; and increased products varieties that possibly result in more outsourcing.

Our results also observe that complexity of food supply chain setting is most likely to be the determinant factor for level of outsourcing. Food companies (Case A1 and AIII) who decided to outsource the execution level seem to operate at a high complexity setting (high perishable, short lead-time, and high delivery frequency, low demand uncertainty). Food companies (Case

AII) who decided to outsource the planning and control seem to operate at a low complexity (medium perishable, less lead-time requirement and less delivery frequency, high demand uncertainty). However these results seem to conflict with Supply Chain Management theory. Supply chain complexity reveals the type of decision problems faced by a supply chain. Supply chain uncertainties are the decision making situations in the supply chain in which the decision-maker lacks effective control actions or is unable to accurately predict the impact of possible control actions on system behaviour because of a lack of information (or understanding) of the environment or current supply chain state. The more uncertainty related to a process, the more waste there will be in the process (Persson, 1995) because the presence of uncertainty stimulates the decision maker to create safety buffers. Differences in the operating complexity could be examined by three measures: volume, variety, and uncertainty (Slack *et al.*, 1995). Low volume of operations results in less repeatability, and it gives high unit costs. High variety results in the need of high flexibility (Slack *et al.*, 1995). Variety causes complexity in planning, for example, a supply chain with the customers located in different countries raise more complications than a supply chain with customers located in the domestic country. Uncertainty, such as demand uncertainty, is one of the complications that raise the problems in the planning, scheduling and control of the products and information flows (Fisher, 1997). A list of source of supply chain complexity is presented at Table 4.

H2: Supply chain complexity is negatively associated with the level of outsourcing.

7.3 Activity-specific investments and choice of outsource versus in-house

Operating the logistics activities requires asset-specific investments. These assets include physical assets (trucks, buildings, refrigerators), human assets (driving skills or planning skills), and information assets (software or hardware). The results show that the food companies tend to outsource logistics activities when they had not invested in physical or human assets. In Case AIII, the company bought transportation services from logistics service providers because they did not own any vehicles. When asset specificity is high (indicating that assets can not be deployed for other purposes, for example, refrigerators), food companies tend to keep the activity in-house. Highly asset-specific investments are also called relationship-specific investments (Grover and Malhotra, 2003). The relevance of asset specificity as a decision consideration has been confirmed across a range of research applications, including, outsourcing decision of salesman (Anderson

Table 4. Source of supply chain complexity.

	Product characteristic	Market characteristic	Process characteristic	Network characteristic
Supply-related			Supply lead-time	Numbers of suppliers
Demand-related	Perishability of end product	Shipment size	Order distribution lead-time	Numbers of customers
		Demand fluctuation		Geographic diversities
		Delivery lead-time		
Production-related		Production volumes		Product varieties

and Schmitteln, 1984) or outsourcing decision of information technology (Aubert *et al.*, 2004). The theory assumed that transactions that are supported by investments in durable, transaction-specific assets experience lock-in effects, on which account autonomous trading will commonly be supplanted by unified ownership (vertical integration) (Williamson, 1985). It can be implied that the greater the activity-specific investments within the firm, the more likely the firm will be eager to engage in internal development (in-house) rather than to engage in cooperation with a logistics service provider.

H3: The higher the asset specificity of a specific logistics activity, the more likely that a food manufacturer will keep this activity internally rather than outsource.

7.4 Transaction uncertainties and choice of outsource versus in-house.

The results reveal that transaction uncertainties were not the main considerations for the three food companies when they outsourced logistics activities. This conflicts with other empirical results. Robertson and Gatignon (1998) confirmed the positive relationship between the technology innovation outsourcing and transaction uncertainties (technological uncertainties and measurement uncertainties). Other empirical researches also suggested that when the difficulty of measuring performance increases, the likelihood of vertical integration increases (Anderson, 1985; Robertson and Gatignon, 1998). It implies that the higher the uncertainty about a transaction, the less likely this transaction will take place. Further, Capgemini (2005) shows that one of the reasons firms do not like to outsource is "time/effort could not be reduced." This can be explained by that time/efforts are possibly high when transaction uncertainties are high.

Measurement uncertainty influences the outsourcing decision. It can be argued that when performance cannot be assessed easily, using markets can be "inefficient" (*i.e.*, less profitable than in-house), the contracting cost is high when writing an incentive compatible contract under a complex performance assessment. Similarly, technology uncertainty also determines an outsourcing decision. When technology changes fast, the contracting costs are high when logistics companies wish detailed contracts for securing their profits, or renew the contracts for installing new technology for clients. In supply chain planning, the efficiency is highly related to technologies changes. Thus, we propose the next hypothesis:

H4: The higher the transaction uncertainties when outsourcing a logistics activity, the more likely that a food company will keep this activity internally rather than outsource.

7.5 Core activities analysis and choice of outsource versus in-house

The observations show that the activities chosen to be outsourced would not be the core activities, such as transportation in Case AII and transportation network management in Case AI. The results are supported by other studies (see Table 2). A firm's activity can be a core activity when it contributes to competitive advantage. It exploits opportunities and/or neutralizes threats in a firm's environment (Barney, 1991). Some companies would not like to outsource planning activities (like Cases AII or BIII), due to the fact that they perceive these activities as essential to their business. Wilding and Juriado (2004) mentioned that information system related logistics activities (tracking and tracking, order entry, forecasting, and logistics information systems) are not outsourced because the systems are essential elements of the business. In brief, strategic outsourcing emphasises a company's core competencies (focus on the core activities), because manager can leverage the firm's skills and resources for increasing competitiveness (Quinn and Hilmer, 1994). Thus we formulate another hypothesis:

H5: A food manufacturer would like to outsource the activities which are not the core business activities

8. A decision making framework for logistics outsourcing in FSCN

These cases suggest a new outsourcing framework (Figure 1) for FSCN with regard to the logistics outsourcing decision. A framework presents a structure for various considerations, which is likely to provide a better understanding to decisive criteria. The decisions for levels of outsourcing inherently include two decision-making processes: the decision to choose the degree of logistics outsourcing is decided by the logistics competitive strategy and supply chain complexity; and the decision to outsource or keep in-house depends on the degree of asset specificity, transaction uncertainty, and core activity analysis. It can be proposed that outsourcing of execution activities is favoured by food companies with a lean operations strategy where the activities are perceived as non-core activity, operating at high complex supply chain setting, low asset specificity and low transaction uncertainty. Similarly, it can be proposed that outsourcing of value-adding activities are favoured by food companies with an agile operations strategy operates where also these activities are perceived as non-core activity, operating at medium complex supply chain setting, low asset specificity and low transaction uncertainty. For outsourcing of planning activities, the hypothesis is that outsourcing of planning activities are favoured by food companies with a leagile operations strategy where the activities are perceived as non-core activity, operating with low complex supply chain setting; low asset specificity and low transaction uncertainty. Performances are expected to be improved through outsourcing.

9. Conclusions

This paper aims to contribute to both theories and practices of logistics outsourcing in Food Supply Chain Networks. A new outsourcing framework is presented that intends to fill the gap identified in literature by capturing factors to be considered form food manufacturers, as well as from the perceptions from the service providers. This paper also presented some preliminary evidence suggestive on the role that SCM theory may play in outsourcing decisions. Results show that the degree of logistics outsourcing in a particular firm depends on the logistics strategy, supply chain complexity, asset specificity, transaction uncertainties, and core activity analysis. The complexity experienced in the supply chain can be viewed as a threat and something that needs to be avoided or reduced; the degree of complexity shows the need for specific degrees of outsourcing. Our results show that degree of complexity is negatively associated to degrees of outsourcing. Different logistics competitive strategies present different degrees of logistics service providers' involvement from basic execution functions to higher planning functions. Degree of transaction uncertainties and asset specificity are assessed to evaluate the likelihood of outsourcing or operating internally. Similarly, analysis of core or non-core activities specifies the conditions that such logistics activities can be a source of competitive advantages.

The limitations of this research are the following. First, our research model is only tested at dairy and veal food sectors. Second, we received more empirical results on decisive factors and execution level of outsourcing, but less evidence was found on planning/control level of activities. Thus, further research is suggested to test and validate the research framework through in other

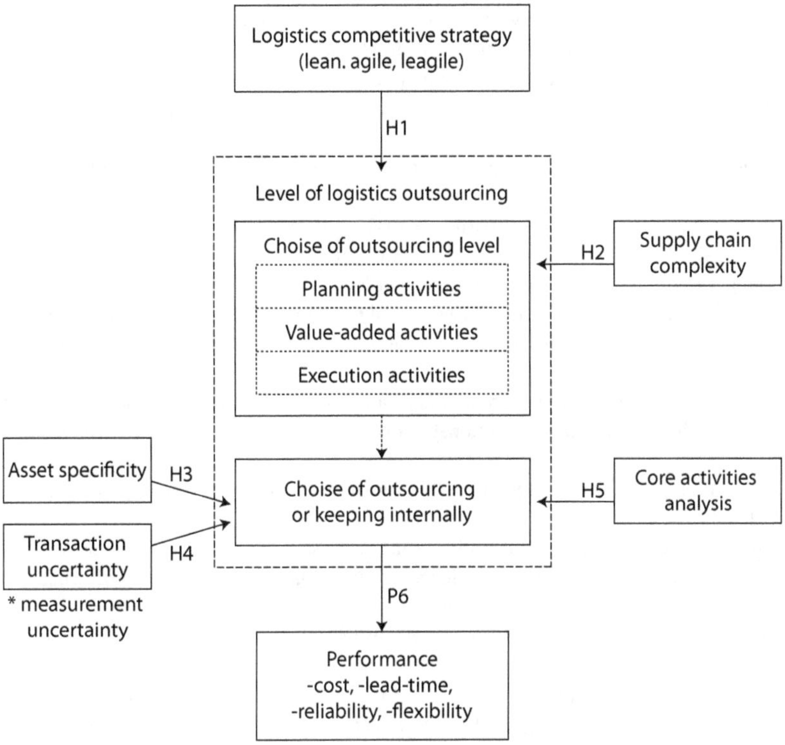

Figure 1. A preliminary decision making framework for level of logistics outsourcing in FSCN.

cases. Moreover, the framework will be tested empirically in a large survey in order to fully understand the practices of logistics outsourcing in FSCN.

References

Anderson, E., 1985. The salesperson as outside agent or employee:a transaction cost analysis. Marketing Science 4, 234-254.

Anderson, E. and D.C. Schmitteln, 1984. Integration of the sales force: an empirical examination. Rand Journal of Economics 15, 245-255.

Aubert, B.A., S. Rivard and M.Patry, 2004. A transaction cost model of IT outsourcing. Information & Management 41, 921-932.

Barney, J., 1991. Firm resources and sustained competitive advantage. Journal of Management 17, 99-121.

Bolumole, Y.A., 2001. The supply chain role of third-party logistics providers. International Journal of Logistics Management 12, 87-102.

Bourlakis, M.A. and P.W.H. Weightman, 2004. Food Supply Chain Management 2004.

Capgemini, 2003. Third-party logistics. http://www.capgemini.com/. Nov 2005.

Capgemini, 2005. Third-party logistics: results and findings of the 10th annual study. http://www.capgemini.com/. Nov 2005.

Damme, D.A.V. and M.J.P. van Amstel, 1996. Outsourcing logistics management activities. International Journal of Logistics Management 7, 85-95.

Fisher, M.L., 1997. What is the right supply chain for your product? Harvard Business Review 75, 105-116.
Grover, V. and M.K. Malhotra, 2003. Transaction cost framework in operations and supply chain management research: theory and measurement. Journal of Operations Management 21, 457-473.
Hertz, S. and M. Afredsson, 2003. Strategic development of the third party logistics providers. Industrial marketing management 32, 139-179.
Insinga, R.C. and M.J. Werle, 2000. Linking outsourcing to business strategy. The Academy of Management Executive 14, 58-70.
Johnson, G. and K. Scholes, 2002. Exploring corporate strategy. Person Education Limited.
Leiblein, M.J. and D.J. Miller, 2003. An empirical examination of transaction-and firm-level influences on the vertical boundaries of the firm. Strategic Management Journal 24, 839-859.
Lieb, R.L., R.A. Millen and L. van Wassenhove, 1993. Third party logistics services: a comparison of experienced American & European Manufactures. International Journal of Physical Distribution & Logistics Management 23, 35-44.
Madhok, A., 2002. Reassessing the fundamentals and beyond: Ronald Coase: the transaction cost and resource-based theories of the firm and the institutional structure of production. Strategic Management Journal 23, 535-550.
Mentzer, J.T., W.D. Witt, J.S. Keebler and S. Min, 2001. Defining supply chain management. Journal of Business Logistics 22, 1-25.
Miles, M.B. and A.M. Huberman, 1994. Qualitative Data Analysis: an expanded sourcebook. Sage Publications.
Norek, C.D. and T.L. Pohlen, 2001. Cost knowledge:a foundation for improving supply chain relationships. International Journal of Logistics Management 12, 37-51.
Persson, G., 1995. Logistics Process Redesign: some useful insights. International Journal of Logistics Management 6, 13-21.
Porter, M.E., 1980. Competitive strategy: Techniques for analyzing industries and competitors. The Free Press.
Quinn, J.B. and F.G. Hilmer, 1994. Strategic outsourcing. Sloan Management Review, 43-55.
Rao and Young, 1994. Global supply chains: factors influencing outsourcing of logistics functions. International Journal of Physical Distribution & Logistics Management 24, 11-19.
Razzaque, M.A. and C.C. Sheng, 1998. Outsourcing of logistics functions: a little literature survey. International Journal of Physical Distribution & Logistics Management 28, 89-107.
Robertson, T.S. and H. Gatignon, 1998. Technology development mode: a transaction cost conceptualization. Strategic Management Journal 19, 515-531.
Slack, N., S. Chambers and R. Johnston, 2001. Operations Management. Pearson Education Limited.
Sohail, M.S. and A.S. Sohal, 2003. The use of third party logistics services: a Malaysian perspective. Technovation 23, 401-408.
Van der Vorst, J.G.A.J., A.J.M. Beulens and P. van. Beek, 2005. Innovations in logistics and ICT in food supply chain networks. In: Innovation in Agri-Food Systems. W.M.F. Jongen and M.T.G. Meulenberg (eds.). Wageningen, Wageningen Adademic Publishers, pp. 245-292.
Van Goor, A.R., M.J.P. van Amstel and W.P. van Amstel, 2003. European distribution and supply chain logistics. Wolters-Noordhoff, Groningen.
Van Hoek, R.I., 2001. The rediscovery of postponement a literature review and directions for research. Journal of Operations Management 19, 161-184.
Voss, C., N. Tsikriktsis and M. Frohlich, 2002. Case research in operations management. International Journal of Operation & Production Management 22, 195-219.
Walther, G. and T. Spengler, 2005. Impact of WEEE-directive on reverse logistics in Germany. International Journal of Physical Distribution & Logistics Management 35, 337-361.
Wheelen, T.L. and J.D. Hunger, 2002. Strategic management and business policy. Pearson Education
Wilding, R. and R. Juriado, 2004. Customer perceptions on logistics outsourcing in the European consumer goods industry. International Journal of Physical Distribution & Logistics Management 34, 628-644.
Williamson, O.E., 1975. Markets and Hierarchies: Analysis and Antitrust Implications. Free Press, New York.
Williamson, O.E., 1985. The economic Institutions of Capitalism. Free Press, New York

About the preparedness of dairy companies for ECR

T. Obersojer and H. Weindlmaier

Abstract

The German food market has undergone tremendous changes within the last years: in the food retail a concentration process has taken place, discount-stores have gained market shares, and consumers have shown a hybrid and smart buying behaviour demanding highest quality at the lowest price. New information technologies have been developed and enable a closer collaboration between the members of the chain. Therefore, the former isolated management of single firms must be replaced by a vertical comprehensive management encompassing each individual member of the supply chain. In this context, Efficient Consumer Response (ECR) is seen to be the successful supply chain strategy for the grocery industry. This paper gives an overview of the ECR concept and its single strategies. Core requirements for a successful implementation of ECR are approved. Thereby, German, Austrian and Swiss dairies were surveyed about their preparedness and implementation ratio of ECR. Seven factors regarding the requirements were identified. But these essential requirements for a successful ECR implementation are rarely fulfilled up to now. In practice, ECR is often only tailored to IT concerns. Distrust in the relationship between dairy companies and retailers and a lack of strategic orientation are the main obstacles for a consequent implementation. According to these findings the implementation ratio of ECR strategies is very low and the polled dairy companies meet only single aspects regarding enablers at the moment. The necessary logistic and marketing instruments are currently not implemented.

Keywords: Supply Chain Management (SCM), Efficient Consumer Response (ECR), dairy industry

1. Introduction and goal

The development of the German food market is characterised, on the one hand, by a rising concentration of wholesalers and increasing market shares of discount-stores. On the other hand, it is characterised by hybrid consumer behaviour and a decline in spending for food (Borchert, 2002). Additional new information technologies enable new forms of collaboration between the trading partners in the supply chain (Kotzab, 2000). In order to cope with these challenges and gain competitiveness and economic growth, new strategies of collaboration between trading partners, namely food companies and retailers, are advisable. The isolated management of single firms must be replaced by a vertical comprehensive management encompassing each single member of the supply chain. In this context, Efficient Consumer Response (ECR) is seen to be the successful supply chain strategy for the grocery industry. The conceptual framework was developed years ago. Several studies indicate high potentials in cost savings and higher sales. In their presentation at the 10th ECR Europe Conference, Holmes and Lonergan (2005) emphasized that ECR yielded € 18 bn in cost savings in the European market for fast moving consumer goods during the last decade. The future potential of ECR was estimated to be € 17 bn. But it is assumed that the implementation ratio of single ECR strategies is still quite low. This can be inferred from the fact that ECR literature often describes empirical findings 'on a single-case-study-basis' (Kotzab, 1999). One could suppose that these cases are used as a strategic 'play-ball by international consulting companies' (Kotzab, 1999) advertising their consultancy services.

This paper investigates to what extent the German dairy industry is ready to implement the ECR concept and which instruments are preferably applied. The article is structured as follows: after a brief description of the ECR concept and its single strategies, core requirements for a successful implementation are deduced from the literature. Section 3 describes the sample, the survey design and evaluation method. In section 4, the results of the investigation are presented. Section 5 summarises the relevant findings and gives an outlook on the future development of ECR in the German dairy industry.

2. Theoretical background

2.1 Concept of Efficient Consumer Response

Although there is not a common definition of Supply Chain Management (SCM) yet, it can be described as a management philosophy that aims to integrate all processes that products, services and information have to undergo from the source of supply to the consumer. ECR is the supply chain strategy developed by the grocery industry. The European ECR Initiative defines ECR as: 'Working together to fulfil consumer wishes better, faster and at less cost.' (www.ecrnet.org; Wildemann, 2003; Alvarado and Kotzab, 2001). It consists of different methods that enable to better coordinate crucial activities along the supply chain (Hoffmann, 2000). By giving up confrontation and building up cooperation in the chain ECR promises less operating costs, higher sales and primarily a higher value to the consumer.

Within the ECR concept, there are three focus areas presented in Figure 1: enabling technologies, the supply side and the demand side (for the focus areas *e.g.*, Corsten, 2004; Weindlmaier, 2003; Alvarado and Kotzab, 2001; Brown and Bukovinsky, 2001; Hoffmann and Mehra, 2000, Zentes and Swobboda, 1999).

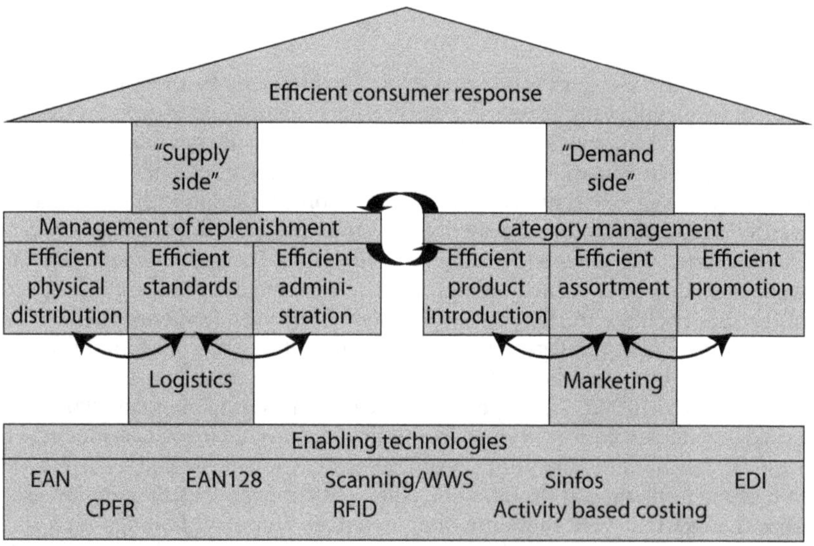

Figure 1. Focus areas of the Efficient Consumer Response concept. Modified from Weindlmaier (2003) and Zentes and Swobboda (2001).

The common basis for collaboration in logistics and marketing are enabling technologies that have been developed over the past years (see Figure 1). These comprise identification standards, standardised communication processes as well as practices to improve supply chain planning and activity based costing (Corsten, 2004). The European article number EAN 13 is used for a distinct identification of every article. The EAN 128 label is affixed on travel units containing a SSCC (Serial Shipping Container Code). This code facilitates tracking and tracing every single travel unit distinctly and optimises logistical processes. Scanning is the key to read and save the information given by these identification standards. Another element of the enablers is the central data pool 'Sinfos' serving to store product attributes. Electronic Data Interchange (EDI) applies the internationally agreed EANCOM standard for a standardised and automatic communication between the members of the chain (Weindlmaier, 2003). Vast amounts of information about sales, orders and inventory levels can be transmitted this way. Fewer errors and lower processing costs result from a broad application of the electronic information flow between enterprises (Brown and Bukovinsky, 2001). Activity-based costing helps to identify this cost reduction on a process level. Collaborative Planning Forecasting and Replenishment (CPFR) is a common process of planning by using an accurate and timely information flow between enterprises (Corsten, 2004). Forecasts of individual chain members and event-based information such as promotions are exchanged with each other. Significant benefits in stock-levels, fill-rates and on-shelf availability are expected by applying these instruments.

The supply side attempts to improve the information and product flow by an automatic order process, standardised shipping units and a reorganization of the physical distribution (Von der Heydt, 1997). A starting point for efficient logistics is the development of appropriate product packaging by using modular sizes for consumer and transportation units (Efficient Standards). On this basis order quantities have to be optimised regarding logistical restrictions aiming to utilise the standardised pallet and lorry capacities in an efficient manner and to avoid inefficiencies and re-supplies (Alvarado and Kotzab, 2001). In a second step, the information flow must be integrated between the trading partners by EDI. This electronic information flow can be applied for orders, despatch advices, receive advices, invoices, and so on. Product movements can thus be tracked chronologically along the whole chain. Better information about storage and demand of products in the chain enable the vendor to carry out the order process (Corsten, 2004). This strategy of Efficient Replenishment can avoid out of stock situations and as well as overstocks of goods in the supply chain. Furthermore, a lot of labour resulting from parallel systems of retailers and manufacturers can be saved. As a result of better transparency and clarity, the Bullwhip Effect (Lee *et al.*, 1997) can be reduced. Depending on the responsibility for collaborative processes in the chain, these concepts are called Vendor-Managed-Inventory or Co-Managed-Inventory. Finally, the physical distribution must be optimised. Goods should be bundled in the production area and distributed in the area of consumption (Zentes and Swobboda, 2001). Cross Docking (CD) aims to unload materials from incoming lorries, to recombine goods and to load these in outbound lorries. CD in its purest form intends minimal or no warehousing. Two types of CD are distinguished: CD I making out the production order for the chain stores by the manufacturer and CD II making out the production order for the chain stores by the retailer in the cross docking centre. Finally, concepts of single enterprises have to be replaced by concepts aiming at reaching the optimum for all members of the chain from the source of supply to the area of consumption.

Category Management is the framework of the three collaborative demand side practices, namely Efficient Store Assortment, Efficient Product Introduction and Efficient Promotion. This means, new categories are developed and managed as strategic business units with the aim to focus on higher consumer value (Von der Heydt, 1997). The store assortment as a starting point is meant to be adjusted to consumer needs by a collaborative effort of manufacturers and retailers (Weindlmaier, 2003). Retailers benefit from an improved efficiency in space utilization and inventory turnover (Hoffmann and Mehra, 2000). All members of the chain can expect higher sales due to fewer out-of-stock events. In another step, manufacturers and retailers can cooperate in developing and introducing new products to meet consumer needs better and to reduce cost by avoiding the development of products that may fail in the market (Brown and Bukovinsky, 2001). Prototypes are evaluated jointly regarding their expected acceptance in the market. Test markets are used for a field perspective. Additionally, the members of the chain coordinate promotion activities in place and time. Forward buying and inefficient promotions should be prevented and replaced by selected promotion activities. A collaborative planning, execution and controlling of promotions help to better meet the consumer and business goals of each category (Corsten, 2004).

2.2 Requirements for a successful Efficient Consumer Response implementation

For a successful implementation of ECR, different requirements have to be met. First, the identification of strategic objectives by top-management and their internalization within the whole company is the starting point for the implementation of ECR and a sustainable development of a company (Mentzer *et al.*, 2000; Homburg *et al.*, 1997). The so-called 80:20-rule states that the soft facts such as human behaviour and, especially, the commitment of the management are relevant for the success rate of cooperation up to 80 percent. Merely 20 percent depend on the hard facts, such as resources and infrastructure (Von der Heydt, 1999). Different empirical studies emphasise the close relationship between the top-management involvement and conviction and the success of cooperation (Mentzer *et al.*, 2000, Homburg *et al.*, 1997).

Second, cooperative concepts like ECR are often threatened by conflicts and opportunistic behaviour of the involved partners. In contrast to that, an intensive and faithful relationship is necessary to stabilise the cooperation (Gerlach *et al.*, 2004). Nevertheless, the current relationship between manufacturers and retailers is characterised by extreme competition. This must be replaced by a 'cooperative spirit of mutual interest' (Gruen and Shah, 2000). Long-term considerations foster fair behaviour in questions of joint aims, allocation of returns and claims to leadership. Private contacts often help to socialise and stabilise business affairs (Borchert, 2002; Belz, 1998). A good relationship is finally characterised by mutual trust, which helps to find ways to solve power games and conflicts (Mentzer *et al.*, 2000). Trust is of great importance as a constitutional success factor of cooperation (Corsten and Gössinger, 2004).

Moreover, implementing ECR and applying new processes requires significant investments in new information technologies, staff and training in the new methods (Lothia *et al.*, 2004; Brown and Bukovinsky, 2001). A lack of these requirements has often been found to be a significant barrier for implementing ECR. A maximum of success can only be reached if all required resources are allocated (Kabel *et al.*, 1999).

The authors' own research shows that the integration of all locations and departments of a company in a consistent IT system are crucial for a successful implementation of ECR enablers. Enterprise resource planning systems actually undergo a wide spread (Großpietsch, 2003). They enable central storage and processing of data. Hence, in consistent management systems these data must be available online (Von der Heydt, 1997). Simultaneously, this internal IT consistency is the prerequisite for communicating efficiently with the entrepreneurial environment.

Interfaces between IT systems of enterprises involving all partners must be installed for coordinated communication with each other (Balling, 1998). By using standardised messages in the frame of Electronic Data Interchange for Administration Commerce and Trade (EDIFACT), bilateral adjustments become redundant (Hansen and Neumann, 2002). Therefore, cost can be saved and new cooperative partners can be won.

Furthermore, the implementation of supply chain management requires the reorganization of the companies. This concerns the organizational structure within the companies, as well as between different companies. Teams with members from different departments have to work together to optimise processes and to achieve a tight coordination between the supply and demand side (Corsten, 2004).

3. Research method

A survey covering three countries to investigate the preparedness and implementation ratio of ECR in dairy companies was conducted in the period from June to July 2005. At first, a sample of 88 German, 15 Austrian and 8 Swiss dairies was generated using a database containing the addresses of several dairies in German-speaking countries. They were predominantly selected in terms of milk processing volume. Additionally, some smaller dairies were selected, which were assumed to be very innovative and open-minded about new technologies.

To develop the survey instrument, previous studies about ECR and cooperation as well as own research results were used. Overall, the eight-page questionnaire consisted of three groups of questions. The first part investigated to which extent the requirements for successful cooperation between manufacturers and retailers were fulfilled. In the second part, the current implementation ratio of the different ECR instruments was investigated. In the third part, additional data concerning the enterprise and the respondent were collected.

As the second and third part consist of questions dealing with facts, their elicitation is quite unproblematic (for the research method see Enneking *et al.*, 2005). However, questions regarding the requirements of successful cooperation had to be paraphrased with statements. We applied a 5-point rating scale that allowed a differentiated evaluation of the statements: -2 meant a total refusal, +2 a total agreement to the respective statement. Those single items were then combined into factors. Their internal consistency was evaluated by Cronbach's Alpha method (Brosius, 2002). This measurement should attain a value of at least 0.7. In sum, 36 questionnaires were returned and evaluated. This corresponds to 32 percent of the sample and represents 58 percent of the turnover and 67 percent of the processing of raw milk of all German dairies. Table 1 shows the success factors and the pertinent statements as well as Cronbach's Alpha values.

Our preliminary research showed that very big enterprises and dairies with an above-average ratio in sales-per-kg of processed milk show higher interest and more advanced implementation of single ECR instruments than their competitors. Therefore, a cluster analysis of the respondents was conducted using the questions describing the implementation ratio of single ECR-instruments. The cluster analysis aims to build groups of cases (respondents) which show very similar values of the variables inside the group, but very dissimilar values of the variables between the groups (Bühl and Zöfel, 2000). The similarity of the cases is deducted from the squared Euclidic Distance. This analysis indicated a similar classification of the single respondents. In the result, the sample was divided into two groups we termed 'precursors' and 'early adopters'. The first group unifies all enterprises with an above-average ratio in sales-per-kg of processed milk (more than 0.64 €/kg)

Table 1. Allocation of items to the success factors.

Success factor	Pertinent Statements	Cronb. Alpha
Internal consistency in IT	Information of different departments is integrated in a consistent IT system. Information of different locations is integrated in a consistent IT system.	0.73
Planned investments in IT	Planned investments in IT	-[1]
Technical and organizational linkage to retailers	Problems are solved by the responsible persons of the involved operational departments. Common ECR-activities undergo an intensive coordination by our and the retailer's stuff. Technical interfaces for an electronic data interchange with retailers exist.	0.78
Intensive relationship to retailers	We cultivate intensive, personal contacts to retailers. Similar occupational qualifications of our and the retailer's staff ease to handle business processes. Retailers are willing to circulate "point of sale" data to their partners.	0.74
Planned investments in staff	We are going to invest above average in stuff for analysing and optimising processes. We are going to invest above average in stuff for implementing and attending cooperative projects.	0.81
Strategic ECR objectives	Our company promotes the development and implementation of vertical cooperation with retailers. We invest in cooperation aiming future profits. All decision-makers of our company promote a commitment in vertical cooperation. Vertical cooperation is not planned by our company[2]. The implementation of vertical cooperation is not an explicit part of our business objectives[2].	0.86
Organizational structures for ECR	For ECR-cooperation clear responsibilities are established in our company. For ECR-cooperation multi-functional teams are established in our company. Our staff has comprehensive knowledge about vertical cooperation in the sense of ECR.	0.81

[1] This aspect was asked in a single statement, as it refers to a single fact which is easily answered.
[2] For further analyses this statement was multiplied with -1.
Source: Own calculations

and additionally all enterprises with a processing of more than 850 m kg milk a year. The early adopters group comprises the rest of the enterprises with a ratio in sales-per-kg of processed milk of less than 0.65 €/kg.

4. Results

4.1 Preparedness for a successful ECR implementation

As described above, the statements concerning the requirements for the implementation of ECR could be reduced to the following seven factors (see Figure 2): Internal consistency in IT (+1.08); planned investments in IT (+1.03); good technical and organizational linkage to retailers (+0.77); intensive relationship to retailers (+0.51); planned investments in staff (+0.25); strategic ECR-objectives (+0.24); and organizational structures for ECR (-0.24). All factors that show a significant difference between the two groups are printed in bold type. The maximum failure allowed for significance is five percent.

The internal consistency in IT is the factor with the highest performance. However there is a wide gap between the precursors and the early adopters. The only factor that shows no significant difference between the two groups is the second one. There are broad investments in IT planned. This shows that all interviewees are currently heavily concentrating on IT concerns with respect to ECR. The third factor shows a lack of technical and organizational linkage especially by the early adopters.

According to our results the relationship to retailers is often seen as intensive. Nevertheless, this does not reflect the real quality of the relationship between the different parties. In additional questions, interviewees characterised the behaviour of retailers more as opportunistic and selfish,

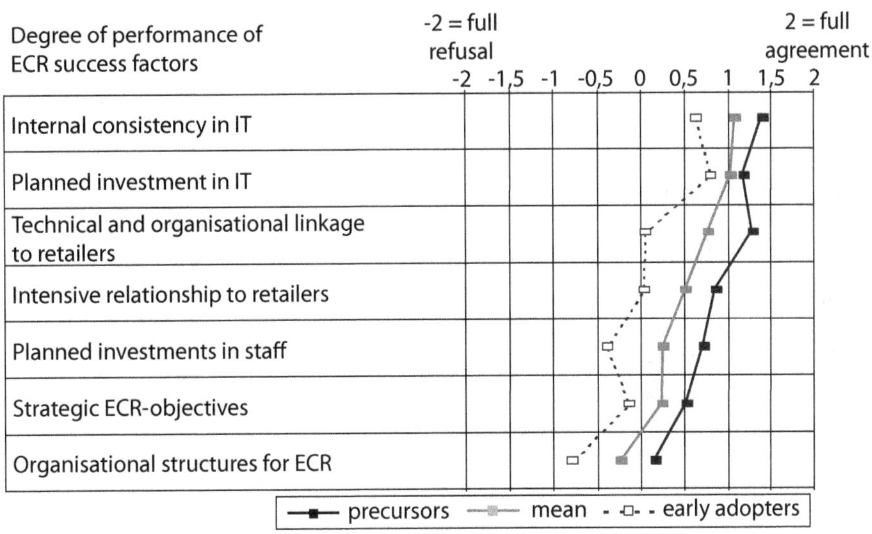

Figure 2. Degree of performance of ECR success factors. Source: Own calculations.

rather than trustworthy. Precursors exhibit a significantly more intensive relationship to retailers than early adopters.

All further factors describing the strategic, organizational or personnel ECR ability are only fulfilled by precursors. But also the arithmetic mean of them shows a quite low agreement. The early adopters indicated a very low performance of these success factors by the refusals of the statements. They are not willing to invest in staff for the purpose of working together in ECR-projects. Cross-functional teams for the tight cooperation within the companies are missing. As a result they have a lack of organizational structures and responsibilities for ECR. In consequence, the interfaces to retailers are not linked in an appropriate manner. Finally, there is a similarly wide gap regarding strategic ECR objectives. While precursors have certain aims, early adopters refrain from setting ECR objectives. In summary, the results show that there are no clear objectives for the development of ECR. Missing objectives inhibit the investment in staff and reorganization.

4.2 Implementation ratio of ECR instruments

According to these findings, the implementation ratio of ECR strategies regarding enablers and supply and demand side instruments is very low (Figures 3-5). The results of the second group of questions are presented in the this section.

Even the basic enablers were given a limited current relevance (see Figure 3). However, it is estimated that their future implementation ratio will be higher. The application of EAN 128 transportation labels on pallet level was currently assessed as most important. This is due to both the commercial pressure and to the regulation 178/2002 of the European Union. The utilization of the 'Sinfos' data pool can be stated as insufficient. The importance was rated as neutral. Up to now, only 43 out of more than 200 German dairy companies are participants of the pool. Furthermore, electronic data exchange is almost exclusively used for the messages INVOICE and ORDER. In case of aiming for an automated order process at least the four EDI messages asked

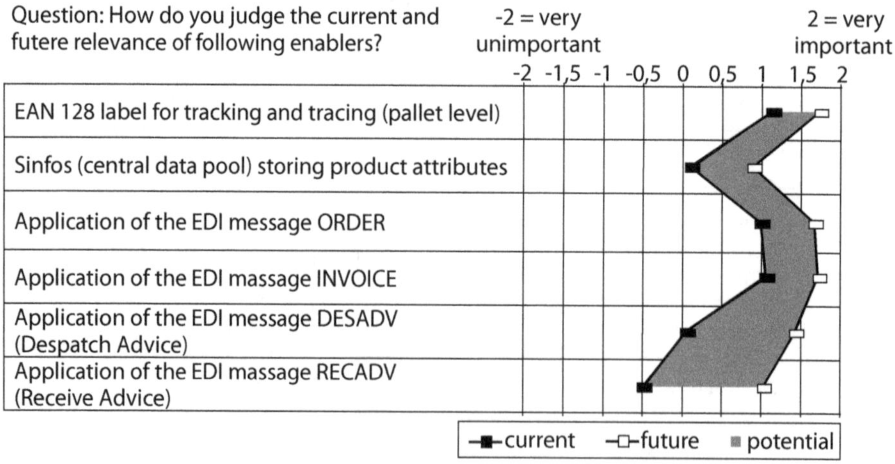

Figure 3. Current and future relevance of ECR enablers. Source: Own calculations

in the questionnaire have to be applied by both of the trading partners. Overall, it can be stated that even the basic enablers have not been implemented thus far. The potential for their future development is stated as high.

Optimising logistic processes in the supply chain in means of ECR has hardly been assessed as important until now (see Figure 4).. However, the future importance is meant to be higher. Transportation standards for the packing of pallets have not even been applied. However, there is a certain linkage to an optimised order process in means of Vendor Managed or Co-Managed Inventory (VMI/CMI). Manufacturers expect that these processes will cost additional manpower and investments in hardware while retailers will benefit from it. On the other hand, case studies from the Austrian dairy industry indicate high benefits in logistic efficiency for both partners by applying VMI. Changing requirements by retailers regarding logistic processes reinforce the implementation of ECR instruments. But there is no way to optimise the flow of goods, other than to cooperate along the whole chain.

ECR cooperation in marketing is considered to be of minor importance at this time see (Figure 5). As a result of high expenditures on products that fail in the first year after introduction, as well as of inefficient spending on promotions, one can assume that these two areas of collaboration are regarded to be of high interest. Dairy managers report that the daily business is rather governed by 'promotion subsidies' and 'listing funds' than by a cooperative atmosphere.

In short, at the moment the interviewed dairy companies meet only single aspects regarding enablers. The necessary logistic and marketing instruments are currently not implemented. Dairies with an above-average ratio in sales-per-kg of processed milk (>0,64 €/kg) or a processing of more than 850 million kg milk a year show a better performance with respect to ECR requirements and more advanced implementation of enablers than the rest of the interviewees. Especially the enormous pricing pressure on dairy products by the highly concentrated and powerful retailers and the surplus production of dairy products are likely reasons for the lack of ECR implementation in this industrial sector.

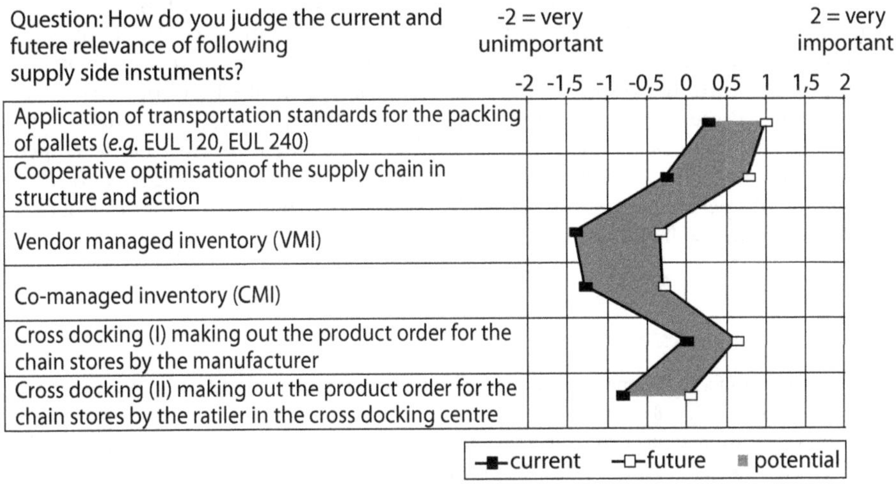

Figure 4. Current and future relevance of supply side instruments. Source: Own calculations

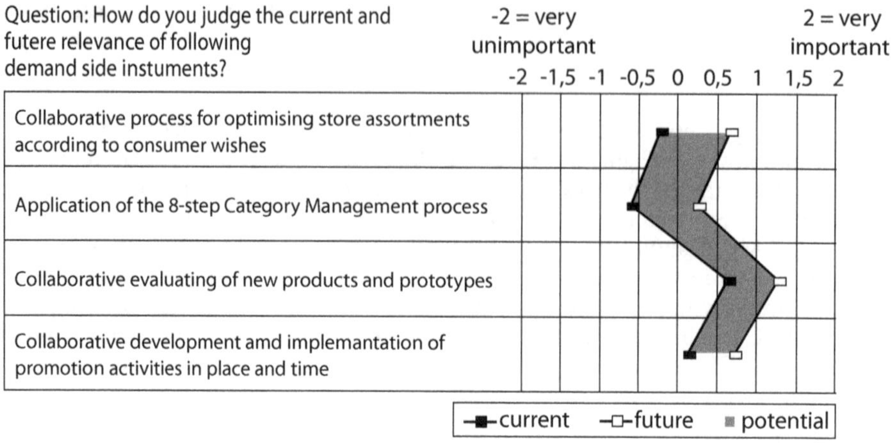

Figure 5. Current and future relevance of demand side instruments. Source: Own calculations

5. Summary and outlook

Joint objectives of all members along the whole supply chain and strategies in accordance with the needs of SCM are crucial for an efficient and effective realization of ECR. However, at least one part of the interviewed dairy companies, namely early adopters, does not fulfil the essential requirements for the successful ECR implementation. The investigation verifies that the implementation of ECR is dominated by IT while the important economic objectives are insufficiently realised. Distrust in the relationship between dairy companies and retailers and a lack of strategic orientation are the main obstacles for its consequent implementation.

However, every single company should analyse which ECR instruments can be implemented to reach a greater benefit for the relevant supply-chain. Therefore, different steps must be considered. First, each individual starting point must be analysed and integrated in the context of the strategic positioning of the company. Second, in case of serious shortfalls, urgent measures have to be taken immediately. Third, the results of the analysis must be solidified and options for action must be developed. These have to undergo a close examination with respect to their economic efficiency. Afterwards, distinct plans for a successful implementation must be developed for promising instruments. Their realization has to respond to the strategic objectives of the top-management. Finally, performance figures have to be defined, which permit the control over the extent at which the aims will be reached.

References

Alvarado, U.Y. and H. Kotzab, 2001. Supply chain management: The integration of logistics in marketing. Industrial Marketing Management 30, 183-198.

Balling, R., 1998. Kooperation: strategische Allianzen Netzwerke, Joint Ventures und andere Organisationsformen zwischenbetrieblicher Zusammenarbeit in Theorie und Praxis. 2nd, rev. edition, Frankfurt am Main, Peter Lang.

Belz, C., 1998. Management von Geschäftsbeziehungen, St. Gallen: Thexis.

Borchert, S., 2002. Implementierung von Category Management im Handel. In: D.F. Frey (ed.), POS-Marketing - integrierte Kommunikation für den Point of Sale. Wiesbaden, Gabler, pp. 253-271.
Brosius, F., 2002. SPSS 11. 1st edition, Bonn, MITP-Verlag.
Brown, T.A. and D.M. Bukovinsky, 2001. ECR and the grocery retailing: an exploratory financial statement. Journal of Business Logistics 22, 77-90.
Bühl, A. and P. Zöfel, 2000. SPSS Version 10: Einführung in die moderne Datenanalyse unter Windows. 7th, rev. edition, München, Addison-Wesley.
Corsten, D., 2004. Efficient consumer response adoption: theory, model and empirical results. Bern, Haupt.
Corsten, H.; R. Gössinger, 2004. Einführung in das Supply Chain Management. München, Wien: Oldenbourg Verlag.
Enneking, U., T. Obersojer, R. Balling, M. Kratzmair and M. Kreitmeir, 2005. Enhancing the acceptance of quality systems by german farmers: The case of quality management and quality assurance. 92nd EAAE-Seminar: Quality Management and Quality Assurance in Food Chains, Göttingen (Germany), http://www.eaae.uni-goettingen.de/Startseite/EAAE-Vortraege/kreitmeir_ enneking_Balling_Obersojer.pdf (24.11.2005).
Gerlach, S., B. Köhler, J. Spiller and C. Wocken, 2004. Supplier Relationship Management im Agribusiness: Ein Konzept zur Messung der Geschäftsbeziehungsqualität. Diskussionsbeitrag 0406, Göttingen.
Großpietsch, J., 2003. Supply Chain Management in der Konsumgüterindustrie. Reihe: Planung, Organisation und Unternehmensführung, Vol. 91, Lohmar - Köln, Josef Eul Verlag.
Gruen, T. and R. Shah, 2000. Determinants and outcomes of plan objectivity and implementation in category management relationships. Journal of Retailing 76, 483-510.
Hansen, H.R. and G. Neumann, 2002. Wirtschaftsinformatik. 8th edition, Stuttgart, Lucius & Lucius.
Hoffmann, J.M. and S. Mehra, 2000. Efficient Consumer Response as a supply chain strategy for grocery business. International Journal of Service Industy Management 11, 365-373.
Holmes, R. and E.L. Lonergan, 2005. The business case for ECR. 10th ECR Europe Conference, http://www.ecrnet.org/conference/files/presentations/Plenary%201%20The%20case%20for%20ECR/1%20The%20case%20for%20ECR.pdf (24.11.2005).
Homburg, Ch., C. Engelbrecht and H. Krohmer, 1997. Handel und Hersteller über Efficient Consumer Response (ECR): Zwei unterschiedliche Perspektiven. Mannheim.
Kabel, D., R. Durst and M. Mühlfelder, 1999. Voraussetzungen für unternehmensübergreifende Kooperationen. In: H. Luczak and M. Schenk (eds.), Kooperationen in Theorie und Praxis: Personale, organisatorische und juristische Aspekte bei Kooperationen industrieller Dienstleistungen im Mittelstand. Düsseldorf, VDI Verlag, pp. 92-105.
Kotzab, H., 1999. Improving supply chain performance by efficient consumer response? A critical comparison of existing ECR approaches. Journal of Business & Industrial Marketing 14, 364-377.
Kotzab, H., 2000. Managing the grocery industry in an efficient consumer response manner. In: Global purchasing & supply chain strategies. December 2000.
Lee, H.L., V. Padmanabhan and S. Whang, 1997. The bullwhip effect in supply chains. Harvard Business Manager 82, 78-87.
Lothia, R., F. Tian Xie and R. Subramamiam, 2004. Efficient consumer response in Japan: Industry concerns, current status, benefits, and barriers to implementation. Journal of Business Research 57, 306-311.
Mentzer, J., S. Min and Z. Zacharia, 2000. The nature of interfirm partnering in supply chain management. Journal of Retailing 76, 549-568.
Von der Heydt, A., 1997. Efficient Consumer Response: Basisstrategien und Grundtechniken, zentrale Erfolgsfaktoren sowie globaler Implementierungsplan. 2nd, rev. edition. Frankfurt, Peter Lang.
Weindlmaier, H., 2003. Die Wertschöpfungskette Milch - Konzepte, Optimierungsmöglichkeiten und Konfliktfelder. Deutsche Milchwirtschaft 54, 109-111.
Wildemann, H., 2003. Supply Chain Management, Leitfaden für ein Unternehmensübergreifendes Wertschöpfungsmanagement. 4th edition, München: TCW Transfer-Centrum-Verlag.
www.ecrnet.org (13.09.2005).
Zentes, J. and B. Swobboda, 2001. Hersteller-Handels-Beziehungen aus markenpolitischer Sicht. In: F.-R. Esch (ed.) Moderne Markenführung, 3rd edition, Wiesbaden, Gabler, pp. 891-911.

Using data mining to improve operations management in food supply networks

Y. Li, M.R. Kramer, A.J.M. Beulens and J.G.A.J. van der Vorst

Abstract

Operations management in food supply networks is especially complex due to the variability of primary inputs and the perishability of products. In order to prevent problems in food quality and operations, early warning and proactive control systems are required. This paper describes a method to identify causal factors for these problems employing data mining techniques. A product of the research will be a reusable knowledge base of important types of variables and relations. In the case study, we investigated the applicability of data mining by applying three data mining techniques in the chicken chain of a large Dutch food company. Relations discovered in this case were confirmed by experts and practice, which demonstrates that data mining could be used to help to improve the efficiency and effectiveness of operations management in food supply networks.

Keywords: data mining, knowledge base, early warning, proactive control

1. Introduction

European consumers are sophisticated and highly conscious of food quality and safety (Beulens, 2005). Recall announcements can be found in newspapers almost weekly due to problems in operations. Operations management in food supply networks is especially complex because of the variability of primary inputs and the perishability of products. To ensure consistent quality of food products and to prevent food safety problems, it is beneficial to set up early warning and proactive control systems in food supply networks.

Box 1. A case for early warning and proactive control in operations management.

December 15, 2005, a Dutch dairy firm had to recall packets of chocolate milk after children turned sick from drinking the milk. The problem was caused by traces of cleaning substance left after cleaning the pipes in the factory. With an early warning system the problem could have been detected before the packets were actually sold and consumed. A proactive control system could even have prevented using the pipes that were not properly cleaned.

In current food supply networks, large amounts of data about business operations and transactions are recorded into ICT systems every day (Petersen *et al.*, 2002). Apart from their day-to-day operational use, the information implicitly present in these data is valuable as a basis for implementing early warning and proactive control. Many unknown causal relations exist between the recorded data about inputs, controls and production means, and operational performance of processes inside food supply networks. Knowledge about these relations provides the possibility to improve operations management by monitoring and proactively influencing the corresponding determinant factors. So both from a scientific point of view and from a practical perspective, it is an interesting topic how to get value from recorded data.

In our research project, we intend to design a generic knowledge based framework to set up early warning and proactive control systems in food supply networks. This paper presents the general research plan and describes a first case study. Ultimately, we will use data mining for discovering causal relations from recorded data, as well as for monitoring determinant factors during actual operations. Here, we present the feasibility of this approach in a practical case study. Therefore we investigated the applicability of three data mining techniques: decision tree, neural networks, and nearest neighbour methods. The data we used were retrieved from the food supply networks of Nutreco, a large Dutch food company.

In the next section, we introduce the key concepts of our problem domain and some terminology of the techniques employed. Then, section 3 presents the design of the research project of which this study is an element. Section 4 shows the results of a first case study in employing Data Mining to find causal relations in actual data. In the last section we discuss the meaning of this case study and give directions for further research.

2. Review of relevant concepts

In this section we introduce the concepts of food supply network, control theory, early warning and proactive control systems, and data mining.

2.1 Food Supply Networks

According to Santoso *et al.* (2005), a supply chain is a network of suppliers, manufacturing plants, warehouses, and distribution channels organized to acquire raw materials, convert these raw materials to finished products, and distribute these products to customers. Van der Vorst *et al.* (2005) extends this to a food supply network, referring to an interconnected system with a large variety of complex relationships such as alliances, horizontal and vertical cooperation, forward and backward integration in supply chains.

Figure 1 depicts a supply chain network with accompanying monitoring and control systems. Processes can be seen at different levels, depending on the kind of problems considered. They may be as large as complete farms, factories or warehouses, or as small as one individual activity. The monitoring system gathers performance data from processes and their inputs and outputs. The control system can influence settings of the processes involved. Together, the monitoring system and control system manage the whole series of processes.

There is much theory on supply chains and networks in general. However, food supply networks are specific in a number of aspects. For example:

- Food products are subject to perishability, so special requirements are imposed on transport and storage.
- Quantity and quality of process yield vary due to biological variations, seasonality, random factors connected with weather, pests, and other biological hazards.
- Strict requirements on homogenous output of processes are harder to maintain due to larger variability of inputs.

Moreover, recent developments in food supply networks make this kind of management even more complicated. For example, product assortments have exploded, product life cycles have

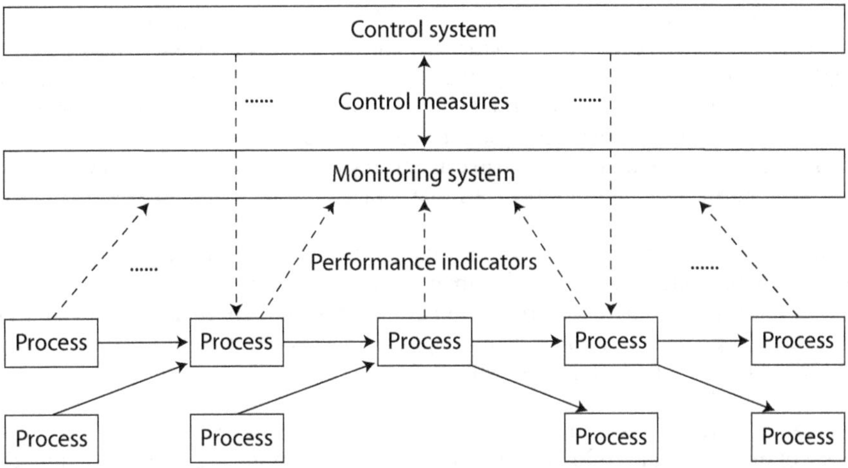

Figure 1. An example supply chain network.

decreased and consumer demands on product freshness and food safety have become more unpredictable (Van der Vorst *et al.*, 2005). Against this background, it becomes a challenging problem how to improve operations management in order to effectively control operational performance, specifically food quality.

2.2 Control systems and performance

Operations management in food supply networks is about the way organizations produce food products and services. Slack *et al.* (1998) distinguish five categories of objectives for operations management: quality, speed, dependability, flexibility and cost. In order to facilitate measuring performance of food supply chains, many authors provided frameworks for constructing performance measurement systems for food supply chains (Beamon, 1999; Gunasekaran *et al.*, 2001; Lohman *et al.*, 2004). A performance measurement system associates measurable performance indicators with important processes and intermediate products in the supply network. For control purposes, the variation of values of certain performance indicators should be restricted. The lower and upper acceptable values for a performance indicator are called its control limits.

We define deviations as any variations that exceed the control limits of performance indicators; a deviation exceeds the normal variation that is inherent to the process. In food supply networks, various kinds of deviations may occur, either from the specifications of product attributes, or from the specifications of logistical processes (*e.g.* time restrictions, excessive use of raw materials or resources). Deviations may cause defective products which may pose a threat to the environment, to the public involved, or to the chain participants.

There are two key mechanisms in control theory to keep variations between limits: feedback and feedforward control (Goodwin *et al.*, 2001). With feedback control, a system maintains its homeostatic action through adjusting the system according to monitored output properties, irrespective of the origins of the input variations that disturb the system. With feedforward control, the process disturbance is forecasted before it actually occurs. The forward cycle anticipates an expected situation based on earlier measurements in the process. A pure feedforward system makes no attempt to monitor and feed back the actual values of controlled variables. Rather it monitors the state of variables that are known to affect performance, and feed forward corresponding control actions to counter the disturbing actions encountered. We define determinant factors as those variables that play an important role in causing any deviations in controlled variables. Thus, a feedforward control system monitors determinant factors to forecast and prevent deviations.

In food supply networks, causes of deviations may reside in many aspects, for example, mistakes by operators, or biological variation of primary products. If feedback control is applied for deviations in food supply networks, production can only be influenced after losses have already been incurred. To prevent deviations from happening, it is necessary to realize feedforward control for deviations in food supply networks.

It follows that, in order to function effectively, feedforward systems require some form of system model, a set of prediction algorithms, and a comprehensive monitoring facility to detect changes in system inputs (Fowler, 1999). Counteractions can then be computed and fed forward to neutralize the effects of external disturbances and keep the system output within bounds.

2.3 Early warning and proactive control

The principle of feedforward control forms the foundation of early warning systems. Early warning systems are well known in natural sciences. These systems, based on historical monitoring, local observation or computer modelling, predict and help to prevent or reduce the impact of natural disasters. They are typically used to monitor potential disasters relating to meteorology (*e.g.* floods, fire and droughts) (Grijsen *et al.*, 1992; Kumar, 1998), geology (*e.g.* earthquakes and volcanoes) or technology (*e.g.* nuclear safety). Early warning is being extended to other application areas

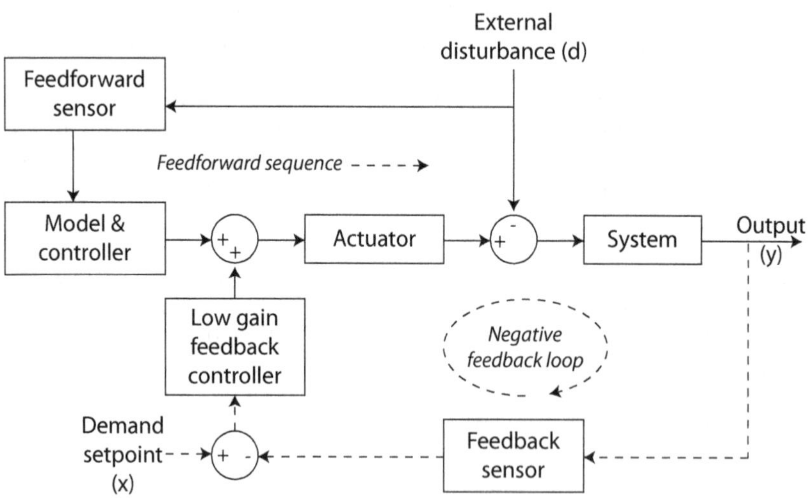

Figure 2. Feedback plus feedforward system (Fowler, 1999).

as well. For example, Costello *et al.* (2003) presented a prototype sensor system for the early detection of microbially linked spoilage in stored wheat grain. Further, the importance of early warning systems is recognized by researchers and professionals in other disciplines to include social, economic and cultural factors. However, to our knowledge there is currently no existing early warning system that is particularly designed to improve operations management in food supply networks.

Our definition for early warning systems in food supply networks is given as follows: Based on the knowledge of relations between deviations in performance of food supply networks and determinant factors, early warning system in food supply networks predict as early as possible potential deviations for decision makers in food supply networks through monitoring measurable determinant factors.

We would like to define the point where determinant factors are monitored. In analogy to Hazard Analysis and Critical Control Points (HACCP) (NACMCF, 1997), we introduce the idea of a *Deviation Analysis Critical Warning Point* (DACWP): the point where an early warning signal about a potential deviation is generated and sent to decision maker before the deviation occurs. It can be a location, practice or procedure where the potential cause for deviation is monitored. As with HACCP, there are two important requirements for DACWP:

1. It is possible to measure at this point
2. The time consumed by measurement should be short enough to take appropriate action

Figure 3 gives a functional architecture for early warning systems in FSCN:

At the critical warning point (DACWP), the monitor system gathers the data about a specific determinant factor, and feeds it to the control system. The control system determines whether to take appropriate action or not. Each of these two steps will take some time, so to realize early warning the total time between DACWP and the warning should be shorter than the time for the product to travel from DACWP to the point where action should take place:

Mathematically, we can define this as Tm + Tc < Ta, where:

- Tm is the time period used by the monitor system.
- Tc is the time period used by the control system.
- Ta is the time period for the product flow from DACWP to the point where action should take place.

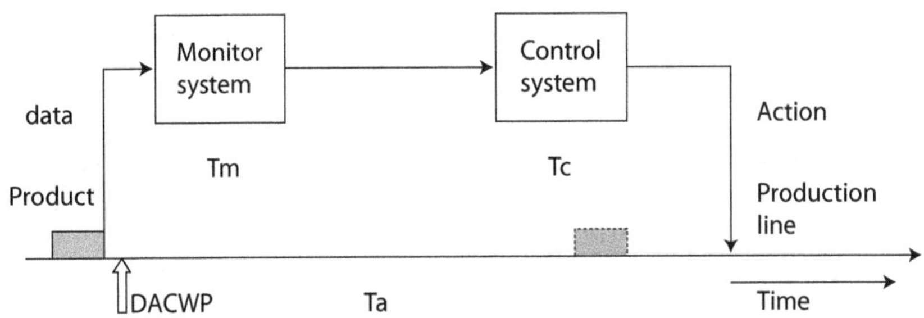

Figure 3. Architecture for early warning system.

This is only a functional architecture to illustrate our idea of early warning system in food supply networks. In practice deviations may have multiple determinant factors, so that we have multiple DACWP to monitor.

After a deviation in performance is early warned, another question appears: how to efficiently and effectively prevent or diminish this deviation? Fortunately with the implementation of early warning system, we already solved the first crucial obstacle. The set of causal relations discovered for setting up early warning systems is a valuable asset. It provides decision makers with information on control points in food supply networks for specific (early warned) deviations. With the knowledge of decision varieties in those control points and their consequences, decision makers could devise a series of actions to proactively control deviations in food supply networks. We define a proactive control system in food supply networks as follows: Based on the knowledge of relations between deviations in food supply networks and determinant factors, proactive control systems in food supply networks proposes appropriate actions in order to prevent potential deviations which have been early warned. The potential actions could be discarding, taking corrective measures, or adapting succeeding processes in order to make amends.

From these definitions we can see that an indispensable component to both early warning and proactive control systems is the set of causal relations between deviations in food supply networks and determinant factors. Ultimately, expert domain knowledge is required to verify those causal relations. However, current advances of ICT systems provide the opportunities to largely automatically disentangle causal relations from huge amounts of data.

2.4 Data mining

The application of early warning and proactive control requires predictive models of the object system (*i.e.* the controlled food supply network). However, to construct such a model, normally would require detailed insight into the processes involved. An alternative approach is to infer a model from available data with the help of data mining. Data mining (DM) is the process of extracting valid, previously unknown, comprehensible and actionable information from large databases and using it to make crucial business decisions (Simoudis, 1996). Data mining has two high-level objectives: prediction of unknown or future values of selected variables, and knowledge discovery (Weiss *et al.*, 1998). The goals of prediction and discovery are achieved by using the following primary data mining tasks (Chen *et al.*, 1996; Fayyad *et al.*, 1996): classification, regression, clustering, summarization, change and deviation detection, and dependency modelling. These tasks coincide with requirements for early warning and proactive control. In practical operation, early warning and proactive control systems in food supply networks have to find infrequent problems in large amounts of operational data (prediction). In our approach, the underlying model of the early warning and proactive control systems is based on causal relations discovered in historical data (knowledge discovery). Table 1 presents a preliminary comparison of some widely used data mining techniques.

There are two approaches for data mining: verification driven, the aim of which is to verify a hypothesis postulated by a user, or discovery driven, which is the automatic discovery of information by the use of appropriate tools. These two aspects of data mining fit well with the observation-hypothesis-experiment cycle of this research. The discovery driven data mining works well for the observation-hypothesis phase, and the verification driven data mining works well for the hypothesis-experiment phase (Crawford *et al.*, 1996).

Table 1. Overview of some data mining techniques.

Name	Features
Nearest Neighbors (Aha, 1992)	Mainly used for prediction, to classify unknown instances into subgroups that have a number of properties in common. Speed is slow relative to other algorithms.
Decision Trees (Breiman et al., 1984)	Mainly used for prediction, to classify unknown instances into subgroups which have a number of properties in common. Can be very sensitive to minute fluctuations in a data set.
Association rule (Agrawal et al., 1996)	Mainly used for knowledge discovery, to find rules such as 'A implies B' in a data set. Results do not necessarily cover the whole data set; they only represent some aspects of dataset.
Neural Networks (Beale and Jackson, 1990)	Mainly used for prediction, to learn (input-output) patterns within data sets, those patterns have a number of properties in common. How to determine the structure of the network depends on the miner's experience.

3. Research design

This research follows the inductive/deductive research cycle in which literatures (about food supply networks, process control, system design and data mining) and multiple case studies are used. In the induction phase, we induce from several cases a systematic framework for applying data mining to set up early warning and proactive control systems in food supply networks. In the deduction phase, we apply this framework to other cases to verify its validity. If the result is not satisfactory, we iterate these two phases again with the problem encountered. By this inductive/deductive cycle we gradually develop a knowledge based framework for setting up early warning and proactive control systems in food supply networks.

3.1 Knowledge base

We could benefit from case studies in several aspects. The major one is the knowledge base. In order to implement early warning and proactive control in food supply network, it is necessary to obtain the following information for every process we intend to control:

- Performance indicators, to describe some aspects of the process's performance and to facilitate monitoring.
- Control limits for all performance indicators to be able to detect deviations.
- Data availability (i.e. whether there is enough data to characterize operations and performance) and the time required to gather data.
- Decision variety (i.e. what are the available decisions) and influence of decisions on subsequent processes.

These kinds of information should be kept in the knowledge base. The knowledge base will be helpful not only for managers to make decisions, but also for researchers to discover new relations in subsequent case studies. By means of the knowledge base we could realize extendable early warning and proactive control systems. Our knowledge base will contain:

- categories of determinant factors;
- categories of deviations;
- categories of relations between determinant factors and performance; and
- specific cases of causal relations in food supply networks.

Relations retrieved by data mining along the way will be categorized and recorded into the knowledge base, together with related determinant factors. Other benefits include the experience in building up these systems, and the merit of various data mining techniques in discovering causal relations. Both knowledge and experiences can be utilized by researchers and practitioners who intend to efficiently and effectively manage food supply networks.

3.2 Design process

Our process for designing early warning and proactive control systems is represented in Figure 4. The process is repeated for each case study; one cycle of the design process consists of these five steps:

1. *Problem description*: With the help of domain experts, descriptions of the problems (or opportunities) and explanations of the variables in the operational database are obtained. The operational database is designed to record data associated with operations and product quality performance. With this information, we specify the requirements of stakeholders involved in food supply networks.

2. *Analysis and matching*: Based on descriptions of problems and the database, we analyze data and, if possible, match them to existing categories in the knowledge base.

3. *Apply DM methods to find relations*: From the knowledge base, we find corresponding data mining techniques that are suitable for these types of data and specific problems (or opportunities). Then we apply these techniques and analyze their results to find relations in these new data. If we find new relations, they will be added to the knowledge base, so the knowledge base grows in the process.

4. *Design control systems*: Based on knowledge of those relations, we investigate the possibility of monitoring corresponding determinant factors and predicting potential deviations in the actual food supply networks, and proactively control deviations with appropriate decision varieties.

5. *Evaluate early warning and proactive control systems (EW&PC systems)*: At the end of a cycle, we evaluate the performance of the early warning and proactive control systems. By interviewing domain experts we investigate whether they regard the effects of these systems as valid in the application domain.

There is a reciprocal relation between domain experts and case studies: first domain experts provide background knowledge on specific problems, corresponding operational databases, and hypotheses of possible relations, *etc.*; then after the case study, they could benefit from relations found in the design process.

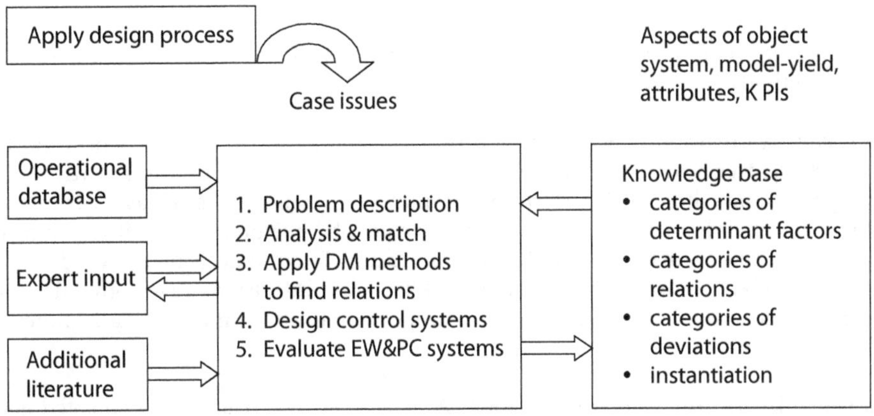

Figure 4. Design process of early warning & proactive control system for one case study.

4. A case study

Here we present an overview of the results of the first case study we have conducted. This case comes from one of Nutreco's food supply networks; it deals with breeding chickens for meat. This supply network was suffering from a high rate of chicken's Death-On-Arrival (DOA). Currently, no early warning or proactive control system is available in this company. However, the information system of Nutreco contains detailed reliable data associated with the processes in the network. These data are gathered from various sources and cover several stages in Nutreco's chicken supply network (see Figure 5). In this case study, we applied data mining techniques to those data. This case study focused on two aspects:

1. identify causal relations for DOA by analyzing data associated with relevant processes; these causal relations are indispensable for setting up early warning and proactive control system;
2. compare the merit and shortcomings of several data mining techniques when applied for discovering causal relations from data.

Further case studies on other data will be conducted as part of ongoing research.

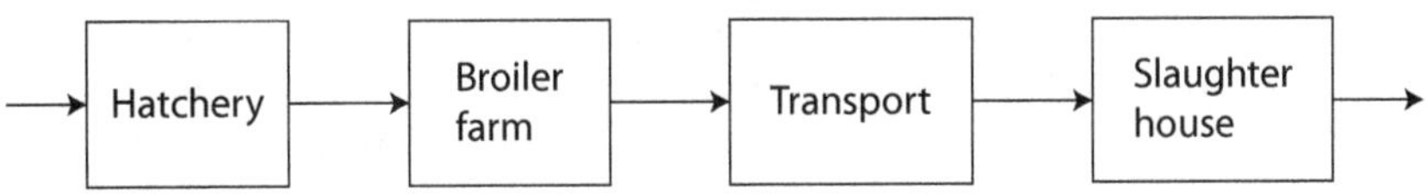

Figure 5. Stages in Nutreco's chicken supply networks, as far as covered in this case study.

4.1 Methods

The data mining techniques we employed include decision tree, neural networks, and nearest neighbour methods.

The Decision tree method is widely used for classification and regression. A decision tree is a flow-chart-like tree structure, where each internal node denotes a test on an attribute, each branch represents an outcome of the test, and leaf nodes represent classes or class distributions. In order to classify an unknown sample, the attribute values of the sample are tested according to the decision tree starting from the root until one of the leaves. To build decision trees, a data mining algorithm recursively inspects the available data set to find decisions that optimally split the data into distinguished subsets. An important property of this technique is that its functioning is easily understood.

Neural networks are used as prediction tools. A neural network consists of simulated neurons and interconnections between them. Each connection has an associated strength, called its connection weight. The neural network has to learn the connection weights from a set of inputs with corresponding optimal outputs. It learns a pattern between input and output by modifying the connection weights between neurons. After learning, a network is used to predict the output for new input of a process.

Nearest-neighbour methods classify a new case based on its 'distance' with stored earlier samples. They assign a new case the same class label as its nearest neighbour(s). In order to get accurate classification, researchers tried several ways to modify the weights of each factors (Wettschereck *et al.*, 1997; Aha, 1998). In this case, we look at each factor's weight as assigned by this technique, in order to find out which factors are important.

4.2 Results

In this section, we focus on methodological aspects of applying DM techniques to find determinant factors and relations that can be included in a knowledge base for the early warning and proactive control systems. We will not go into details regarding the specific determinant factors and relations that we identified for the DOA performance.

Decision tree

Figure 6 shows the result obtained through the decision tree technique. Each node in the tree compares the number of flocks with DOA (in black) to those without DOA (in grey). Take the node in the black circle for example. It indicates that if rejection (PRCAFK) is less than 0.010587, then almost no flocks have DOA (whereas with PRCAFK above 0.010587 significantly more flocks have DOA). The absolute number of instances in each leaf is not shown here, to reduce space. In practical use, we have to take the number of instances into consideration as well.

This graph shows a number of decision rules to determine whether a flock has a higher chance of DOA. For example, as the arrows indicate, if rejection (PRCAFK) is larger than 0.01, breed (KKRAS) is Cobb, transportation time (TITRN) is larger than 1.84, the amount of DOA is relatively large. Some care is needed when interpreting the results. For example, rejection (PRCAFK) has a very high correlation with DOA. However, it is not a determinant factor because both are dependent on other factors.

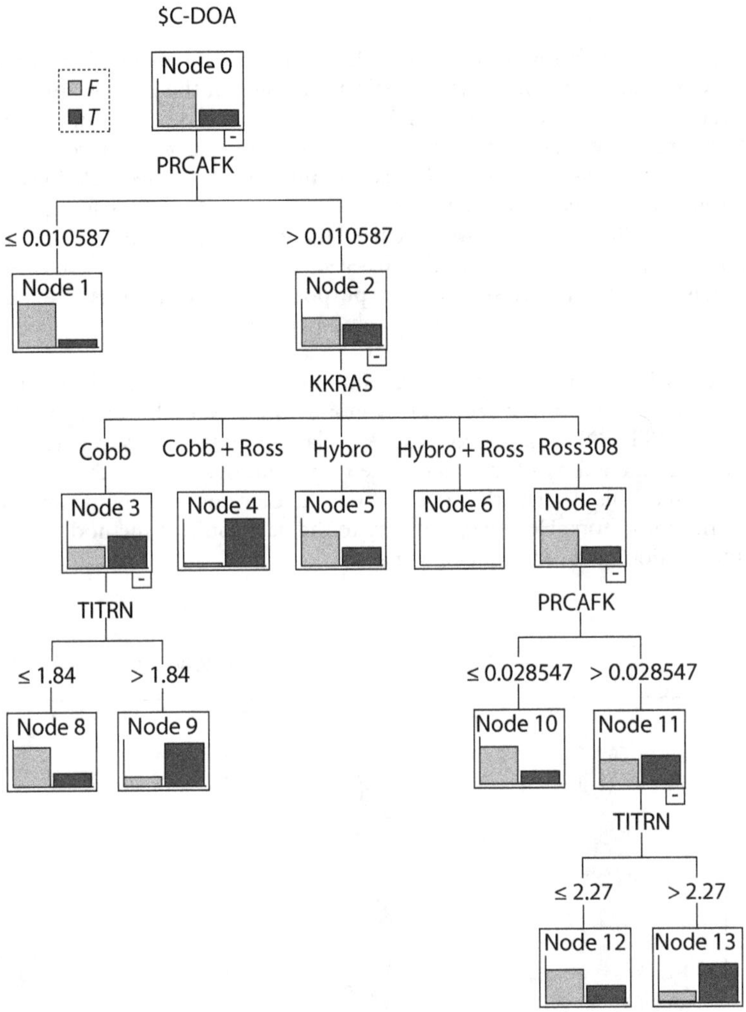

Figure 6. Result obtained through decision tree.

This example of a decision tree shows that the decision tree method could discover relations between determinant factors. These relations provide suggestions for operations management. For example, from this decision tree we can find that when the breed 'Cobb' is subjected to long transportation times, DOA will increase. So managers are suggested to limiting the transport time for 'Cobb'. Sometimes there is only a small set of instances within one leaf, so the decision tree method does not only find large groups of objects having some properties in common, but also tiny groups, that might correspond to exceptional patterns. This feature is especially helpful when dealing with a quality problem that is insignificant in quantity but potentially has serious impact.

Neural networks

Although neural networks are mainly used as prediction tools, data obtained during training the network may provide additional insights. The effect that each of the network inputs has on the network output provides feedback as to which input channels (*i.e.* determinant factors) are most significant. With sensitivity analysis, we retrieve valuable information from the learned network structure. Sensitivity analysis is a method for extracting cause and effect relations between the inputs and outputs of the network. There are several methods to do sensitivity analysis with neural networks (Yao, 2003), of which we have used the variable perturbation method. The first run of sensitivity analysis shows that rejection has a strong correlation with DOA. As discussed above, this result is obvious but useless for control purpose. So we deleted this factor and analyzed again. The graph in Figure 7 shows the results of this analysis.

The magnitude of each bar represents the corresponding change in output (DOA) after applying some change in that specific input factor. This graph shows that chicken's growth (groei/d), transportation time (TITRN), and breed (KKRAS) play important roles in determining DOA. This result illustrates the ability of this technique to give an overview of the influence of each factor in the data set. This information could direct further analysis about how these factors affect performance indicators. However, contrary to the decision tree method, this method can not discover the relation between influencing factors.

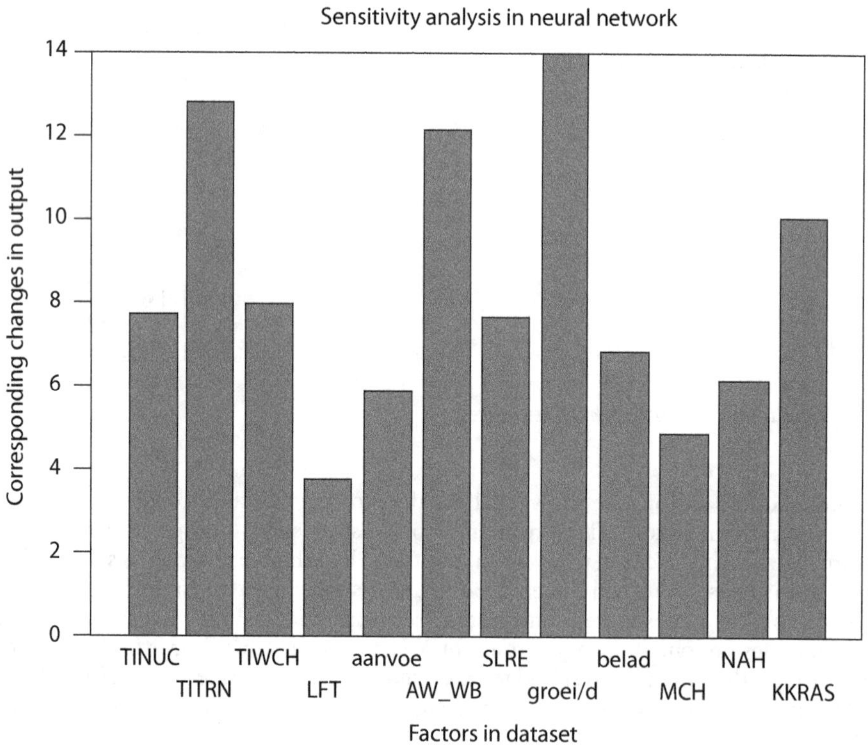

Figure 7. Result from sensitivity analysis on neural networks.

Nearest neighbours

Normally nearest neighbours methods are used for prediction instead of exploring. However, we investigated the weights of factors involved in order to find the importance of each factor in predicting DOA. The factor weighting algorithms we used here is IB4 (Aha 1992). The results obtained show that this method has similar representation power as neural networks:

1. This method gives an overview of the importance of each factor.
2. This method also can not discover the relation between influencing factors.

Figure 8 shows the result obtained with this technique. Most factors' relative importance is similar with those obtained through the neural network. However, some factors' relative importance allocated by this technique is different from those assigned by the neural network. For example, chicken's breed (KKRAS) plays an important role in the result obtained through the neural network, but it is relatively unimportant in the result of the nearest neighbours method. So it is necessary to compare the results of different DM techniques. In this case, the importance of KKRAS has already been demonstrated by two techniques.

By applying data mining techniques, we found a number of important relations in the data set. Most of our findings were confirmed by experts in the company, either because the relations

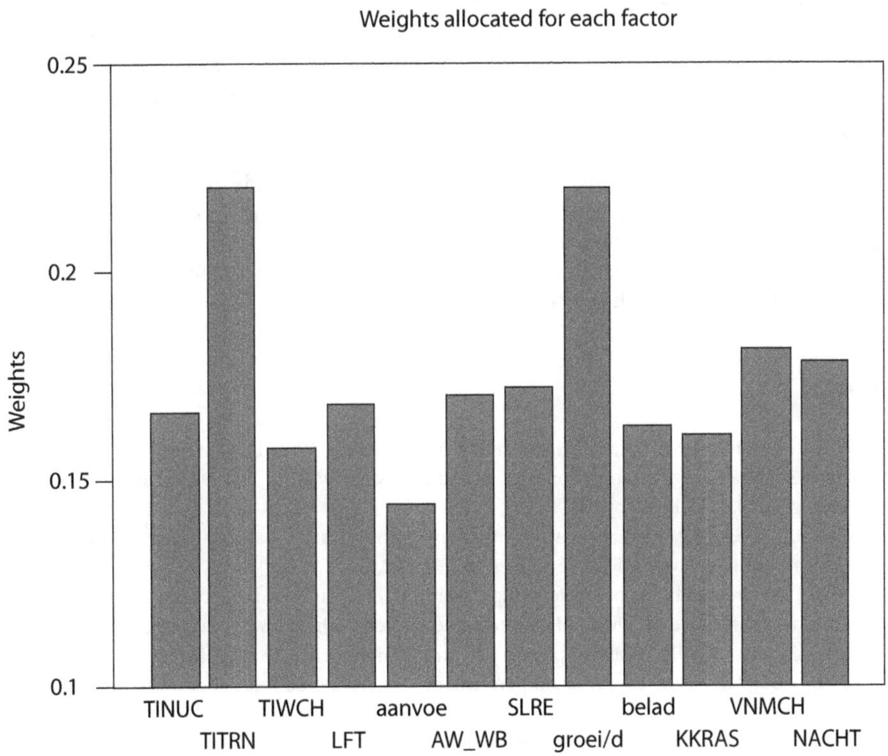

Figure 8. Result from Nearest Neighbors.

were apparent from domain knowledge, or because the experts recognized the relations after reviewing them. The former confirms the capability of data mining techniques to discover domain knowledge. The latter means that the data mining techniques employed are capable of finding new relations that enhance or confirm expert knowledge.

Some relations found turned out to be causal relations actually. Relations found in the earliest part of the data set predicted the effects of measures effectuated in the later parts of the data set. Those relations pointed out possible measures to be taken to decrease losses and improve performance in the operations management of food supply networks.

5. Conclusions

This paper presents the first results in an ongoing research project aimed at constructing early warning and proactive control systems in food supply networks. Such systems can help domain experts improve the efficiency and effectiveness of operations management in food supply networks. In this ongoing research project, we will conduct multiple case studies to gradually build up a knowledge base that can be used for constructing these early warning and proactive control systems. The knowledge base will contain generic information about operations management in food supply networks, especially about types of determinant factors and their influence on performance indicators.

The first case study has illustrated that data mining techniques can help to efficiently find relations between performance indicators and variables related with operational processes in food supply networks. Still, there is a need to improve current data mining techniques to predict which relations are actually causal relations, *i.e.* that some attributes of products may be influenced by changing attributes of designated processes.

Although we gained some promising experience by the case study, we have not yet acquired systematic insight into the typology of relations that may exist in food supply networks and related data mining techniques which are suitable to discover them. This knowledge is also crucial for implementation of early warning and proactive control systems in food supply networks. We are conducting further research in this aspect.

References

Agrawal, R., H. Mannila, R. Srikant, H. Toivonen and A.I. Verkamo, 1996. Fast discovery of association rules. In: Advances in knowledge discovery and data mining. U.M. Fayyad, G. Piatetsky-Shapiro, P. Smyth and R. Uthurusamy (eds.), MIT Press, pp. 307-328.

Aha, D.W., 1992. Tolerating noisy, irrelevant, and novel attributes in instance-based learning algorithms. International Journal Of Man-Machine Studies 36, 267-287.

Aha, D.W., 1998. Feature weighting for lazy learning algorithms. In: Feature extraction, construction and selection: A data mining perspective. H. Liu and H. Motoda (eds.), Dordrecht, Kluwer Academic Publisher, pp. 13-32.

Beale, R. and T. Jackson, 1990. Neural computing: An introduction. Bristol, Adam Hilger.

Beamon, B.M., 1999. Measuring supply chain performance. International Journal of Operations & Production Management 19, 275-292.

Beulens, A.J.M., 2005. Food safety and transparency in food chains and networks: relationships and challenges. Food Control 16, 481-486.

Breiman, L., J.H. Friedman, R.A. Olshen and C.J. Stone, 1984. Classification and regression trees. California, Wadsworth International Group.

Chen, M. S., J.W. Han and P.S. Yu, 1996. Data mining: An overview from a database perspective. IEEE Transactions On Knowledge and Data Engineering 8, 866-883.

Costello, B., R.J. Ewen, H. Gunson, N.M. Ratcliffe, P.S. Sivanand and P.T.N. Spencer-Phillips, 2003. A prototype sensor system for the early detection of microbially linked spoilage in stored wheat grain. Measurement Science & Technology 14, 397-409.

Crawford, J. and F. Crawford, 1996. Data mining in a scientific environment. AUUG 96 & Asia Pacific World Wide Web 2nd Joint Conference, Australia.

Fayyad, U.M., G. Piatesky-Shapiro and P. Smyth, 1996. From data mining to knowledge discovery: An overview. In: Advances in knowledge discovery and data mining. U.M. Fayyad, G. Piatesky-Shapiro, P. Smyth and R. Uthurusamy. Menlo Park (Ca), AAAI-Press, pp. 1-37.

Fowler, A., 1999. Feedback and feedforward as systemic frameworks for operations control. International Journal of Operations & Production Management 19,182-204.

Goodwin, G.C., S.F. Graebe and M.E. Salgado, 2001. Control system design, Prentice Hall.

Grijsen, J.G., X.C. Snoeker and C.J.M. Vermeulen, 1992. An information system for flood early warning. 3rd international conference on floods and flood management, Florence, Italy, Delft Hydraulics Publication.

Gunasekaran, A., C. Patel and E. Tirtiroglu, 2001. Performance measures and metrics in a supply chain environment. International Journal of Operations & Production Management 21, 71-87.

Kumar, V. (1998). "An early warning system for agricultural drought in an arid region using limited data." Journal of Arid Environments 40: 199-209.

Lohman, C., L. Fortuin and M. Wouters, 2004. Designing a performance measurement system: A case study. European Journal Of Operational Research 156, 267-286.

NACMCF, 1997. Hazard analysis and critical control point - principle and application guidelines. National Advisory Committee on Microbiological Criteria for Foods.

Petersen, B., S. Knura-Deszczka, E. Ponsgen-Schmidt and S. Gymnich, 2002. Computerised food safety monitoring in animal production. Livestock Production Science 76, 207-213.

Santoso, T., S. Ahmed, M. Goetschalckx and A. Shapiro, 2005. A stochastic programming approach for supply chain network design under uncertainty. European Journal of Operational Research 167, 96-115.

Simoudis, E., 1996. Reality check for data mining. IEEE EXPERT 11, 26-33.

Slack, N., S. Chambers, C. Harland, A. Harrison and R. Johnston, 1998. Operation managements. London, Pitman Publishing.

Van der Vorst, J.G.A.J., A.J.M. Beulens and P. van Beek, 2005. Innovations in logistics and ICT in food supply chain networks. In: Innovation in agri-food systems - product quality and consumer acceptance. W.M.F. Jongen and M.T.G. Meulenberg, Wageningen Academic Publishers, The Netherlands, pp. 245-292.

Weiss, S.M. and N. Indurkhya, 1998. Predictive data mining: a practical guide. San Francisco, Morgan Kaufmann Publishers.

Wettschereck, D., D.W. Aha and T. Mohri, 1997. A review and empirical evaluation of feature weighting methods for a class of lazy learning algorithms. Artificial Intelligence Review 11, 273 - 314.

Yao, J.T., 2003. Sensitivity analysis for data mining. Proceedings of the 22nd International Conference of the North American Fuzzy Information Processing Society NAFIPS'2003, Chicago, Illinois.

Tailored improvement of supply chain processes in agri-food networks

M. Fritz and T. Hausen

Abstract

Agri-food production is realized in complex network comprising large multinationals and small and medium sized enterprises producing and processing inputs for food production as well as food products. Production and processing in this network involves complex supply chain and logistics processes with complex material and information flows. These processes are often inefficient due to the special infrastructure and conditions in the agri-food production network. Electronic supply chain management is an option to improve efficiency and coordination in complex supply networks. Prerequisite for the realization of improvements with electronic supply chain functionalities in complex markets is the appropriate development and embeddedment of a tailored solution according to the supply network scenario. The paper shows a five-stage procedure model for the analysis, tailored improvement, development and testing of a tailored improvement solution for the agri-food supply network. It shows results from the application of the procedure model to the agri-food production network. With the help of a tailored and embedded supply chain management platform for the network, several improvements could be realised. Empirical evidence from laboratory experiments and expert evaluations clearly showed possible improvements with tailored supply chain management solutions. In this sense complexity could be reduced, process efficiency improved and change efforts be minimized. The tailored improvement is realized with a customized supply chain management platform for the network offering a variety of process support options for members of the supply chain network, ranging from functionalities to speed up processes to the reduction of inventories.

Keywords: supply chain management, logistics, agri-food production network

1. Introduction and objective

Agri-food production is realized in complex network comprising large multinationals and small and medium sized enterprises producing and processing inputs for food production as well as food products. Production and processing in this network involves complex supply chain and logistics processes with complex material and information flows. These processes are often inefficient due to the special infrastructure and conditions in the agri-food production network. Web technology offers support potentials for the coordination of supply chain and logistics processes. The potentials provide particular opportunities for supply networks rather than single firms. ICT support potentials regard coordination and efficiency improvements and excellence in a broad sense, not only speedy operations. However, how to proceed to exploit these potentials to streamline supply chain processes in complex production networks? Prerequisites for the realization of coordination improvements and logistics efficiency in supply networks are a customized ICT support for supply chain and logistic processes in a given production network and a tight collaboration between network firms in order to achieve a global optimum.

Purpose of the paper is to present a procedural model for the tailored efficiency improvement of coordination and processes in complex supply networks. The paper explains the five phases of the procedural model and links it with suitable analysis and design methods (Section 2). It presents results from its application to the part of the agri-food supply and production network ranging from the input production, the distribution of these inputs, the agricultural production, and the distribution of the agricultural products to the food industry for further processing (Sections 3-7).

2. Procedure model for the tailored improvement of supply networks

It is the objective of the procedure model to offer a stepwise procedure to systematically lead to tailored improvements of the coordination of supply chains and networks by reducing complexity, improving process efficiency and minimizing change efforts. The procedure model takes a holistic perspective on the entire network, iteratively supports the improvement of the network processes in five phases (see Figure 1) and provides suitable analysis and design methods and tools for each phase. The combination of different methods and tools supports the analysis from different focal points.

2.1. As-is analysis of value network and benchmarking

The first phase of the procedure model regards the as-is analysis of the supply network scenario to be improved. For the as-is analysis, a flexible palette of methods and instruments is provided for the take-up and analysis of the supply network. An important aspect is the holistic approach to capture the supply chain structures and process organization in an encompassing way. The holistic approach is a prerequisite for tailored improvements and is of uttermost importance for the development and design of improvement options. The as-is analysis of the supply network scenario uses an analysis framework specifically developed for process and organizational

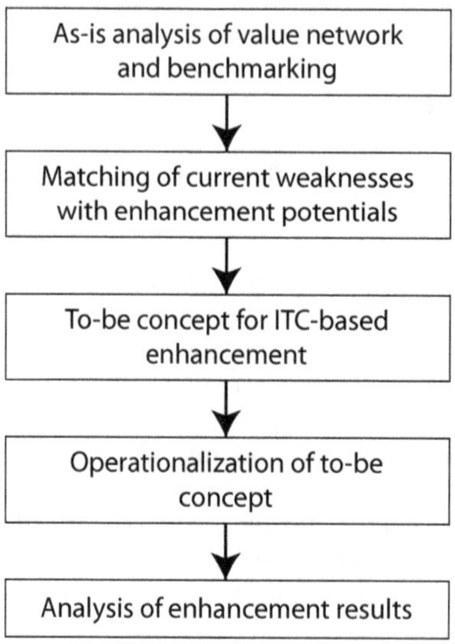

Figure 1. Procedure model for tailored improvement of supply networks.

transformations. Main dimensions are the value network configuration, the products (inputs, outputs), and supply chain and logistics processes (see Table 1). Knowledge about the value network and its products allows for understandings its supply chain and logistics processes and is a key to the identification of their problems. The as-is analysis may also includes a benchmark analysis. The benchmark analysis has an important role to determine potentials for the current situation and possible development paths for supply chain functions by taking best practices form competitors or companies or value chains from different sectors into account. It accounts also to determine implications for the operationalization in a later stage of the procedure model.

2.2 Matching of inefficiencies with potentials

The second phase of the procedure model regards the matching of inefficiencies in the supply network emanating from the as-is analysis with potentials of web technology for supply chain management (see Figure 2). This matching should overcome current inefficiencies and derive opportunities for further improvements from ICT potentials. The matching is an analysis and estimation of potentials with the objective of an improvement of the supply and logistics process coordination in the entire network. The matching takes logistic goals regarding performance (level of availability, flexibility, time and timeliness of delivery or quality of delivery) and costs (reduction of inventory costs, costs of order processing) as a starting point to discuss potentials for the tailored improvement of supplier-buyer-relationships, the customized organization of

Table 1. Dimensions and criteria for as-is analysis.

Dimensions	Criteria
Value network configuration	Firm sizes, number of chain levels, fragmentation degree on levels, relationships
Products	Number and kind of product, product characteristics, demand structure and forecasting, production structure
Supply chain and logistics processes	Number of processes, core processes, interdependencies between processes, process phases, process instabilities

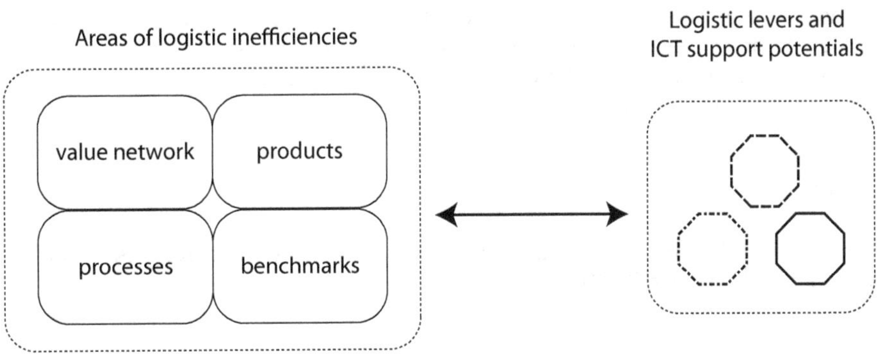

Figure 2. Matching of logistic inefficiencies with ICT potentials.

related processes on all chain levels, and the identification of suitable logistic levers. Typical potentials from ICT include process automation, information exchange and transparency, or new coordination forms.

2.3 To-be concept

The third phase of the procedure model regards the derivation and definition of concrete ICT-based measures to implement the logistic levers and support potentials identified before. For the development of the to-be concept for the tailored improvement of the supply and production network, the procedure model provides a comprehensive and well-tested guideline to define shape, functions and services of the electronic process support according to the demands emanating from the process environment (see Hausen, 2005). As a consequence, tailored and customized improvement in line with demands emanating from the value network, product and processes characteristics through functionalities of electronic support and automation such as transaction facilities can be realized.

2.4 Operationalization of to-be concept

The operationalization of the to-be concept describes the development of IT based solutions to enhance supply chain management. The operationalization depends on the results of the as-is analysis and the developed to-be concept. The operationalization has to consider good practice of software development and is often a trade-off of functionalities, timeliness of development and costs. To achieve an embedded solution, the prototype development has to follow an iterative approach with permanent feedback from experts of the targeted field. Early user involvement and experiments and pilot studies are critical factors for successful development.

2.5 Evaluation of improvement results

As information and communication systems such as supply chain management systems are human-machine-task-environment-systems (*e.g.* Alter, 2002; Heinrich, 1993), the evaluation procedure has to take not only the system itself, but also the system environment (enterprise, value chain network) into account. Especially the execution of processes between enterprises comprises the interaction of humans with a process system (machine) directed towards the process task in a given scenario (environment). Therefore, the evaluation should consider the different objectives of the different target groups of supply chain systems (*e.g.* user as process owner, company's strategic goals, IT maintenance, *etc.*). Reference for improvement are the processes and situation as analyzed in the as-is analysis.

3. As-is analysis

The as-is analysis is based on expert interviews and literature studies (see Bassermann, 1999; Schulze-Düllo, 1995; Müller, 2000).

3.1 Value network environment

The agri-food production network (see Figure 3) ranging from the input production, the distribution of these inputs, the agricultural production, and the distribution of the agricultural products to the food industry for further processing pictures a complex value creation network with a large variety of differently sized companies, ranging from micro enterprises, small and medium sized enterprises to multinationals. Differences in size and owner structure lead to different company cultures and business policies. Companies in the network have a wide variety of different IT systems for process support. The IT systems supporting supply chain processes range from ERP systems to simple spreadsheet based programs and relating breaks in the information flow. Both traders and wholesale traders distribute many product categories. An important success factor for the traders and wholesale traders is logistic flexibility. Stocks therefore play an important role for these companies.

On the value chain levels, different degrees of fragmentation are prevalent. As a consequence, suppliers have a large number of customers and vice versa resulting in a large number of interfaces on the suppliers' distribution side. The different fragmentation degrees require a high level of coordination of suppliers and customers. Logistic interfaces increase from input suppliers to agricultural production linked with the disaggregation of input lots. From agricultural production to the food industry, often product aggregation takes place. The network in general is characterized by either one-tier or two-tier distribution. The large number of interfaces creates high complexity and makes coordination of the information and material flows difficult and inefficient.

Relationships in the agri-food production network vary from spot market relationships to long term contracting. The number of small scale orders increases. Agricultural producers often have only one or two regular suppliers on a regional basis. Larger buyers in the network often show a higher price sensibility. Wholesale traders and traders act as both sellers and buyers in the network. Two opposed flows of material and information regarding input and output of the agricultural production are the consequence. Especially on the distributing levels of the network, the logistic complexity is very high. As a consequence, the chain levels between wholesale traders, traders and agricultural production and vice versa have a bottleneck character in the agri-food supply and production network. Due to this bottleneck character, efficiency improvements on these levels are a prerequisite and starting point for improvements in order to reach a global optimum in the network. The main focus of the subsequent as-is analysis therefore is on this bottleneck in the network and the prevalent products and processes.

3.2 Products

Products distributed and further processed on the bottleneck chain levels are inputs for agricultural production such as crop protection products, seeds, feedstuff or fertilizers and agricultural output including crops and animal products. Inputs for agricultural production have a high need of explanation by specialists, a limited shelf life, and are under a strict legislation. High costs for warehouses and transport makes distribution capital intensive. Agricultural output is a natural product characterized by perishability and changing qualities.

The demand for agricultural input is discontinuous as highly dependent on ad hoc weather conditions and seasonality. As a consequence, forecasting and production planning is extremely difficult for input producers and distributors. The input industry has only rough estimates on total stocks in the supply network due to information asymmetry caused by their traditionally

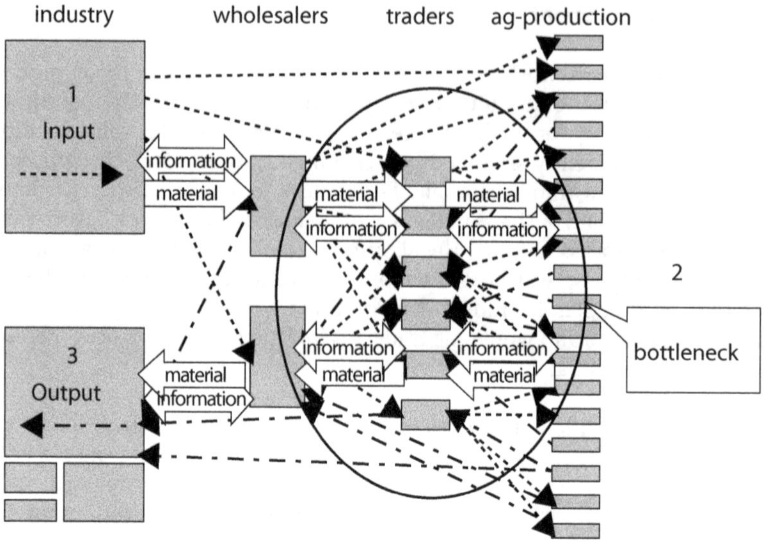

Figure 3. Agri-food supply and production network.

high margins. In the input industry, production inputs are principally produced in campaign production for the entire season to avoid seasonal stock outs. This results in high inventory costs throughout the distributing chain. Agricultural customers expect a high flexibility in delivery and service. As a consequence, suppliers need to realize a high service level. High stock capacities and safety stock levels exist. As supply flexibility is an important element of logistics quality, suppliers should be flexibly integrated in the supply chain. The demand for agricultural output is fairly continuous. Also downstream levels have high inventories levels and safety stocks and expect flexibility of delivery. High quality demands regarding traceability and innocuousness exist and quality management plays a crucial role.

The offer for agriculture input is characterized by frequent new product launches, which are often substitutes to existing products. Highly qualified personnel are required for consulting and product advice for agriculture input. The offer for agriculture output is discontinuous and seasonal and of changing qualities.

3.3 Processes

In general, complex and antidromic information and material streams exist in the bottleneck of the agri-food network entailing a large variety of processes and interfaces exist in different corporate structures (*e.g.* culture, size, IT).

Core process of the traders of the bottleneck is order processing and related subprocesses. The process of order processing is seen as a critical success factor for wholesalers and traders and highly influences the distribution service. It is essential for creating relationships with customers. Process efficiency of the order processing is considered as essential. Order processing consists of the subprocesses order reception, processing, fulfilment, documentation, and customer communication - including specialists' advice in sales - and related processes coordinated by order processing such as storage and transportation. Especially the order processing for the sales of seeds and plant protection products in the high season is time-critical.

Processes are error prone mainly in high season at demand peaks. Process breaks at some point of the process flow are the consequence. Main errors in the order processing process are mistakes in order taking, insufficient specialist advice, inefficient route planning, and mistakes in commissioning.

To select processes in the agri-food network to improve, process portfolio technique is a prioritizing method (Helbig, 2003, see Figure 4). The process portfolio pictures the value for customers on the Y-axis and the web technology improvement potential on the and X-axis. The result shows the general suitability of processes to be improved by web technology and allows for identifying where to quickly achieve results. For the bottleneck of the agri-food network, order reception was determined as most suitable for improvement and analyzed in more detail. The analyses were conducted with several enterprises in order to determine similarities and common problems and inefficiencies. The order reception was in most cases characterized by manual activities with several breaks of the information flow due to media changes.

3.4 Benchmarking

A benchmarking study has identified the state-of-the-art of successful solutions for supply chain management in the food and life science industry (Fritz *et al.*, 2004). The study revealed networking activities of e-marketplaces by cooperation and strategic partnerships and joint activities for the development of functionalities and standardization efforts. The network shows a structure of a kernel of e-marketplaces surrounded by satellite e-marketplaces.

The kernel of this networked infrastructure is formed by major industry driven e-marketplaces like Transora, GNX or CPGmarket. By forming partnerships and focusing on joint standards in both product description and process technology, these e-marketplaces allow for sharing trade functionalities and transferring transaction requests. This kernel of e-marketplaces is expanded by partnerships with smaller, more specialized e-marketplaces surrounding the core network like satellites.

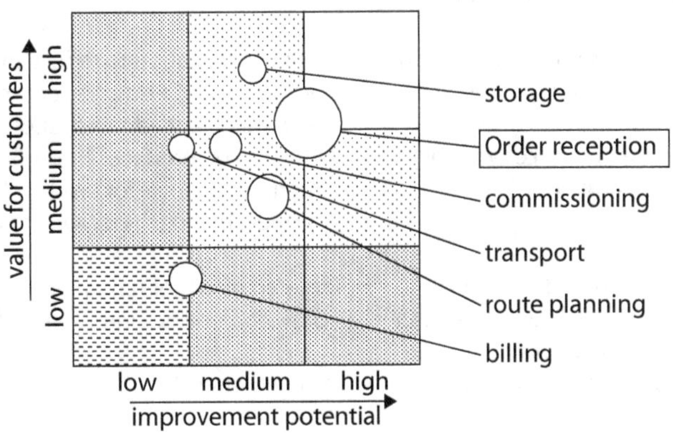

Figure 4. Process portfolio for improvement suitability.

Key findings of success of the networked platforms are acquisition of key players of the sector, tailored and embedded processes and expansion of additional services like collaborative planning.

5. Matching of inefficiencies with potentials

For the logistic inefficiencies in the bottleneck of the agri-food network, ICT-based improvement potentials regarding logistic performance and cost objectives need to be identified. For the main inefficiencies resulting from the as-is analysis, the potentials stated in Table 2 exist to improve the situation. Potentials involve automation, bundling, virtual stocks and targeted offers as logistic levers.

Due to the results of the process portfolio, the improvement of coordination between companies is focused. Web-based support of coordination processes facilitates efficient matching of business partners on different chain levels as well as efficient matching of product lines on one chain level. Improvement between chain levels can be realized through automation. The reduction of customer interfaces improves the organization on the distributing chain levels.

5. To-be concept

The to-be concept derives and defines concrete ICT-based measures to implement the logistic levers automated order receiving, bundling or group buying according to geographic and product demand settings, use of a virtual stock system, and active targeted offers. It follows the objective of a tailored improvement of the agri-food production network and its processes. The to-be concept focuses a web-based platform for supply chain management for the entire network. A web-based platform for supply chain management for the network improves efficiency by combining the advantages of the different logistic levers for synergies. Especially the combination of process automation with bundling or targeted offers creates synergies. The supply chain management

Table 2. Inefficiencies and potentials.

Inefficiencies in processes and organization	Potentials from web technology and logistic levers
Inefficiency of order processing, particularly in reception and order placement processes	Process automation to reduce inefficiency, process cycle time and errors, reduce logistic interfaces, realize efficient administration
High number of customer interfaces with high order frequency in demand peaks	Reduction of interfaces through bundling of information flows for product lines and business partners by matching, adds to improvement of order processing
Need for high safety stocks due to seasonal varying demand patterns on input side often leading to large remaining stocks	Targeted offers for remainder of stock and provision of stock information (virtual stock)
Varying qualities and perishability with the need for aggregation and fast turnover, weak information exchange between levels of network/chain	Targeted offers for remainder of stock and provision of stock information (virtual stock)

platform is a basis for further improvements for the supply chain including planning, the sharing of quality management related information, or tracking and tracing.

The supply chain management platform as a solution for the entire network rather than for a single stage or enterprise needs to be a flexible instrument for supply network management and implemented as an open platform with various functionalities. Companies can use the functionalities independent of each other and picture their business relationships within the supply network. An essential prerequisite for a network solution valid to support and automate the network processes is the availability and identification of reference processes. The reference processes capture and unite process variations existing in different network companies. A reference process builds on these variations and creates a joint basis for the further improvement of the bottleneck in the agri-food network. The process analysis of the order processing in the as-is analysis offers process data of several different companies for the extraction of a reference process for order processing (see Figure 5). This reference process is the starting point for the further definition of the web-based platform for supply chain management in the agri-food network.

5.1 Automation order reception process

The automation of the order reception process regards the order processing of wholesalers and traders in the agri-food network in the distribution of input such as plant protection to the agricultural production (see Figure 6). It builds on the reference process valid for the entire network and has the objective of the automated order reception to enhance efficiency of the order process especially at the seller's side and reduce the process errors in high season. In addition, the reduction of process breaks in high season opens time for customer communication and advising. Figure 6 shows where in the network the platform supports the order processing. In addition, it schematically shows the process improvement by the elimination of process steps and media breaks. For the automation of the order reception process, the platform for supply chain management requires a shop based system with and electronic catalogue of products and prices. The electronic catalogue supports information on available articles, amounts, possible delivery date and varieties of trading units to reduce communication errors.

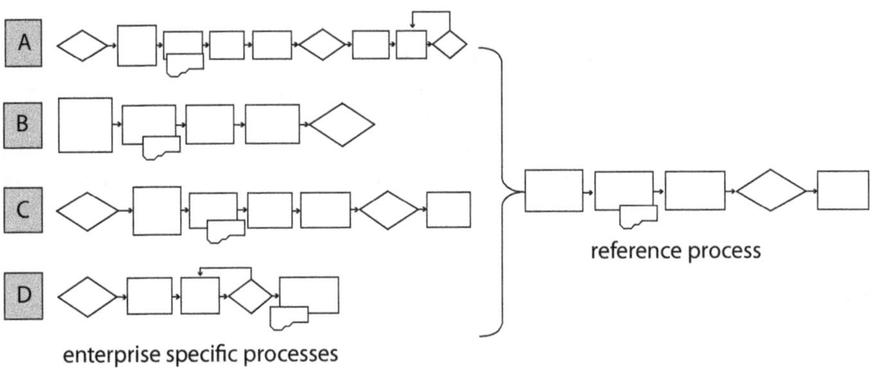

Figure 5. Definition of reference process.

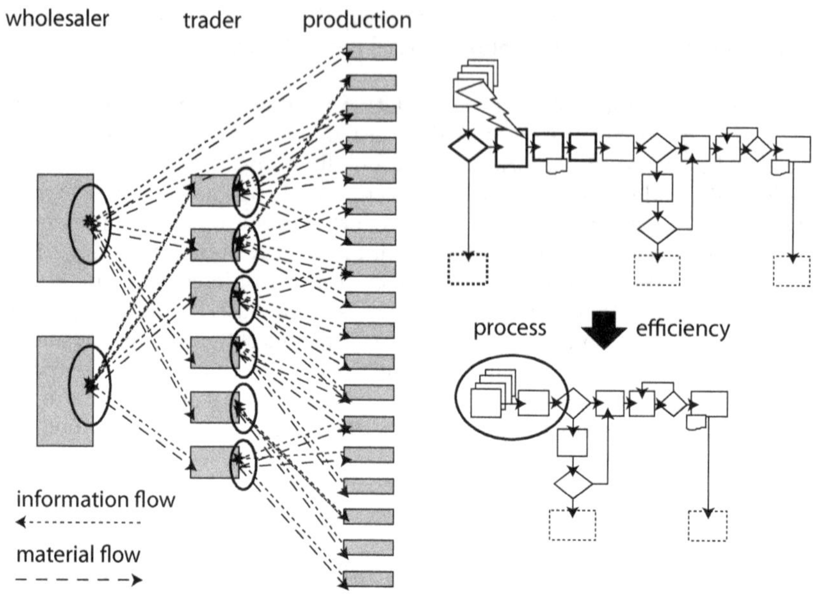

Figure 6. Automation of order processing.

5.2 Group buying

Bundling or group buying as second logistical lever is implemented on the supply chain management platform to reduce the logistical interfaces in the bottleneck of the agri-food supply network. Group buying bundles buyers according to region, product, and delivery day. It builds on the automation of the order processing based on the reference process creating synergies. The group buying mechanism bundles on a common chain level product lines horizontally. For the seller, the group buying mechanism effectively reduces customer interfaces by bundling the customers' information flows.

Figure 7 shows the bundling principle and the reduction of customer interfaces regarding the information flows. The group buying functionality on the supply chain management platform offers several options, ranging from bundling volumes of a same product or bundling different products and variations according to the transport mode. This means that it groups or pools logistics for the same route to optimize transportation or groups loads with flexible delivery dates.

5.3 Virtual warehouse

The virtual warehouse as third logistical lever is implemented to realize an exchange of information on inventories and available products on stock between suppliers and buyers throughout the chain. It has the objective to improve the planning and efficiency of supply chain management processes like procurement, inventories, production or sales in terms of reduction of safety stocks and traceability. The virtual warehouse enables information exchange on available quantities

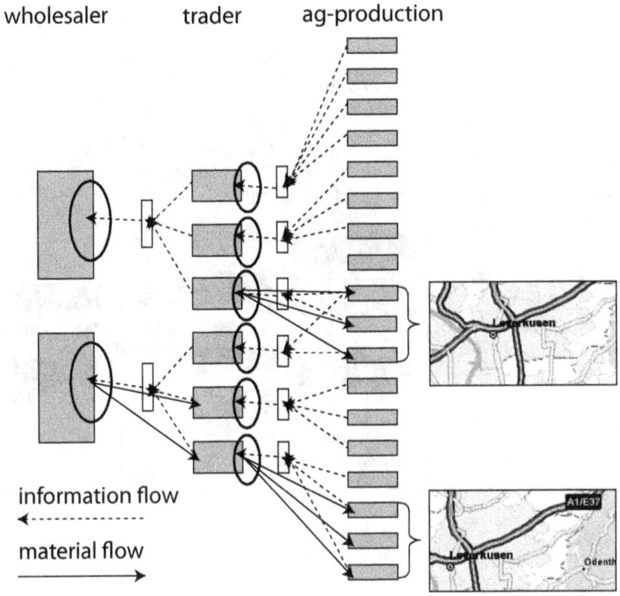

Figure 7. Principles of bundling or group buying for transportation optimization.

of products and an exchange of detailed information on qualities. Information on qualities is particularly important for the levels of the agri-food production network on the output side of agricultural production.

The improved information exchange on stock availability and product qualities in inventories is important for the aggregation of specific product qualities to serve specific supply needs throughout the chain. Figure 8 gives on the left side an overview on the principle of the virtual warehouse for the exchange of inventory information between the chain levels. On the right side, the figure explains in more detail a possible process involving the virtual warehouse on the supply chain management platform for the exchange of inventory information between the ag production, the trade level and the processing food industry. Ag producers (1) as inventory holders provide access to information about the availability and quality of their output to traders (2). Traders combine this information about the (future) availability of ag output (a + b) with the inventory they hold on stock (c) to a virtual warehouse (a + b + c). Information about the inventory in the virtual warehouse is provided to food processing companies (3) when they request a quote over the platform. This shows the synergies between the first logistic lever on the supply chain management platform, the automated order processing, and the virtual warehouse. It is important to note that the provision of access to inventory information between companies requires a high amount of trust between the firms and ability to cooperate.

5.4 Targeted offers

The fourth logistical lever for improvement of the supply chain processes in the bottleneck of the agri-food network regards the option to place targeted offers. Targeted offers are an immediate solution for the purposeful reduction of inventories by offering and selling remainders of stock to

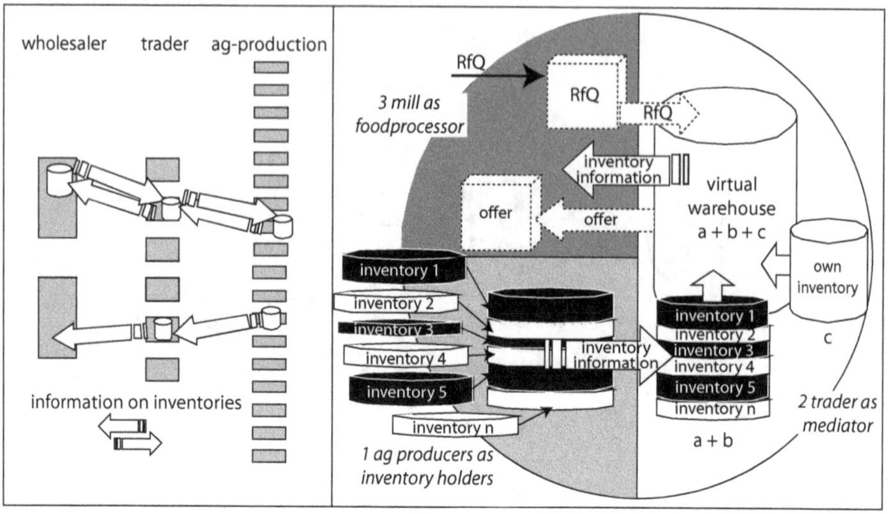

Figure 8. Virtual warehouse.

specific, interested customers. In particular, targeted offers are interesting for perishable products. The targeted offers functionality implemented on the supply chain management platform uses a routing mechanism for matching suppliers' targeted offers with potential demands of buyers. The routing mechanism builds on profiles of companies in the agri-food production network. In addition the reduction of remainders of stock, targeted offers can be used as tool for customer relationship management as it allows for targeting specific customers with specific information or product offers. Figure 9 explains the principles of targeted offers.

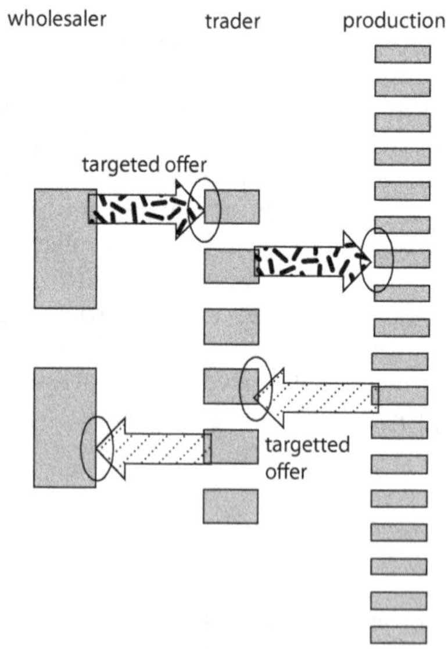

Figure 9. Principles of targeted offers.

6. Operationalization of to-be concept

The to-be concept of the supply chain management platform for tailored improvement of processes in the agri-food production network is operationalised as a customized software application. To realize an enterprise independent software application, the supply chain management platform is accessible over the Internet and has a flexible modular structure integrating functionalities for the four logistical levers mapping the rules, processes, and product demands in the bottleneck of the agri-food network. Companies can use the functionalities for process support independent from other companies. Its modular structure and flexibility enables the future integration of additional logistical levers and functionalities such as tracking and tracing, quality communication, or additional process improvements. From the technical point of view, the software application is designed as a database driven web application using Macromedia ColdFusion as application server and CFML as programming language.

To ensure the customization of the supply chain management platform, an iterative development process in collaboration with future users from network companies (see Figure 10) is necessary. The iterative development and evaluation by experts and the redesign according to their feedback supports the customization.

Figure 11 shows examples for the web based supply chain management platform for the agri-food network. The supply chain management platform is the technological basis that integrates the information infrastructure of the entire network.

7. Analysis of improvement results

After pre-tests to validate functions and processes of the system, the system was configured according to common scenarios and used for laboratory experiments. Several experiments and pilot studies were conducted in order to validate the results of the procedure model for the tailored improvement of supply chain processes in production networks.

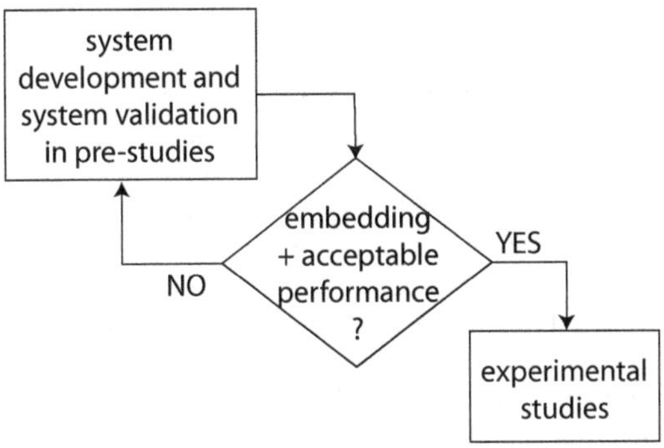

Figure 10. Iterative development procedure.

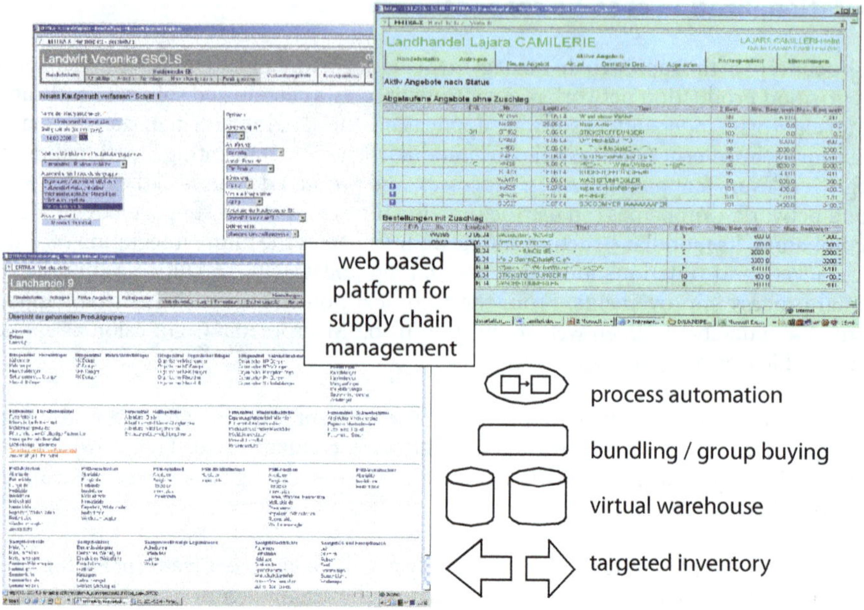

Figure 11. Web based supply chain management platform for the agri-food network.

7.1 Automation of order processing

In laboratory studies, the order processing process via the electronic supply chain management platform was compared with the traditional order processing in the sector. The results showed for both buyer and seller efficiency improvements for a basic direct order of a required input and a request for quote (RfQ) for a required input (see Calosso *et al.*, 2003). These process types build on the extracted reference processes and are the most common order processing variants in the bottleneck of the agri-food network.

The results showed efficiency improvements at the sellers' side by 57% in the direct order process and by 43% in the request for quote processes. As sellers face a high number of transactions per year, with special peaks high seasons, efficiency gains can result in competitive advantages. In the agri-food network, the importance of efficiency gains for sellers of agricultural input to the ag production is influenced by the market structure, as the number of producers is approximately 250 times higher than the number of traders. The competitiveness raises not only due to the streamlined transactions itself, but due to the fact that sales personal can put their efforts into specific sales advice, which is a critical success factor in the selling of ag input. Common errors like wrong product sizes and amounts by manual note taking are also reduced with the electronic order processing system leading to higher customer satisfaction.

For the buyers of agricultural input, the use of the supply chain management system also showed efficiency gains in order processes, 49% in direct ordering and 26% in the RfQ process. The efficiency gain for buyers compared to the traditional RfQ processes would even more increase with a larger number of input sellers to request a quote from. However, as most buyers in the given scenario situation have a low order frequency, the buyers' efficiency improvements may not have a large effect on their competitiveness.

7.2 Group buying

The group buying mechanism to reduces the number of customer interfaces for sellers by up to 80%, depending on the group size formed as a bundling group. In laboratory studies was shown that customers can receive a price reduction on their orders depending on the volume purchased.

Transport routes could be improved and streamlined due to the geographical matching of buyers of ag input. But as coordination efforts decline for the supplier side, some experts mentioned increased coordination efforts on the buyer side. This issue may need further clarifications and empirical evidence by additional laboratory studies and experts tests. As group buying is a sophisticated feature for buyers, it is important to start with low additional efforts regarding coordination.

7.3 Virtual warehouse

An important aspect for the operationalization and implementation of the virtual storage is a stepwise implementation of functionalities starting with functionalities needing only a minimum of additional amount of coordination. These results are similar to the experiences with the group buying feature. An important aspect for the virtual warehouse is power within the buyer-seller relationship. Virtual warehouses build on trust as the prerequisite for cooperation. Experiments have also shown that information asymmetry leads to more inefficient transactions in the described scenario and so further strengthen the principle of the virtual warehouse (Hausen *et al.*, 2006). A prerequisite for the virtual warehouse is a good cooperation of the network members, since the reduction of information barriers makes companies more vulnerable to aggressive competitive behaviour of linked suppliers and customers. Especially the trade levels are reluctant and characterized by protective behaviour for sales margins. The virtual warehouse has an important role in the area of tracking and tracing and was further developed to a system in itself with additional functionalities (Poignée *et al.*, 2004).

7.4 Targeted offers

The targeted offers option to reduce remainder of stock sis an intermediary option to flexibly reduce inventories. Nevertheless, the long term goal should lead to a reduction of safety stocks by improved planning and forecasting throughout the chain.

As targeted offers can also be used as an instrument for sales, customers can be specifically targeted like with customer relationship management (CRM) tools. Experiments in this area have shown that the targeted offer mechanism results in higher prices compared to direct order or RfQs as buyer tend to often to be price takers (Hausen *et al.*, 2006). In general the results implicated a first-mover advantage for either customers or suppliers who initiated a transaction.

8. Conclusions and discussion

Supply networks with small and medium sized enterprises are complex in many industry sectors and, as a consequence, often inefficient. This paper has taken the complex supply chain situation in the agri-food sector as example for complex supply chains in general. The agri-food sector is an example for an industry sector with complex supply chains due to its market structure, with a majority of SMEs, the specifics of the agri-food products, and the transaction processes. Electronic supply chain management is an option to improve efficiency and coordination in complex supply chains. Prerequisite for the realization of improvements with electronic supply chain functionalities in complex markets is the appropriate development and embeddedment of the solution according to the supply chain scenario.

The paper showed and used a five-stage procedure model for the analysis, tailored improvement, development and testing of a solution for a specific supply network. With the help of the tailored and embedded supply chain management platform several improvements could be realized. Empirical evidence from laboratory experiments and expert evaluations have clearly shown the possible improvements of tailored supply chain management solutions. In this sense complexity could be reduced, process efficiency improved and change efforts be minimized. The platform developed with the procedure model offers a large variety of solutions for members of the supply chain network, ranging from functionalities to speed up processes to the reduction of inventories.

Especially the process efficiency was raised by electronic process automation and support and may lead to competitive advantages like improved customer services as was stated by experts. The efficiency gains also opposed the opinion of many company leaders who stated electronic transaction as cumbersome due to personal efforts of using an electronic supply chain system, in this way, the results of the paper may be considered as incentive for companies in complex supply chains to engage in electronic commerce, as the processes of direct ordering and request for quote are also common processes in other areas. Electronic supported supply chain transactions have the strongest impact on operational effectiveness in scenarios where transactions are characterized by manual activities and a high transaction frequency, as the paper has shown. Fragmented value chain levels are commonly found in various industry sectors like machinery, wholesale, transport, pharmaceuticals or SMEs in general, especially on the buyer's side (see Naslin *et al.*, 2000).

Transport routes could be improved and streamlined due to geographical matching on the demand level. Stock visibility could be enhanced and remainder of stocks be reduced. The demonstrated benefits clearly show an impact on operational effectiveness. These benefits should be the driver for organizational change. To realize the potentials of electronic supply chain management, a close collaboration of the supply network partners is necessary. This was also indicated by the asymmetric realization of efficiency benefits from electronic supply chain transactions. As a consequence, trust within the network is an important driver of improvements and for the network companies to achieve global optimum. Starting point for improvement should be the point of the network where largest possible gains may be reached with smallest possible organizational and structural changes. Closer relationships and collaboration, *e.g.* supported by and realized in virtual networks (see Khalil and Wang, 2002; Manthou *et al.*, 2004), may allow for sharing the efficiency benefits and – as results from current experiments show – an even higher efficiency of the entire vertical supply chain (see Hausen and Schiefer, 2004) and realize the full advantage of group buying and virtual warehouses. The experience in system implementation has shown to work with key clients in order to demonstrate the benefits with minor organizational change disturbances in the first place.

References

Alter, S., 2002. Information systems. The foundation of e-business. 4th ed. Upper Saddle River.

Bassermann, K., 1999. Der Markt für Pflanzenschutzmittel. Bestimmungsgründe und Marktanalyse. Agrimedia, Bergen.

Calosso, T., M. Cantamessa, D. Vu and A. Villa, 2003. Production planning and order acceptance in business to business electronic commerce. International Journal of Production Economics 85, 233-249.

Fritz, M., T. Hausen and G. Schiefer, 2004. Developments and development directions of electronic trade platforms in US and European agri-food markets: Impact on sector organization. International Food and Agribusiness Management Review 7, 1-20.

Hausen, T. and G. Schiefer, 2004. Experiments in the electronic market space: First-mover-advantage, information asymmetry, and trust. In: G. Schiefer, P. Wagner, M. Morgenstern and U. Rickert (eds.), Reports of the 25. GIL Conference 8–10 September, Bonn, Germany. Lecture Notes in Informatics (LNI) – Proceedings, Series of the Gesellschaft für Informatik, vol. 49, pp. 245–248 (in German).

Hausen, T., Fritz, M. and G. Schiefer, 2006. Potential of electronic trading in complex supply Chains: An experimental study. International Journal of Production Economics (In press)

Heinrich, L.J., 1993. Wirtschaftsinformatik. Einführung und Grundlegung, Oldenbourg, München, Wien.

Helbig, R., 2003. Prozessorientierte Unternehmensführung. Eine Konzeption mit Konsequenzen für Unternehmen und Branchen dargestellt an Beispielen aus Dienstleistung und Handel. Heidelberg.

Khalil, O. and S. Wang, 2002. Information technology enabled meta-management for virtual organizations. International Journal of Production Economics 75, 127-134.

Manthou, V., M. Vlachopoulou and D. Folinas, 2004. Virtual e-Chain (VeC) model for supply chain collaboration. International Journal of Production Economics 97, 241-250.

Müller, F. (ed.), 2000. Agrochemicals. Composition, Toxicology, Applications. Weinheim, Wiley-VCH.

Naslin, S., A. Wong, K. Majdalany, P. Jensen, J. Molina, A. Roskill and K. Carey, 2000. The A–Z of B2B. UBS Warburg Global Equity Research Study, September 2000.

Poignée, O., Th. Hannus and G. Schiefer, 2004. Quality Management in the Grain Chain - Requirements, Organizational Alternatives and Implementation Priorities. In: G. Schiefer and U. Rickert (eds.), Quality Assurance, Risk Management and Environmental Control in Agriculture and Food Supply Networks, University of Bonn – ILB-Press, Bonn, pp. 129-138.

Schulze-Düllo, M., 1995. Logistik-Controlling im Landhandel: Analyse und Gestaltung eines Managementunterstützungssystems. Verlag M. Wehle, Witterschlick/Bonn.

Part 5. Power

A power-based approach to the analysis of the food system

V. Sodano

Abstract

This paper stresses that in order to understand the current restructuring processes in the food system it is necessary to take explicitly into account the role of power as a driving organizational force. Agricultural economics, drawing pervasively on the Walrasian model, has mainly analysed power in the form of market and bargaining power. Stemming from different definitions of power, the paper focuses on some definitions suggested by the new institutional economics and the network theory, showing their relevance to the analysis of the food market.

Keywords: power, scale-free networks, new institutional economics

1. Introduction

During the last two decades a good deal of theoretical and empirical literature have tried to model and estimate market power in the food sector, especially with the aim of supporting and guiding antitrust legislation. This literature has mainly been based on models of imperfect competition offered by the standard economic theory (Sheldon and Sperling, 2001). Only a few studies (Magdoff *et al.*, 2000; Goodman and Watts, 1997) have tried to use non-orthodox approaches to explain the ways in which the occurring concentration and centralization of capital in the food sector could affect national economies and societies well beyond the general accounted "market power effects".

In this paper I stress that in order to understand the current restructuring processes in the food system it is necessary to take explicitly into account the role of power as a driving organizational force in the system.

In the next section, I introduce the concept of power, confronting some different definitions of power from different realms of social science. Section 3 surveys economic literature on power, stemming from the view of the standard model to the newest theories developed by the new institutional economics and to power definition and concepts suggested by different strands of social network analysis. Section 4 uses different concepts of power to address some organizational problems in the food system.

2. A power definition: power as a relational behaviour

A great deal of social and political literature has attempted to give accurate definitions of power and it would be a challenging task to even give only a taste of it. Some classical monographs (Lukes, 1986, 2005; Popitz, 1986; Clegg, 1989) offer a wide survey of the issue. My aim in this section is to sketch a short "systematic of power" limiting the attention to those definitions of power that better fit the economics of the food system.

Russel (1986) defines power as the production of intended effects, that is the capability of somebody of achieving her own interest. This definition is a wide one and states the first attribute of a workable concept of power: the behavioural attribute.

Power is mainly a behaviour, involving a rational human being, who acts to pursue some aim. That which gives the actors the possibility of deciding (defining goals and means) and acting (actually performing their task) is their endowment of material and immaterial assets. The magnitude of this endowment also determines the degree to which the actor is able to achieve his desires (that is the success of his behaviour). In this general form power can be labelled "power to act" (Palermo, 2000), or "power to do" (Holloway, 2002).

In order to achieve her aims the actor can also command somebody else to do something and/or control the behaviour of somebody. In this case the power to do something somehow implies the power over somebody. Accordingly, power is defined in a relationship, and becomes a strategic behaviour (this means that the behaviour of one actor takes into account the behaviour of other actors; in other words the choice domains of two or more actors are in someway overlapped). The relational dimension of power is underlined by the well-known definition by Dhal, thoroughly quoted in the literature on power: *"At the most general level, power terms in modern social science refer to subsets of relations among social units such that the behaviours of one or more units (the responsive units, R) depend in some circumstances on the behaviour of other units (the controlling units, C)"* (Dhal, 1986).

The difference between these two power characterizations, power to act and power over, is the first step in my power systematic, as illustrated in Figure 1. Power to do refers to the "brigth" side of the concept, *i.e.* power as possibilities (can-ness), freedom, opportunities. Power-over refers to the "dark side" of the concept, *i.e.* power as control, command, coercion.

The second step in my power systematic is the characterization of power-over stemming from the very famous weberian definitions of power.

Weber defines generically power (macht) as *"the probability that one actor within a social relationship will be in a position to carry out his own will despite resistance, regardless of the basis on which this probability rests."* (this sentence, taken from the English edition of Wirtschaft und Gesellschaft, is quoted in Dhal, 1986).

Besides this primary form of power, that indicates a relationship de facto, regardless of its legitimacy (Ferrarotti, 2004), Weber dedicates his analytical effort to the legitimate power (Herrschaft), *i.e.* to that power which identifies a legitimate relationship of subordination. The latter takes three ideal forms: 1) bureaucracy, when legitimacy lies in the legal system; 2) the traditional power, when legitimacy lies in culture and tradition, that is in those informal contracts (common values, conventions) that integrate societies besides the state; 3) charismatic power, when legitimacy lies in the peculiar characteristics of the leader that make the subordinates 'instinctively' be willing to obey.

The weberian definitions of power (see again Figure 1) are consistent with the basic definition of power as a relationship (power-over). The absolute form of power (macht) does not need legitimacy to be exercised and includes the extreme form of coercive power, violence. The legitimate power (herrschaft) is based on consensus obtained through a formal contract (bureaucracy), or by an informal contract (traditional power) or by faith. These definitions are useful to restrict the analysis of power to those kinds of power that assume the form of a rational

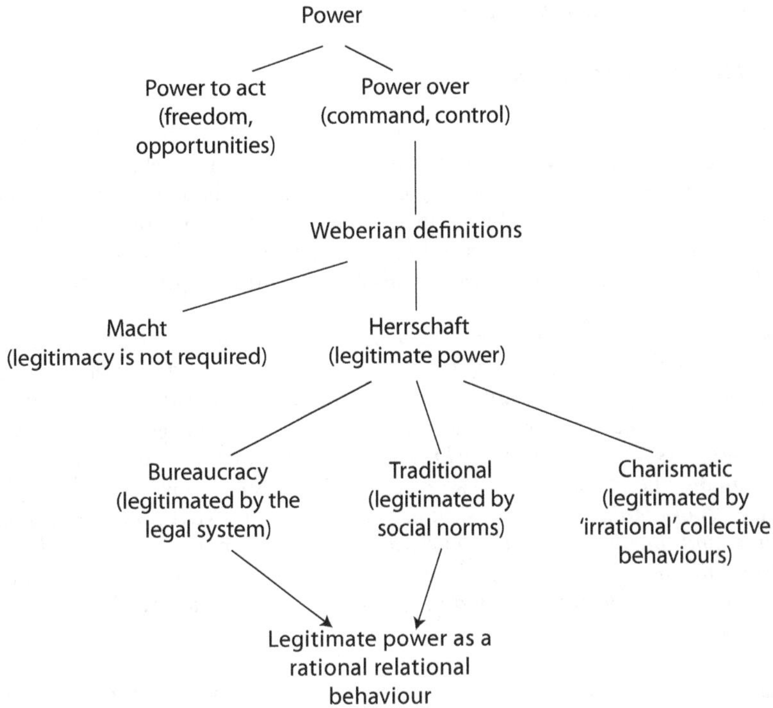

Figure 1. General power definitions.

behaviour in a given institutional constraint. In this way it is not taken explicitly into account the problem of the role of power in determining a specific institutional framework (*i.e.* the problem of the genesis of the constitutions), and the kind of power linked to non-rational behaviours (where a rational behaviour is defined as a purposive behaviour, *i.e.* a behaviour guided by goals known with awareness by the actors). Power-over legitimated through contracts, either formal or informal, can be referred to as "legitimate power as rational relational behaviour", as reported in the last row of Figure 1. In its relational dimension power can be viewed as an actors' behaviour (determining the structure of the relationship) or as a type of structure of the relationship (determining the actors' behaviour). The first view is baked by social theories based on the assumption of methodological individualism (basically all approaches that make a wide use of the theory of rational choice), while the second view is baked by those theories based on the assumption of methodological collectivism (Archer, 1995) (basically functionalism and structuralism, but someway also the theory of conflict). Both approaches are essential for a full understanding of the power phenomenon.

In the next sections, in surveying different approaches to the study of power in economics, I first present approaches stemmed from the standard model that preserve hypothesis of methodological individualism and rational choice. Subsequently I introduce three structural approaches stemmed from literature on sociological economics based on the hypothesis of methodological collectivism. Finally, I integrate these different approaches in a general framework drawn on the perspective of the New Institutional Economics.

3. Power and the economic theory

3.1 Power and the standard model

Methodological individualism, and the assumption of rational optimizing actors, make Coleman's social theory based on the analysis of social exchange, be a sort of mirror of the standard economic theory (based on the analysis of economic exchanges). Given this theoretical and methodological similarity, one question is pertinent: why has the analysis of authority relations, and of the sources of power held by actors in the relation, not been studied with equal emphasis in Economics? Coleman (1990) gives an exhaustive answer to this basic question at page 782 of his book:

> *"If the system of action is a perfectly competitive market and all transactions are voluntary exchanges within that market, it is possible to characterize the power of each actor very simply. It is the amount of wealth he brings to the market, either in the form of money (which is the measure of wealth) or in the form of goods and services desired by others who have wealth (which is measured by the extent of that demand). In such an individualistic, voluntary, and competitive system, power is well specified and consists of nothing more or less than wealth. In such a system money will buy anything, and thus the amount of power an actor can exert in the system (that is, the degree to which he can realize the outcomes he desires) is merely the amount of money (or other resources measured in terms of the money they can be exchanged for) he holds."*

Coleman's statements clarify that even when we assume 1-that the political problem (who decides to whom rights are originally distributed, *i.e.* before the exchange system starts working) is already solved, and 2- that there exists a complete system of perfectly competitive markets, even then power is present in the system. In this case power assumes explicitly the form of power to act, but it can also assume the form of power-over, when among goods being exchanged are rights limiting personal freedom or basic human rights. Furthermore, in the real extreme competitive case authority relations do not occur because they would entail long-term contract, while an ideal competitive system foresees only short-term contracts (spot markets). In the enlarged model long-term contracts are foreseen, but in the form of complete contracts, where power-over occurs in the form of legitimate voluntary exchange relations drawn by Coleman.

Concluding, mainstream economics explains power to act (in the form of purchasing power) as freedom to act and deals explicitly with power only in the form of market power and bargaining power (Rothschild, 2002). Market power refers to the possibility to set a price exceeding the competitive one. Also in the case of bargaining power, the power is connected to the possibility for the parties to make the price, with the party with more bargaining power being able to "command" the exchange terms to the other party.

Market power and bargaining power are analysed as cases of market imperfection, *i.e.* cases in which the market (intended as the basic institution of capitalist economies) fails to perform its function of coordination. It has been the literature on market failure together with the modern theory of firm that have opened the way to the power issue in economics. There are at least two important strands of literature that directly address the issue: the contested exchange approach within the neo-behavioural economics; the literature on the internal politics of firms, within the managerial theories of the firm.

3.2 The contested exchange approach and the influence cost theory

Through the contested exchange approach Bowles and Gintis (1999) show that in capitalist economies voluntary market exchange engenders a structure of power relations among economic agents, even when a competitive equilibrium is attained. This view contrasts with the assertion (generally sustained by orthodox economists) that power relations are produced by deviations from competitive conditions, such as the pervasiveness of monopoly, the autonomy of management, corporate influence over government policy and consumer demand, the ubiquity of disequilibrium.

By a 'competitive capitalist economy' Bowles an Gintis (1999: 14) mean "*one in which productive assets are privately owned, production is carried out by employees, and all markets are characterized by free entry and large number of buyers and sellers*". In such an economy power relations can occur whenever contracts defining exchange's terms are incomplete and the Walrasian 'exogenous enforcement assumption' does not hold.

Incomplete contracts give rise to competing claims, that make transactions need an adequate enforcement system to be viable. When a third party enforcement is not possible (for instance when the contested attribute can be measured only imperfectly or at considerable cost; typical examples are work effort, quality in case of experience and credence goods, the degree of risk assumed by a firm management), an endogenous enforcement system is required. Bowles and Gintis (1993, 1999) refer to "contested exchange" as an exchange setting where the benefit the parties derive from the transaction depends on their own capacities to enforce competing claims. The party in the relationship who exercises power is the one able to institute monitoring and sanctioning mechanisms by which to induce the other party to provide the desired level of the contested attribute. The authors take as an example an employment relationship where the contested resource is the work effort. In the example the employer (A) gets the desired effort from the employee (B) offering B a higher wage than the reservation wage (reservation wage equates the equilibrium wage in the Walrasian model), at the same time using an endogenous enforcement mechanism, called 'contingent renewal'. A contingent renewal obtains when A elicits performance from B by promising to renew the contract in future period if satisfied, and to terminate the contract if not. The difference between the reservation wage and the offered wage is the enforcement rent. The existence of the enforcement rent allows the equilibrium wage to be higher than the reservation wage and consequently makes the labour market not clear in competitive equilibrium, with a part of workers rationed out of the market. Defining the short side of a market as the side for which the quantity of desired transaction is the least, it happens that "*short-side agents have power over the long-side agents with whom they transact, since they may at little or no cost to themselves impose significant sanction by terminating the contract*" (Bowles and Gintis, 1993: 90). In this case A who can purchase any desired amount of labour (given the unemployment associated with an equilibrium wage higher than the reservation one) is on the short side of the market. On the contrary B is on the long-side.

Business and management literature has approached the power issue with the aim to analyse the political dimension of firms. Political activities can be largely defined as those activities carried out by different members of an organization in order to influence the organization's decisions in a way useful to achieve their own interests, even when the latter are not consistent with the goals of the organization. Political activities are carried out by firms by trying to influence the decisions of outside organizations in a way to reduce constraints (or raise opportunities) for the betterment of their own internal strategies. Political activities are also carried out within the firms by different members of the organization (owners, managers, workers, stakeholders) in order to exploit possible sources of rents. Literature on rent-seeking behaviour and influence cost theory (Milgrom and Roberts, 1999) have analysed the effects of these activities on firms' performance, and have suggested organizational "mechanism designs" adequate to mitigating the negative ones (such as distortions in decision making processes). Actors who succeed in their influence activities are those who exert power over the actors whose actions and decisions affect their interest. Power is here associated with the fact that, contrary to the prescriptions of the standard model, *"the outcomes of firm decisions create winners and losers, where winners enjoy quasi-rents"* (Milgrom and Roberts, 1999: 47). The reference is to the general definition of rents and quasi rents: rents are payments beyond those required to attract a resource in an activity; quasi-rents are payments beyond those that a resource could expect to get by quitting the activity where it is engaged and entering into another activity. Therefore power can arise whenever firm decisions have non trivial distributional consequences.

3.3 Structural approaches

As structural approaches to the analysis of power I consider three theories that can be widely inscribed in the social network analysis literature (Wasserman and Faust, 1994): the power dependence theory; the theory of structural holes, and network theories inspired by the small world phenomenon. According to these approaches power, as other actors' behaviors and characteristics, arises from the structure of the system, where structure refers to the "web" (with its proper scale, scope and shape) of relational ties among actors in the system.

The concept of power featured by the power-dependence theory (Emerson, 1962) derives from exchange theory emphasis on the ties of mutual dependence that underlie all social structures. Social exchange theory depicts many forms of social interactions as exchanges of benefits. The underlying idea is that people strongly depend on one another for the different resources they value and they provide them to one another through the process of exchange, also outside the economic marketplace. To the extent that dependence is mutual, actors in social relations have power over each other. Moreover to the extent that their dependencies are unequal, their relations are unequal, with more powerful (less dependent) actors enjoying greater benefits at lower costs.

In spite of its very "structuralist assumption" (the web of relations matters "in primis", actors' behaviours depending on it), power-dependence theory allows for a better understanding of at least two important features of actors' behaviour that affect the role of power in determining exchange outcomes: normative concerns and strategic use of power.

Normative concerns are taken into account by Cook and Emerson (1978) in addressing the problem of indeterminacy in bilateral monopoly. Specifically they stress the role of power and equity, both considered as social structural phenomena, in determining the location of the outcome along the contract curve, this latter defined as the cluster of points of different distributive equilibria of the maximum outcome available from the exchange. They demonstrate that the equitable solution of a dyadic exchange (*i.e.* the equal distribution of the total maximum exchange value) occurs when neither party has alternative sources and when the behaviour is driven by normative concerns about equity (*i.e.* the parties will refuse any outcome that unequally distributes the total profit, in the same way as in the ultimatum game the responder will refuse low offers). When one agent has alternative sources and equity concerns are weak, the exchange outcome will be chosen by the agent with more power, with the power associated with the position of the agents in the network, *i.e.* with the number of alternative available sources. In other terms power reaches its highest "organizational performance" when actors actually behave like the homo economicus, while giving way to reciprocal norms and trust as organizational devices when homo reciprocans and homo egualis (in the terminology used by Gintis, 2000) enter the stage.

The strategic use of power is dealt with by Molm (1997) in her research on coercive power in social exchange. Molm offers experimental and theoretical evidence that the basic assumptions and concepts of exchange theory can be extended to include punishment and coercive power, besides the traditionally assumed reward power. Stemming from this evidence Molm features a strategic use of power that differs from the structurally induced power in an important behavioural principle. Actors who use power strategically respond to others' behaviours (as in the case of principle of reciprocity of the "standard" power-dependence theory), but they also attempt to influence them. *"Rather than giving rewards to obtain benefits in return, actors administer rewards and punishment contingently; in particular they impose cost (by administering punishment or withholding rewards) on partners who fail to provide sufficient benefits."* (Molm, 1997: 161).

Social exchange theory focuses on the analysis of singular dyadic relations taking as its subject matter the more or less enduring relations that form between specific partners (or in terms of contract theory incomplete resource-specific long term contracts). Accordingly it offers a structure-based solution to the classical bargaining problem of bilateral monopoly, leaving out the issue of structural determinants of competition, this latter specifically addressed by Burt's theory of structural holes. In the introduction of his book Burt explicitly states that the subject of the book is "how competition works when players have established relations with others" (Burt, 1992: 1), *i.e.* when the competitive arena is investigated as a network whose nodes are the players in the competition game. Network structure is responsible for differences in competitive advantage among the players, with structural holes, defined as disconnections or nonequivalencies between players, being the core structural element. Burt's main argument is that players with networks rich in structural holes (and therefore with high structural autonomy) enjoy high rates of return on investments because *"they know about, take part in, and exercise control over more rewarding opportunities"* (Burt, 1992: 2). Burt explains the competitive advantage associated with structural holes in terms of freedom (freedom to exploit market opportunities), explicitly excluding the power argument: *"This is the focus of the structural hole argument, a theory of freedom instead of power, of negotiated instead of absolute control. It is a description of the extent to which the social structure of a competitive arena creates entrepreneurial opportunities for certain players to affect the terms of relationships."* (Burt, 1992: 7). With the previous definitions of power in mind, one can contend that Burt's statement holds only when power is defined as illegitimate and/or coercive. When power is considered in the form of power to act, or, as power over, in the form of legitimate reward power (bargaining power is an example), then the competitive advantages of actors with networks rich in structural holes stem from the exercise of power. The asymmetry in the social capital endowment by actors and the structurally induced consequences of this asymmetry on their market opportunities (also in the form of exercising market power) and bargaining power is the very cause of high performance associated with structural holes. Burt fails to recognize the power source of competitive advantages; this for the same reason that in the standard model, the power to act of wealthier actors is termed freedom, whilst bargaining and market power are considered as mere results of market imperfections, instead of peculiar behaviour able to influence in their turn market structures.

When an actor provides bridges across structural holes (for instance a manager who creates interdivisional links in a large organization), combining information from disparate groups that would not otherwise communicate, she relocates herself in a "higher–power" position. Networks with an architecture characterized by a certain number of actors located in bridging nodes across structural holes exhibit a small world structure. A small world structure is intermediate between regular (each node with exactly the same number of links) and random (as links placed randomly in a graph, nodes have a random number of links). Regular and random networks display a low degree of clustering. While this property is evident for regular networks, in the case of random networks it is predicted by Erdos and Reny random network theory, one finding of which is that if the network is large, despite the links being placed completely at random, almost all nodes will have approximately the same number of links. As pointed out by Barabasi (2003: 22-23) both of these networks have a very democratic nature, *i.e.* the histogram plotting nodes and associated links follows a Poisson distribution, meaning that, for example, in a given society most people have roughly the same number of friends and acquaintances. Erdos' random model dominated research on complex network modelling until recent developments made by Watts and Strogatz (1998) on clustering. Studying communication and biological network architecture (while trying to solve challenging problems on chaos and synchronization) they found that these displayed a high clustering coefficient, in contrast with the small one predicted by the random theory. The clustering coefficient (with a value ranging between 0 and 1) is obtained by dividing the number of actual links between a set of nodes linked to a given node, and the number of potential links they could have if they were all linked to each other. If the clustering coefficient is zero, then the "central" node is the only element that allows for an (indirect) relationship between other nodes. Watts and Strogatz found that the architecture of randomly created networks can be depicted as a regular lattice (where each node is connected to its immediate and next-nearest neighbours) to which a few extra links are randomly added, making the regular turn in a small world network. This finding formalizes Granovetter's model on the role of weak ties in social networks (Granovetter's society includes many highly connected clusters, linked to each other by weak ties, where weak ties correspond to the extra link added to the regular network) and offers a general model for the small world phenomenon (small world networks can emerge from a minimal random rewiring of a regular network). Nevertheless, since it forbids nodes with significantly more links than average, it does not account for the actual shape of real social networks.

As informally introduced by Gladwell's "tipping point argument" and as formalized by Barnabasi's research into power law, real social networks usually display hubs, *i.e.* nodes with an extraordinarily large number of links. Hubs tell us that in the social and economic system, in order to succeed not only does it matter who, as opposed to what you know, but that we lack the same opportunities (as in egalitarian random models), with actors in hub positions controlling most opportunities. Inequality and power (opportunity) asymmetries are ubiquitous in real networks, where the 80/20 rule applies. According to this rule, known also as Pareto's principle (since Pareto observed that 80% of Italy's land was owned by 20% of the population), in a network a few hubs grab most of the links, meaning that in the network the degree distribution follows a power law. "*While a bell curve has an exponentially decaying tail (consistent with the absence of hubs), a power law distribution exhibits a far slower decay allowing for "rare events" such as the hubs*" (Barnabasi, 2003: 69). Power laws dominate the architectures of complex networks and indicate that these systems, being far from random (in physics power laws describe systems in transition from chaos to order), are instead governed in their evolution by the law of preferential attachment. Networks with power-law degree distribution are referred to as scale-free networks. These networks are the result of a growth process guided by the principle of preferential attachment: "*in the expanding process of real networks, when deciding where to link, new nodes prefer to attach to the more connected nodes; thanks to growth and preferential attachment, a few highly connected hubs emerge*" (Barnabasi, 2003: 87).

In economic exchange networks a scale-free structure predicts power whether because hub actors (firms) have a negotiating power higher than that of their attached nodes where customers and suppliers are located (according to the prediction of power-dependence theory), or because they have access to more sources of competitive advantages (according to the prediction of structural holes theory). Moreover, in the measure to which hub actors can forbid the attachment of given nodes, a strategic use of power can also be assumed.

Industries characterized by a network form of organization, exhibiting a scale-free structure, can be thought of as inherently non-competitive, and should be closely monitored by antitrust authorities.

3.4 A general framework from the perspective of the New Institutional Economics

In a sense, New Institutional Economics can be considered as an extension of the standard model that takes explicitly into account the power issue. The famous Robertson's definition of firms as "islands of conscious power in this ocean of unconscious cooperation like lumps of butter coagulating in a pail of buttermilk", was used by Coase (1937) to launch the question of how is that in capitalist economies firms (and thus power) do substitute the market as a means to allocate resources (this issue has been recently addressed by Rajan and Zingales, 1998).

When examining the role of firms in economic organization we can think of two aspects: 1) the internal organizational environment (*i.e.* authority relations that define employment relationships) and 2) the external organizational problem (*i.e.* the way through which the firm exchanges resources with the external environment, as in the case of procurement relationships), *i.e.* the vertical integration issue. The ways in which New Institutional Economics addresses the two issues are useful to integrate different approaches in the analysis of power in economics.

The vertical integration problem

New Institutional Economics points out that certain characteristics of procurement relationship (for instance repeated purchases, with complex contract specification) can lead the parties to switch from short-term contracts (spot markets) to long term contracts that entail power in terms of Coleman's voluntary authority disjoint relationships.

When these contracts are complete and enforceable by a third party, it is even possible, according to Coleman, separate the economic realm from the political one because exchange of rights can be thought in this case to occur in a competitive market, where, following Coleman's reasoning, the only source of power is "money" (the power-to-act type that is in any way in a capitalist economy). It does exist a coercive power, that makes the authority relation work, but it resides outsides the exchange relationship, in the state's legal system, *i.e.* in the political realm.

On the contrary when long-term contracts are incomplete and not enforceable by a third party, power arises in the form foreseen by the contested exchange approach and it is no more possible to separate the economic realm from the political realm. The outcome of the exchange and its distribution will depend on the behaviour of the more powerful party and any grant of equity and efficiency assured by the standard model will vanish. The incomplete contract theory of the firm (Hart, 1988; Grossman and Hart, 1986) states that vertical integration (and hence the firm) occurs when contracts fail because of their incompleteness, then the party more able to exploit the residual claims (the party in the terminology of Crozier and Friedberg more "puissance"-endowed, or in the lexicon of Coleman, with more power) will acquire the residual rights of control. In the model of Grossman and Hart any process of vertical integration entails a redistribution process that will favour the more powerful party, with contradictory equity and efficiency effects. This result sharply contrasts with the Williamson explanation of vertical integration that accounts for the shift from one organizational form (the long-term contract) to another (the firm) exclusively in terms of transaction costs minimizing, implying an efficiency improvement. According to Williamson (1985) high resource specificity (along with opportunism, bounded rationality and an imperfect legal system) turns a competitive relationship into a bilateral monopoly; when the bargaining entails unbearable transaction cost, a contractual agreement fails and vertical integration occurs. In this way power is intended only as bargaining power, linked to a lack of competition. In other terms while in Williamson's explanation firms integrate to save on transaction costs, the incomplete contract theory of firm claims that firms integrate to exploit residual rights of control. What Grossman and Hart fail to underline, is that incomplete long-term contracts do not necessarily lead to vertical integration (and thus to power in the form of voluntary legitimate relations that define firms) or to exchange failure. They can instead survive thanks to endogenous enforcement mechanisms (on the type of 'contingent renewal' suggested by Gintis) that secure the power to the party on the short side of the exchange relationship, or thanks to trust-based or power-based self-enforcing agreements. In the case of power-based self-enforcing agreement power refers to the capability of one party to "informally" enforce the agreement by obtaining obedience by the other party (through promise, threat, retaliation, reward, punishment). Source of power is either structural or resides in the "power-endowment" (or the "atouts" endowment in the terminology of Crozier and Friedberg, 1977) of the powerful actor.

The internal organizational problem

Neo-institutionalists are used to thinking of firms as "nexus of contracts". In this way power issues related to the firms' internal organizational problems take the same form as those related to the external organizational problems. Here again contracts defining relationships among various members of the organization can be complete and-or enforceable by a third party or incomplete and not externally enforceable. In the first case power assumes the form of authority relations (and specifically, disjoint authority relations in Coleman's terminology; Coleman, 1990) whose ex-ante and ex-post problems, such as those related to shirking and monitoring, can be solved defining the kind of incentive contracts that correct adverse selection and moral hazard problems in agency models. In the second case a form of power, that could be labelled 'politically-based power' (this is the kind of power foreseen by Crozier and Friedberg, in their analysis of power in formal organization, based on concepts of "atouts" and "marge de liberté", Crozier and Friedberg, 1977) must be accounted for. In this latter case self-enforcing or self-policing agreements (Furubton and Richter, 2000), are involved. Once again when constitutional problems arise (that is the problem of whom to give the power to to enforce rights, or, in Coleman's lexicon, the "choice" of relevant others), economic exchanges, such as social exchanges, are supported by political other than economic behaviours.

Results attained so far (Figure 2 summarizes relationships between power and economic organization) allow for the following four general statements on power and economics:

1. A distinction can be made between power-to act and power over.

2. Standard economics explicitly considers only three kind of power: purchasing power, market power and bargaining power. The first is clearly a form of power to act, but is also an "atouts" that gives actors the possibility to exert power-over. The second is a form of power over. The third is a form of power-over, even if standard economics fails to recognise it, because it vaguely

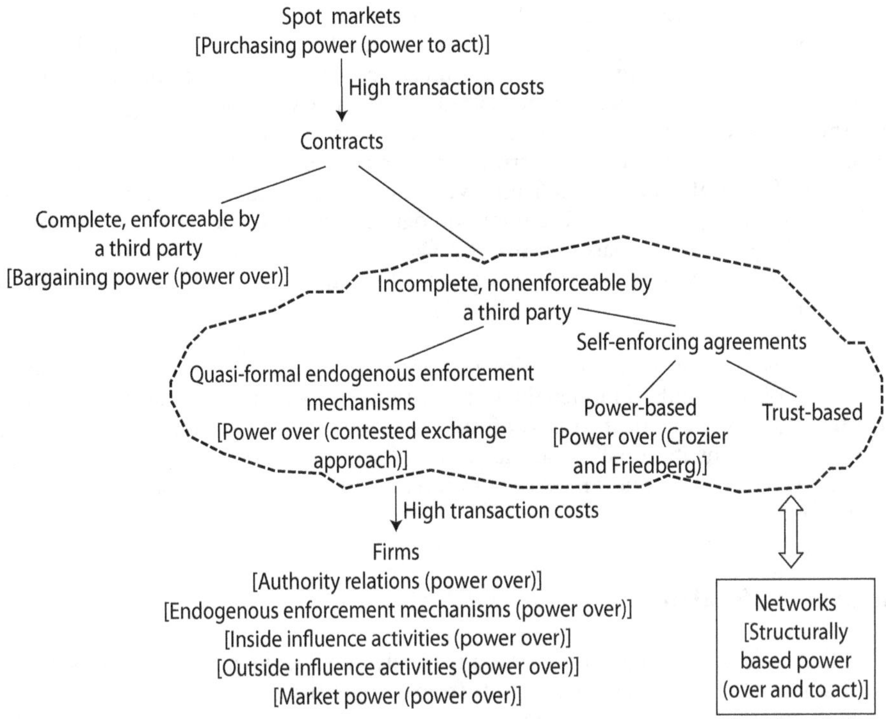

Figure 2. Organizational forms and entailed forms of power.

defines bargaining power as the share of the surplus that one gains in a bargain, rather than to consider the means of attaining it (*i.e.* through a power-based behaviour).

3. The contested exchange approach clearly indicates contract incompleteness as a source of power in the system.

4. The new institutional economics implicitly puts the issue of power at the core of the organizational problem, demonstrating that, even stemming from a methodological individualism approach, and deploying the analytical techniques of the rational choice theory in a liberal context (actors act voluntarily), power emerges as a fundamental organizational force in capitalistic economic systems.

In the presence of positive transaction costs resource allocation can no more be attained exclusively through the competitive exchange on spot markets and other organizational forms take place, contracts and firms. Complete contracts enforceable by a third party entail power in the form of voluntary authority relations. Incomplete contracts not enforceable by a third party involves another form of power, stemming either from the use of quasi-formal endogenous enforcement mechanisms (as in the case of the contested exchange approach) or from informal power-based or trust-based self-enforcing agreements. Firms (vertical integration) occur when contract incompleteness is not corrected by power (in either the two forms), or alternatively by trust (in the case of trust-based self enforcing agreement). Once the exchanges are organized inside the firm, then again power takes the form of Coleman's disjoint authority relations, or, in case of contract incompleteness, the two forms associated with endogenous formal and informal enforcement mechanisms; furthermore, market power, and power associated with the influence activities inside and outside the organization must be considered. To complete the picture I circled the cases of contract incompleteness, jointing the circle with a fourth organizational form entering the framework, the network form. This means that when in an industry a good deal of incomplete contracts survives, thanks to trust or power, networks of manufacturers and suppliers take the place of vertically integrating firms. When networks are observed, besides the different forms of power investigated through a methodological individualism approach, we must take into account also structurally-based forms of power described by structural theories analysed in the third paragraph of the previous section. Concluding, we can add a fifth statement to the four just mentioned:

5. Although new institutional economics is grounded on the very same epistemological assumptions of the standard model, the organizational issues it raises draw a bridge between this latter and new theoretical strands of radical economics and economic sociology. In the case of power new institutional economics allows also orthodox researchers for a dive into the sea of structuralism and of new models of rationality introduced by Kahneman and Tversky's prospect theory.

4. Power in the food market

It is a fact that currently the agro-food system is widely populated by imperfect markets. Figure 2 shows that when the economic system shifts from competitive (*i.e.* spot markets and complete contracts) to imperfectly competitive environments made up of incomplete contracts and large firms, power affects the system through a variety of forms beyond the "traditional" market and bargaining power forms. Therefore, in studying food system organization, whether for descriptive or normative aims, the analysis of power, carried out using the different approaches surveyed in the previous sections, should not be neglected. In this section I make three examples of power-related issues in the analysis of the food market: 1) welfare-assessment of vertical contractual arrangements; 2) the analysis of recent processes of networking; 3) the multiform power of large firms and the role of corporate social responsibility.

4.1 Welfare assessment of vertical contractual arrangements

New institutional economics gives an efficiency-based explanation of organizational dynamics in the economic system. Every time a new hybrid arises moving along the continuum of organizational forms that range from spot markets to contracts and hierarchies, it entails an efficiency improvement in terms of transaction and/or production cost reduction. Using the power argument I stress that a contractual arrangement with a low exchange value can last in the system even when other arrangements exist with a higher exchange value. Let us assume a high contractual incompleteness, with transaction costs associated with problems of post-contractual enforcement and ex-post incentive alignments. According to the Williamsonian view, either a hierarchy emerges (with one party assuming via legitimate formal contracts the control of residual claims), or the incomplete contract survives because a high level of trust reduces risks of opportunism and reduces monitoring and assurance. On the contrary, according to the power argument, the incomplete contract can survive also in presence of high transaction costs. That happens when the party on the *short side* of the market (as predicted by the contested exchange approach) or less dependent on the exchange (as predicted by the power-dependency theory) "forces" the other party to bear all monitoring and insurance costs.

More generally power argument linked to contract incompleteness helps to explain the low level of vertical contractual integration in the food system. Traditionally, the shift from a weak contractual relationship to a tight contractual integration is deemed to occur when it can lower transaction and /or production costs (for examples those costs associated with a high degree of resource specificity or with scope economies) and/or raise market power. Looking at the problem in term of power associated with contract incompleteness helps to explain why a weak incomplete contract is used even when a tighter contract (it does not matter if more or less incomplete) could yield a marketing surplus. This is the case when the pre-existent contract is incomplete in a way such that the party with more power (with more atouts) can control sources of incertitude in such a way as to raise her own interest. For example, let us be the case of a new opportunity to raise the marketing surplus by making investments whose specificity needs more contractual integration (this can be the case of a product innovation that needs vertically coordinated quality assurance activities). If in the new contract the party in high power position in the pre-existent contract cannot get a gain higher than the current one, he will not switch to the new contract. Therefore power motivations (other than market power) will stick the channel to an inefficient organizational arrangement.

The policy implication is that to promote the supply chain efficiency it is not always appropriate to rely on competition policy and market deregulation. A higher market concentration could lead to a better exploitation of market and technological opportunities. Moreover appropriate market regulations (intended here in a broad sense as legal enforceable constraints that raise the completeness of contracts) could change the "structure" of the "contract-game" in a way such as to prevent parties from exerting power, whenever power harms more than benefits organizational efficiency.

The above considerations suggest that difficulties in assessing welfare effects of vertical restraints (Dobson and Waterson, 1996) partially arise from the limited attention given to power. As argued by Hamilton (2003), vertical restraints, as retail slotting allowances, more than being associated with retailers' market and buying power, seem to be associated with various strategic practices carried out by retailers as well as processors in order to raise their gain opportunities in a broader way than simply exercising market power. From a power perspective we could say that vertical restraints change the size and the distribution of the economic surplus by changing the organizational structure of the chain. Specifically, actors who promote vertical restraints try to change the institutional framework of the exchange in order to increase their "marge de liberté" in exploiting possible sources of rents.

4.2 Understanding recent processes of networking

In the 1960s and 1970s the key players in world economy were large vertically integrated transnational corporations (TNCs), whose use and abuse of power has been often deemed to jeopardize sovereignty and effectiveness of national governments. In the 1980s and 1990s radical changes took place. In this period *"the rising integration of world markets through trade brought with it a disintegration of the production process of multinational firms, since companies found it profitable to outsource (domestically and abroad) an increasing share of their non-core manufacturing and service activities"* (Gereffi, 2005: 166). As a consequence, the whole production structure in the global economy changed, with the emergence of international trade and production networks.

During the last two decades in the food sector many TNCs have progressively reduced their involvement in first-tier processing and have engaged in policies of outsourcing together with high investments in brand equity. Brand and financial management are currently the core activities of these companies and the coordination of their suppliers' networks is central to their corporate strategies. (A good example of this behaviour is the implementation of Project Millennia and Project Excel by Heinz in the processing tomato industry – as reported by Pritchard, 2003). The shift to those globally networking is likely to affect at least three general issues: 1) the process of industrial upgrading in developing countries, 2) the efficiency of industrial structures, and 3) how the new transnational capitalist class will rule the global system. The structurally based power argument suggests that distributive and efficiency effects of networking processes will depend on network structure and on power imbalances characterizing linkages in networks. When network structures shift from regular or random types to scale-free types it is very likely that the growing networking process will lead to an exacerbation of disproportion in income distribution (both at intra and inter country level), due to the 'rich get richer' effect. As a result, when a network exhibits a scale-free structure, anticompetitive effects can occur (requiring the intervention of antitrust authorities), even in presence of a not excessively high concentration index.

Literature on industrial organization has compared networks to vertical integration as alternative supply structures for specialized inputs, where networks turn dedicated into flexible assets. The general wisdom is that flexible assets and networks involve social benefits when there is high demand uncertainty. When relationships between buyers and suppliers entail incomplete long-term contracts, payments are often determined or adjusted ex-post through informal contracts. In this case networks' welfare effects will depend on how the exchange surplus in ex-post negotiation is shared among firms in the network. Kranton and Minnehart (2000) demonstrate in their model (using game theory) that when the Shapley value is used instead of the Nash bargaining solution, players' investment incentives are no more aligned with economic welfare. In terms of power argument this means that when relations in networks are affected by power imbalances (considering bargaining as well as structurally based power) causes of vertical disintegration do not need to reside in efficiency seeking (or "simply" market power enhancing) behaviour.

4.3 The multiform power of large firms

As Coleman's theory points out, power is a trivial issue only in the case of a perfectly competitive market system (where the only power is the purchasing power), otherwise power is at the core of the socio-economic organizational puzzle. The fundamental trait of market imperfection is not the chance to get positive price-cost margins; instead it is the switching from the market institution to the firm institution as a new exchange government form. Firms, unlike the ideal economic actors in the Walrasian design, besides purchasing power, exert at least five other forms of power, as summarised Figure 2. While the first form of power (authority relations within the firm) can be thought to have positive efficiency effects (it is supposed to make the system save on transaction costs), the other four forms are associated with monopoly inefficiencies and rising costs, due to influence costs and rent-seeking behaviours. Moreover, considering the political activity of firms calls for a deeper attention to all the activities they carry out (especially multinational corporations, as recently documented by various authors, such as Bakan, 2004; Korten, 2001; Nace, 2003; Rosefield, 2002) by persuading, bribing or threatening public regulators in order to obtain "personal" advantages. "Power abuse" by firms also includes practices that corporations habitually carry out in order to externalise (more or less legally) their costs, thus causing harm to "others" (workers, consumers, communities, the environment), and other practices such as unfair presentation of financial statements, deceptive advertising, breach of trust and so on.

The power argument upheld so far leads to at least four basic tenets:

1. When valuing negative consequences of market imperfections, distortions stemming from forms of power other than market power should be seriously taken into account.

2. While deregulation policies are generally deemed to improve efficiency by boosting private enterprise, they should instead be considered as a way of facilitating firms' exercise of power, with all the possible negative welfare consequences.

3. More generally, when comparing different institutional frameworks (in order for instance to choose the "right" agricultural or food safety policy instrument) power-based considerations should "accompany" efficiency-based considerations.

4. In many cases power abuses by firms cannot be prevented by ad hoc regulation policies (when these latter induce excessive distortions of private incentives). In such circumstances, the hope for a "better world" eventually resides in corporate social responsibility, and public intervention in the socio-cultural sphere can turn out to be much more effective than intervention in the economic sphere.

5. Concluding remarks

The paper is an attempt to introduce power, as an autonomous analytical dimension, for the analysis of food chains and networks. To that aim many theoretical perspectives of the study of power have been explored and have been eventually incorporated in a unified framework, keeping together standard economics, institutional perspectives, and network and managerial sciences. One result is that power is a kind of topic that fits the "encompassing" scope and methods quoted in the early research agenda drawn for the chain and network science (Omta *et al.*, 2001).

While quite successful on a theoretical ground the paper is very incomplete on a more empirical ground, giving only a few examples of "real food issues" related to the topic of power. This lack is due both to the limited length of the paper and to the scant available literature. To partially fill the gap between theory and practice, it is worth ending the paper with a list of some topics of the current research on food chains and networks "sensitive" to the outlined concepts of power: transparency, strategic alliance, and global sourcing.

While transparency is generally deemed to improve food safety and organizational efficiency in the supply chain (Hofstede *et al.*, 2005), there is still controversy about institutional features able to create and sustain transparency. It has been argued that the more intense the division of labor, the more process interdependencies result and the lower transparency is (Theuvsen, 2004). The power argument suggests that the more incomplete and non enforceable contracts are, the more the relationships must rely either on trust or on power to succeed; while trust positively affects transparency, power tends to erode it.

The rise and fall of strategic alliances are still "obscure" phenomena in business and organizational literature. Recently, Kumar *et al.* (2005), suggested tackling alliances as open social systems in order to face their multidimensional and complex features. Power can add new insights to this issue at least in three ways: 1) power can be considered as a conflict-solving means, alternative to trust; 2) the hidden power (Cross and Parker, 2004) of social networks linking CEOs and managements of "allied organizations" emerging as a result of alliances, can be the ultimate arbiter of alliance fortunes; 3) the parties empowerment of structural-based power emerging from the alliance can be the main strategic motivation of the agreement.

Recently, a strain of literature (Feenstra and Hanson, 2005; Antras and Helpman, 2004) has tried to tackle the issue of global sourcing by integrating lessons from incomplete contract theory, general equilibrium models of international trade, and industrial organization models of imperfect competition. While power enters these model only in the form of bargaining models (equilibria are very sensitive to the kind of solutions assumed for the bargaining games at the level of buyer-supplier relationship, with the Nash bargaining solution being the favourite to allow for analytical tractability) it would be challenging to consider power in the form of control asymmetry due to the unequal information and legal enforcement systems among the parties, as well as in the structural form emerging as a result of the outsourcing processes, with firms that consolidate their hub positions being the ultimate exploiters of outsourcing advantages.

Concluding, the paper has barely touched upon the "hot" issue of power and food system, to deal with which a large amount of theoretical and empirical research is required.

References

Antras, P. and E. Helpman, 2004. Global sourcing. J. of Pol. Ec. 3, 552-580.

Archer, M.S., 1995. Realist social theory. Cambridge University Press.

Bakan, J., 2004. The corporation: The pathological pursuit of profit and power. Free press.

Barabasi, A.L., 2003. Linked. Penguin Book.

Bowles, S. and H. Gintis, 1999. Power in competitive exchange. In: Bowles, S., Franzini, M. and Pagano, U. (eds.). The politcs and economics of power. Routledge, London.

Bowles, S. and H. Gintis, 1993. The revenge of homo economicus: contested exchange and the revival of political economy. J. of Ec. Perspectives 7, 83-102.

Burt, R.S., 1992. Structural Holes. Harvard University Press.

Clegg, S.R., 1989. Frameworks of power. Sage Publications.

Coase, R., 1937. The Nature of the Firm. Economica 4, 386-405

Coleman, J.S., 1990. Foundations of social theory. Harvard University Press.

Cook, K. and R. Emerson, 1978. Power, equity and commitment in exchange networks. Amer. Soc. Rev. 43, 721-739.

Cross, R. and A. Parker, 2004. The hidden power of social networks. Harv. Bus. Press.

Crozier, M. and E. Friedberg, 1977. L'acteur et le système. Edition du Seuil.

Dahl, R., 1986. Power as the control of behavior. In: S. Lukes (ed.). Power. New York University Press. pp. 94-143.

Dobson, P.W. and M. Waterson, 1996. Vertical restraints and competition policy. Research Paper 12, Office of Fair Trading, London.

Emerson, R.M., 1962. Power-dependence relations. Am. Sociological Review 27 31-41.

Feenstra, C. and G. Hanson, 2005. Ownership and control in outsourcing to China. The Quart. J. of Ec. May, 729-761.

Ferrarotti, F., 2004. Il potere. Newton Compton editori.

Furubotn, E. and R. Richter, 2000. Institution and economic theory. The Un. Michigan Press.

Gereffi, G., 2005. The global economy. In: N.J. Smelser, R. Swedberg (eds.), The handbook of economic sociology. Princeton University Press.

Gintis, H., 2000. Game theory evolving. Princeton University Press.

Goodman, D. and M. Watts (eds.), 1997. Globalising food. Routledge.

Grossman, S. and O. Hart, 1986. The costs and benefits of ownership: a theory of vertical and lateral integration. Journal of Political Economy 94, 691-719.
Hamilton, S.F., 2003. Slotting allowances as a facilitating practice by processors in wholesale grocery markets: Profitability and welfare. Amer. J. Agr. Econ. 85, 797-813.
Hart, O., 1988. Incomplete contracts and the theory of the firm. J. Of Law, Economics and Organization, Spring.
Hofstede, G.-J.; L. Spaans, H. Schepers, J. Trienekens and A. Beulens (eds.), 2004. Hide or confide: the dilemma of transparency. Reed Business Information.
Holloway, J., 2002. Change the world without taking power. Pluto Press.
Korten, D.C., 2001. When Corporations rule the world. Kumarian Press.
Kranton, R. and F. Minehart, 2000. Networks vs vertical integration. Rand J.of Ec., 570-601.
Kumar, R., G. Duysters and T.K. Das, 2005. Understanding alliance evolution: establishing a research agenda. Journal on Chain and Network Science 5, 51-63.
Lukes, S., 2005. Power: a radical view. Palgrage McMillan, second edition, first edition 1974.
Lukes, S. (ed.), 1986. Power. New York University Press. pp. 94-143.
Magdoff, F., J.B. Foster and F. Buttel, 2000. Hungry for profit. Monthly Review Press.
Milgrom, P. and J. Roberts, 1999. The internal politcs of the firm. In: S. Bowles, M. Franzini and U. Pagano (eds.). The politics and economics of power. Routledge, London.
Molm, L.D., 1997. Coercive power in social exchange. Cambridge University Press.
Nace, T., 2003. Gangs of America. Berret-Koehler Publishers.
Omta, O., J. Trienekens, and G. Beers, 2001. The knowledge domain of chain and network science. Journal on Chain and Network Science, 77-85.
Palermo, G., 2000. Economic power and firm in new Institutional Economics: Two conflicting problems. Journal of Economic Issues, 34, 573-601.
Popitz, H., 1990. Fenomenologia del potere. il Mulino.
Pritchard, B., 2003. Agri-food globalization in perspective. Ashgate Publishing Company
Rajan, R.G., and L. Zingales, 1998. Power in a theory of the firm. The Quarterly Journal of Economics 1998(May), 387-432.
Rosefield, S., 2002. Comparative economic systems. Blackwell publishers.
Rothschild, K.W., 2002. The absence of power in contemporary economic theory. The Journal of Socio-Econ. 31, 433-442.
Russel, B., 1986. The forms of power. In: S. Lukes (ed.). Power. New York Un. Press.19-27.
Sheldon, I. and R. Sperling, 2001. Estimating the extent of imperfect competition in the food Industry: What have we learned?. The Ohio State Univ., Work. Paper: AEDE-WP-0018-01.
Theuvsen, L., 2004. Transparency in netchains as an organizational phenomenon. Journal on Chain and Network Science 4, 125-138.
Wasserman, S. and K. Faust, 1994. Social network analysis. Cambridge Univ. Press.
Watts, J. and S. Strogatz, 1998. Coll. dynamics of small-world networks. Nat. 393, 440-442.
Williamson, O., 1985. The Economic Institution of Capitalism. Free Press, New York.

Chain management challenges and market power in the Chinese fresh produce sector

B.J. Revell and X. Liu

Abstract

The paper provides both a descriptive and an analytical understanding of how the possibilities for chain management and logistics are linked to the distribution of market power within the supply chain for fresh produce in China. The article is structured around four parts. The first outlines the concept of power in supply chains and the types of chain associated with dyadic buyer-seller transactions. The second part provides some typologies of supply chains in the Chinese fresh produce sector based on the key functional stages. The third section examines in detail the main characteristics of the chains and the power relationships in dyadic exchange, and the types of chain management structures emerging in relation to the power distribution through the chains. The final section examines the logistical and developmental challenges facing China's fresh produce sector.

Keywords: China, fresh produce, market power, logistics

1. Introduction

China is the major producer of fresh produce (fresh fruit and vegetable) in the world. In 2004, the area in China planted to fruit was 20% of the world total, and for vegetables (including melons), 43% of world's total. China supplied 16% of world production of fruit, and 50% of world vegetable production. The area planted to fruit and vegetables was over 32 m ha, 20% of the total area of planted crops, fibre and tree crops in China, of which vegetables accounted for 13% of the total area (FAO, 2005). Fresh produce is a major source of commercial income for farmers, and about 16% of farmers' per capita net income came from fresh produce in 2002. Around 78 million farm workers are engaged in vegetable growing alone, and a further 80 million in vegetable processing and marketing. Per capita consumption of vegetables in 2002 was almost 111 kg, not far below the EU 15 average of about 128 kg. Some 11% of household food expenditure in China was on vegetables (National Bureau of Statistics, 2003). China's comparative advantage in the production of non-tree fruits and vegetables is in part due to the high labour intensity in production and low labour costs and China is highly competitive in the international fresh produce market, exporting over 6% of the world's fresh produce worth some US$ 5,429 million in 2003, of which fresh-unprocessed vegetables accounted for 8% of world exports by value (Liu *et al.*, 2003).

In order to be competitive in the global market, one of the key challenges facing the Chinese fresh produce sector is to deliver its produce efficiently to final consumers both domestically and worldwide. Distribution has been recognized as a major problem for companies within the chain and the management of relationships with distributors ranks the highest amongst all of the problem areas. The regionally-fragmented nature of the Chinese food market also inhibits the development of efficient chain management structures

However, although China's food chain is experiencing great changes, research on distribution channels in China has been limited (Liu and Wang, 1999; Fang, 2001; Xian, 2003; Bean, 2004,

2005; Hu *et al.*, 2004) and scant attention has been paid to fresh produce supply chains (Chang *et al.*, 2003). This paper therefore aims to help fill a gap in the literature, by examining the structure of the fresh produce supply chain in China and the resulting changes from shifts in the power relationships in the sector. It develops the arguments presented in a recent paper by the authors (Liu *et al.*, 2005) and examines the challenges the fresh produce logistics is now facing and how future developments will relate to shifting power structures in the sector driven increasingly by retailers.

2. Power in supply chains

An interesting theoretical concept developed in the area of supply chain management in recent years is that of the power regime. Hitherto, the concept of power has rarely been discussed in supply chain management, except to deny its importance, or to argue that power should not be exploited in chain management because lean approaches should be based on equity, trust and openness (Cox, 1999). More recently, these latter views, which have become embedded in the concept of "relationship marketing", have come into question, and a school of thought has arisen that maintains that power-imbalanced business relationships are equally important in understanding the mechanisms and dynamics of business exchanges and chain management. (Hingley, 2005).

Much research focus since the 1970s has been centred on the most appropriate power source. Richardson stated that coercive (punishment) and non-coercive (reward) are two basic sources of power (Richardson *et al., 1995*). It can be argued that companies are only successful if they possess power over something or someone. This is because only by having the ability to extract value from relationships with others, whether these are with customers, employees or suppliers, can business success be sustained (Cox, 1997). There must, therefore, be inherent objective conflicts of interest between vertical participants in supply chains, just as there are between those competing horizontally in the markets that form around specific supply chain resources. This arises because participants in the chain try to appropriate value for themselves and, if individuals seek to maximize their own profit rather than that of the whole chain, they will attempt to gain more of the value-chain for themselves if they are able so to do. Because certain participants in the chain may also recognise that they have limited power to appropriate value from others, this is not the same as asserting that they would not try to acquire greater value for themselves if circumstances permitted. This is important in developing an appreciation of how supply chains might be managed both strategically and operationally, for which a necessary condition is that participants understand the power structures that exist (Cox, 1997).

The analysis presented here must also take account of a number of important political, economic and environmental variables that have impacted on both the demand and supply sides of the fresh produce supply chain over the last decades. Key factors on the demand side driving change in China's fresh produce markets are urbanization, consumers' growing awareness of the importance of vegetables in a healthy diet, rising incomes, and government policy on health, nutrition and environmental protection. Furthermore these factors are interrelated. Most importantly, stakeholders within the fresh produce supply chain now face the demand for assurances on the quality and safety of fresh produce, as well for minimal impact on the environment and ecology. Therefore, at each functional stage within the fresh produce supply chain there are certain resources that provide an operational basis to meet the demand for quality and safety effectively. If these key resources can be owned or controlled in a way that is unique or difficult to imitate, and they are vital to customers or suppliers, then they become what Cox has called *"critical supply chain assets"* (Cox, 1997). These assets are the foundations of supply chain and market power. They give their owners or controllers the potential to appropriate value on a sustainable, long-term basis, because they create a market structure characterised by low or non-existent competition. We might therefore expect a firm that possesses a critical or unique supply chain asset to be achieving above average profit margins (rents) on a sustained basis (Cox *et al.*, 2000).

Globally many perishable food supply companies have transformed themselves from supply-driven to demand-driven procurement agents and value chain managers. With regard to the best way for buyers to manage business relationships with suppliers, two polarised approaches have been propounded. One argues for the adoption of a transparent, win-win approach associated with partnering (Carlisle and Parker, 1989; Sako, 1992) and the other, the diametrically opposite zero-sum win-lose outcomes (Cox, 2004). Retailers' power over food manufacturers is positively related to the importance of economies of scale in manufacturing, retail concentration and own brand penetration of the market and it increases as product shelf-life increases and as manufacturers' specific investments increase (Collins, 2002). In simple terms, a firm operating under conditions of open competition will have little or no power over its customers, suppliers and competitors. Conversely, firm operating under conditions of near monopoly/monopsony or oligopoly/oligopsony will potentially have immense power. A further factor in affecting power distribution in the supply chain is the role of regulatory intervention. If regulation (*e.g.* health or sanitary restrictions or requirements) is imposed at those stages either where effective market competition cannot be introduced, or where compliance costs for some chain participants are made prohibitive, then others may be able to gain advantage and appropriate greater chain value to themselves (Sanderson, 1999).

Even though power in supply chain management has been the subject of some recent studies there are still gaps in the literature. Most existing work has been focused on either retailer–manufacturer relationships or retailer-consumer relationships, which is thus partial and inadequate for a whole chain perspective. Furthermore the fresh produce sector has been neglected, though it has been of growing importance globally, nationally and regionally and undergoing some of the most rapid transformation of all the various food distribution systems.

A power matrix can be constructed according to the relative power attributes of buyers and suppliers involved directly in bilateral exchange with each other in a chain. These in turn derive from market structure, competitive conditions, product attributes and uniqueness, transaction costs and information asymmetry (Cox, 2004). Table 1 displays such a matrix and identifies four dyadic/bilateral trading situations between buyers and sellers that can be defined in relation to these factors. Exchange at each point in the chain can be characterised in terms of the degree of dominance either buyer or seller is able to exert over the other and these are represented symbolically below:

- Buyer dominance (>).
- Interdependence (=).
- Independence (0).
- Supplier/seller dominance (<).

In this paper, we aim to explore the power regimes in China's fresh produce supply chain from a dynamic perspective.

3. Mapping fresh produce supply chains in China

The generic fresh produce chain can be defined as the process of delivering fresh produce from producers to the end consumers. The stakeholders in the chain traditionally include farm-input suppliers, farmers, collectors/transporters, wholesalers and retailers. Suppliers of farm inputs and services include fresh produce seed companies, chemical suppliers and extension service stations throughout China, but at present there is little or no integration between fresh produce-input suppliers and farmers or farmer groups and hence the farm-input supply tier of the chain is excluded in discussion.

The Chinese fresh produce supply system has undergone considerable development from a government-planned sector to a more market-oriented one. In general, the organization of current fresh produce supplies in China is complex, characterized by diverse chains, often small in scale, with a low access threshold to the market, many participants and fluidity of personnel. The availability of refrigerated transport, storage depots and warehousing is also limited. This is

Table 1. The Power Matrix: Buyer and Seller Attributes (after Cox, 2004).

	Supplier power relatively low	Supplier power relatively high
	Buyer dominance (>)	Interdependence (=)
Buyer power relatively high	Few buyers/many suppliers Buyer has high % share of supplier's market Supplier highly dependent on buyer Supplier's switching costs high Buyers switching costs low Buyer's account attractive to supplier Supplier offers standardised product Buyer's search costs relatively low No supplier information asymmetry advantages over buyer	Few buyers/few suppliers Buyer has high % share of supplier's market Supplier highly dependent on buyer Supplier's switching costs high Buyers switching costs high Buyer's account attractive to supplier Supplier offers unique/differentiated product Buyer's search costs relatively high Supplier has some information asymmetry advantages over buyer
	Independence (0)	Supplier dominance (<)
Buyer power relatively low	Many buyers/many suppliers Buyer has low % share of supplier's market Supplier low dependence on buyer Supplier's switching costs low Buyers switching costs low Buyer's account not very attractive to supplier Supplier offers standardised product Buyer's search costs low Limited supplier information asymmetry advantages over buyer	Many buyers/few suppliers Buyer has low % share of supplier's market Supplier not dependent on buyer Supplier's switching costs low Buyers switching costs high Buyer's account unattractive to supplier Supplier offers relatively unique product Buyer's search costs high Supplier has substantial information asymmetry advantages over buyer

however changing. Apart from the demand factors mentioned earlier, hypermarket development associated with Foreign Direct Investment has seen the entry of multi-national retailers such as Carrefour, Walmart, Pricesmart, Metro and Tesco into key East coast cities (Bean, 2005), that are introducing new practices in chain organization, logistics and structure to service their hypermarkets.

We identify 3 broad chain typologies normally found in the fresh produce sector (Figure 1), though within these there are considerable sub-types:

- Chain 1: Direct traditional chains in the rural area where producers sell their fresh produce directly to consumers.
- Chain 2: Traditional wholesale/wet markets chains, and
- Chain 3: Modern chains to retail supermarkets, hotels and catering.

However, it will become clear that significant modifications are now taking place to these broad chain types, particularly in servicing the supermarket sector. The bulk of fresh produce sold through inter-province markets passes through all tiers in the chain, but in local markets, fresh produce may go straight to wholesalers, retailers or directly to consumers. The traditional wholesale/wetmarket chain is still important in general food retailing, but supermarket and hypermarkets are taking an increasing share of the grocery market in larger cities and towns, and had over an 11% share by 2002 (Hu *et al.*, 2004).

4. Analysis of power regimes in China's fresh produce sector

We now outline the characteristics of the various tiers in China's fresh produce chains and the dyadic buyer – seller power relationships implicit within each.

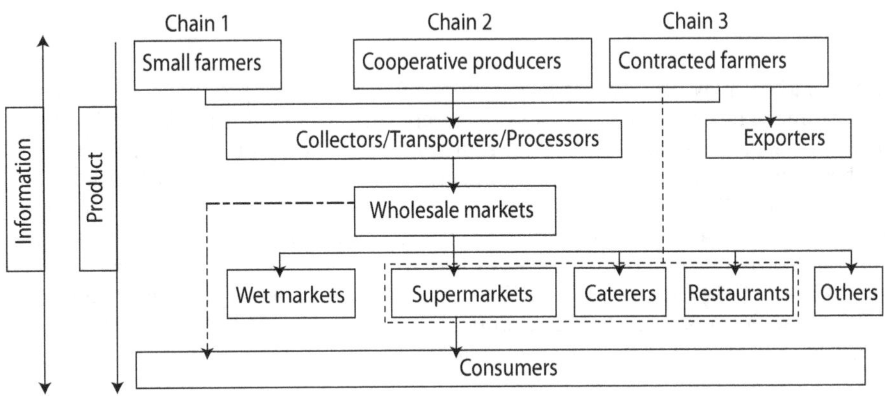

Figure 1. Typical fresh produce supply chains in China.

4.1 Farmers

Traditional fresh produce growers in China have little if any bargaining power over buyers, collectors or wholesalers. One of the main limitations is the absence of accessible cool-chain storage, which forces farmers to sell their perishable products at the earliest opportunity after harvest. During the one-month peak season, there is much competition by farmers in the same villages to sell their produce, rather than any collaborative/ co-operative action to exercise any influence on the prices buyers will pay for their vegetables. We therefore characterise their overall power situation as weak and hence in a buyer dominant (>) situation except where co-operative action may increase supplier relative strength, but not necessarily to a dominant position (0). The only exception to this is where farming companies, often state owned or under joint state-foreign ownership have recently begun integrating downstream into direct retailing through supermarkets (Hu *et al.*, 2004), or into contractual relationships with supermarkets (=).

4.2 Agents/collectors/transporters

There are some professional middlemen who help wholesalers to collect fresh produce from growers and charge per kg of transaction between producers and wholesalers. Mutual trust, rather than contracts, links middlemen to producers and wholesalers. Both farmers and wholesalers consider it as an efficient way of doing business with the middlemen. The middlemen act as an information conduit in which wholesalers inform the middlemen on what products they need, what prices they offer and how much they need and the middlemen provide information to wholesalers on when and who will have harvests and is ready to sell. After collecting products at the farm gate, long distance traders (some are specialised transporters and sometimes wholesalers as well) may transport these products from one wholesale market to another. If the middlemen are just trading on behalf of others at higher tiers in the chain, their role is one of facilitating transactions between a buyer and a seller. In such circumstances both the buyer and supplier power is probably quite low, which we characterise as independence in the power matrix (0).

4.3 Wholesalers

Wholesale markets play an important role in the Chinese fresh produce chain. They either developed from the more than 80,000 traditional wet markets, or established by local government, state owned enterprises, collectives and even the private sector (Han, 2000; Zhang *et al.*, 2001). The numbers expanded at a rate of 20-23% annually during the period 1990-93, but slowed rapidly thereafter to around 2% annual growth at the end of the decade and have subsequently gone into decline. However, about 2000 have developed into large scale wholesale markets without much government intervention and regulation, especially where there are good transportation facilities or in the vicinity of big cities (Peng and Xiao, 2000) and these have an established permanent infrastructure, such as cool facilities, distribution centres, trade halls and storage (Han, 2000). In 2002, there were more than 4,150 wholesale markets in total (Figure 2).

Given the larger trade volumes, sellers in wholesale markets may carry out a greater range of activities than those in wetmarkets. There are professional traders (long-distance and local), producers, state owned enterprises, agricultural processors, supply agents, and franchisers. Wholesale markets (other than standardised ones for bulk commodities such as grains and oilseeds) do not have strict entrance qualification for traders.

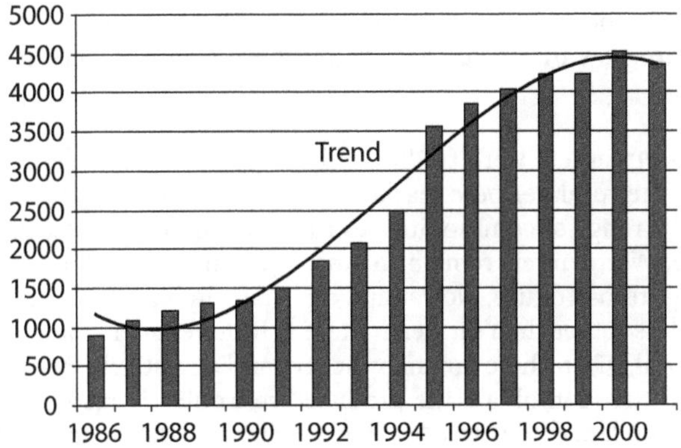

Figure 2. Trend in wholesale market numbers in China 1986-2001.

Management of wholesale markets varies among different types of markets . Taxes are collected either by transaction values, or as a lump-sum amount, similar to at wetmarkets. In the Shouguang vegetable wholesale centre in Shandong province, traders must close their accounts at an accounting centre and 6% of trade value is charged as taxes and market fees (author's survey). In most agricultural wholesale markets however. Long term contracts between traders were not observed in the survey, although "I owe you" notes can be accepted between long term business partners.

Wholesalers usually take a more positive approach to marketing by improving their services to attract both more suppliers and buyers. In Shandong province, over 80% of wholesale market turnover is now from fresh produce, compared with 60% nationally (Ministry of Commerce, 2003). Nowadays, most wholesale markets have an established permanent infrastructure, such as cool facilities, distribution centres, trade halls and storage. Shouguang fresh-produce wholesale market, the biggest in the country, is the only wholesale market with an electronic auction. Given the larger trade flow, sellers in wholesale markets may carry out a greater range of activities than those in wet-markets. There is also evidence of the establishment of specialised wholesalers dealing in fresh produce aiming to service the supermarket sector through co-ordination of procurement, grading and packaging.

We would characterise wholesale market agents as having relatively low power in general, but in having a wide customer base, their trading position in the power matrix is largely independent (0). However specialised wholesalers are more likely to have an interdependent relationship with their buyers (=) and a buyer dominant relationship with many of their suppliers (>).

4.4 Wet and street markets

China has seen the existence of wetmarkets for hundreds years (Peng and Xiao, 2000). The activities of these traditional markets were restricted in the mid 1950s and even subject to closure in the early 1960s, when procurement products could only be sold to the government and not to private buyers. Only after 1978, wet markets were open again. After 1985, extra products other than procurement contracts could be traded directly on the markets, which boosted the revival of traditional markets in rural and urban areas. National statistics show a static situation in rural wet-market numbers in the early 1990's, and a decline from 1998 onwards. Urban wet market numbers continued to grow in the last decade, but have begun to decline also in urban areas (Figure 3).

Most sellers at the wetmarkets were traditionally producers with whom consumers directly negotiated prices. In the rural periodic markets, itinerant traders also travelled from market to market and supplied a range of commodities to rural consumers. Transactions are by cash and prices are negotiable. Wet-markets remain the most common place for consumers to purchase fresh foods including fresh produce. Nowadays, especially in big cities, the majority of sellers in wet-markets are traders rather than farmers. There is little regulation and market management of wet-markets. Overall, given there are many sellers and a relatively large potential number of suppliers from whom they can obtain the produce, we would characterise this dyadic element largely as one of independence (0) with a tendency to supplier dominance (<) where supplies are derived from wholesalers.

4.5 Supermarkets

The supermarket revolution is spreading faster in China than anywhere else in the world. The number of supermarkets has soared in China, especially since economic reforms of 1998, with an average annual growth rate of 20% between 1998 and 2002 (Figure 4) and an estimated 53,100 retail outlets in 2002, accounting for 30% of the urban food retail market sales in 2002 (Hu *et al.*, 2004). The annual growth rate in supermarket sales over the same period, estimated at 40%, outstripped the expansion in sales outlets. Although becoming a significant force in urban food retailing, they still do not yet have the complete market dominance that they enjoy within many Western countries. In the fresh produce sector, vegetable sales in supermarkets were almost unknown in 1995, but by 2002 in the large and medium cities, some 20% of fresh produce was being purchased in supermarket outlets (Fang, 2002).

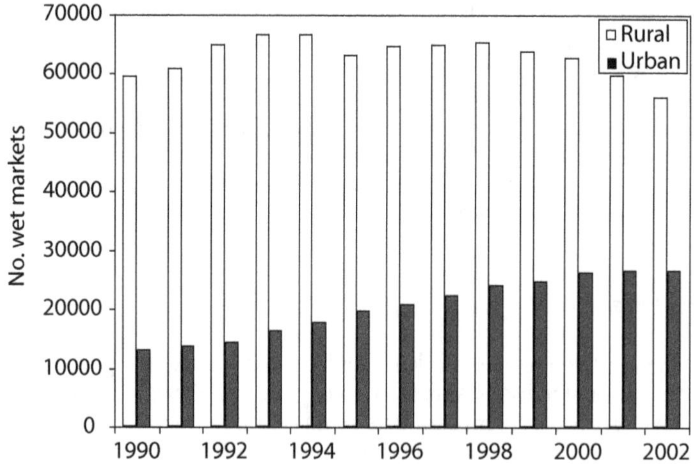

Figure 3. Trends in wet market numbers in China 1990-2002.

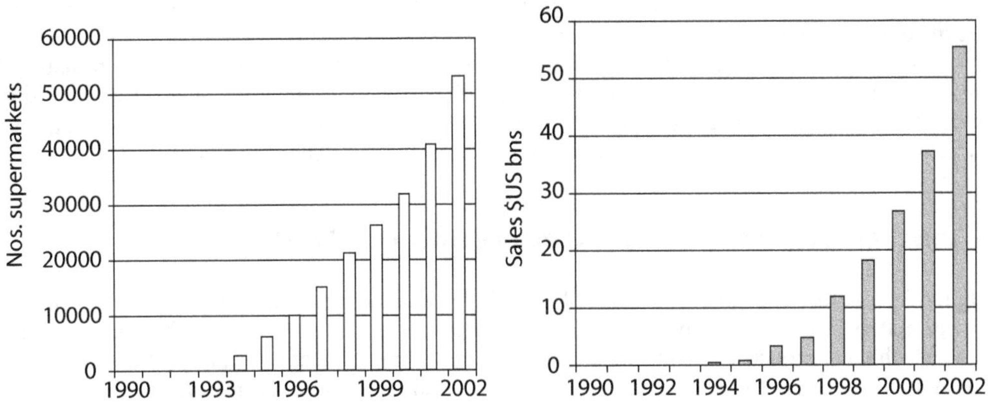

Figure 4. Developments in supermarket numbers and sales in China 1990-2002.

Furthermore, the supermarkets are no longer just serving a niche population of the rich or middle class in the major cities – they have spread well beyond to penetrate the food markets of the poor, from cities of coastal China to inland smaller cities, and from small towns into rural areas. The opening of China's retail sector to FDI and subsequent establishment of international food retail companies is a major factor contributing to the rapid development of supermarkets in China. Having rapidly penetrated the Chinese food retail sector, supermarkets have slowly begun an upstream transformation of the food marketing system all the way to the farmers.

Generally, fresh produce via supermarkets is sourced locally. Those fresh products that can be stored and more easily transported, such as Chinese cabbages, onions, potatoes and winter melons etc can be sourced inter-regionally. Because fresh produce distribution requires investing in facilities for timely-delivery, quality control and temperature control, most supermarkets in China rely on specialised companies or dedicated wholesalers and logistics companies with cool chain storage and transport facilities to maintain product freshness. Some supermarket chains have their own regional procurement and distribution centres. When fresh produce is retailed via supermarkets, they may use their market power to enhance, assure and standardize the quality and safety of the products they buy and have regular residue and checks carried out on the produce procured.

Supermarkets mainly operate within three chain structures –direct from farming companies or farming co-operatives, and increasingly through specialised fresh product wholesalers. Where they buy direct from producer co-operatives or larger farming companies (which may also be multi-nationally owned farming companies or joint ventures such the Glorious Land Agricultural Company in Beijing), and where product specification, quality and volume supplied are high, then there will tend to be an interdependent relationship (=), because the transactions costs for both supplier and buyer of switching to alternatives sources or users are high. Where supermarkets procure substantial proportions of their fresh produce through the general wholesale chain structure, then there may be a tendency to buyer dominance (>). However, between 2000-2003, two of the largest domestic chains - Lianhua and Hualian - more than halved the proportion of fresh produce procured from wholesale markets with consequent expansion in favour of specialised wholesalers (Hu *et al.*, 2004). Overall however, the top-10 firm concentration ratio of the supermarket sector still remains low at some 25-30%, and estimates of the supermarkets' share of total fresh produce retail sales as been put at 30% nationally, although as high as 60% in the major cities. Hence although supermarkets may be able to exert their buying power in certain dyadic buyer-seller transactions, the sector as a whole is probably still insufficiently dominant in fresh product retailing at present to dominate and dictate transactional practices throughout the whole fresh food sector. Clearly however, as their share of fresh food retailing continues to grow, there will be further shifts in the distribution of market power within the chains.

5. The chain power structures and management implications

Figure 5 presents the power structures of the whole chains on the basis of the previous dyadic power matrix characterizations for the different chains and tiers within them. This in turn will suggest which approaches to sourcing and chain management are appropriate and possible. The chain management options will depend on the power and leverage circumstances between tiers, and whether these are enforceable or transmissible up or down the chain.

Positive pro-active chain management is only enforceable or likely to emerge where there is consistent direction in dominance or interdependence between buyers and sellers longitudinally throughout the chain. Chains characterised by independence in their overall power regimes are unlikely to possess the initial conditions to encourage pro-active chain management by either buyers or sellers. Chains in which there is generally interdependence and neither a tendency to buyer or seller dominance will lend themselves to pro-active non-adversarial collaborative management approaches in developing the chain quality standards, in sharing chain value and in investment. Chains with overall buyer dominance will be more adversarial with buyers selecting amongst suppliers; testing the market for better offers and appropriating as much value share as possible; suppliers can only acquiesce, having little alternative but to comply. Supplier dominant chains will exhibit similar adversarial characteristics, but with suppliers extracting greater chain value, and buyers compliant.

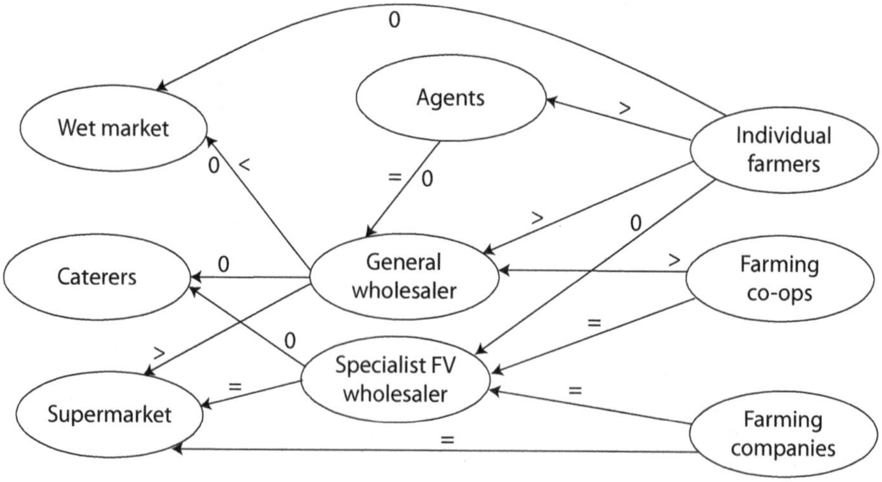

Figure 5. Fresh produce supply chain power structures.

The only chain with clear unidirectional transitive buyer dominance in Figure 5 is that from supermarkets through traditional wholesalers or their agents to individual farmers. The fact that supermarkets still do not wholly dominate fresh produce retailing in China has thus far restricted the extent to which they have been able to select traditional wholesale suppliers and drive rationalization in this particular supply chain type, as has happened notably in the UK, Latin America and the USA. Only where whole chain management is feasible are integrated logistics solutions likely to emerge.

6. Chain development challenges facing the Chinese fresh produce sector

Market forces will inexorably continue to stimulate structural change in fresh produce production, marketing, processing, retailing and distribution in China. However, as a recent report observes *"one of the key challenges for retailers and suppliers in China is overcoming the problems that the underdeveloped supply chain presents ..and ..a fragmented and immature logistics industry"* (IGD, 2005). Regional distribution centres for foreign supermarkets are still in the early stages of development (except by Wal-Mart), and until these overseas chains have established a wider network of outlets and scale, it will limit their willingness to invest in centralised warehousing and distribution. Increasing transport costs, manual warehousing operations, poor inventory management, congestion at delivery points and inefficient back-hauling operations are also problems reducing chain efficiency. This makes it more likely that retailers will initially focus on securing improved service levels from their suppliers, reducing their supplier base and squeeze out wholesalers as they increase direct deliveries. In the medium term, they will begin to develop automated distribution centres, employing internet technologies and real-time inventory and ordering systems. The development of such directed chain management will depend on foreign and domestic supermarkets securing an even greater share of the fresh produce retail market. This in turn will inevitably lead to greater concentration of power and in chain structure within the fresh produce sector as the driver of cost reduction throughout the chain necessitates a greater scale of operations in production, packing, processing, warehousing, transport and distribution.

Fresh produce is perishable in nature, which means that efficient distributing and handling under controlled conditions are critically important to maintain quality. Cold warehousing facilities are one important link in the fruit and vegetable chain; temperature-controlled transportation is another. However, the long history of state ownership has not equipped the system to handle, transport and track the fruit and vegetable produce. Cold storage capacity is believed to be only 20-30% of growing cargo demand. Spoilage losses of up to 33% of perishable freight and illnesses caused by poor handling are common. Apart from a lack of refrigerated transport infrastructures, the lack of tracking systems and geographic fragmentation of networks are other barriers to the efficient distribution of fresh produce.

The growing road network in China has been well developed in the country and is a major option for moving fruit and vegetables from scattered production sites to urban market and ports, thereby raising rural incomes. But several common issues have been identified in the trucking industry. Firstly, trucking firms are extremely fragmented and no single trucking firm can offer truly national coverage (Wu, 2003). Secondly, trucking costs are increasing due to more strict emission standards, road usage fees, rising fuel prices and the need for advanced technology, as well as truck maintenance. Furthermore, the limited opportunities to back-haul goods increases transport costs and often discourages long distance transport. Thirdly, overloading of fruits and vegetables is quite common, which results in damage and spoilage of the goods, delays and higher cost if fines are charged.

Even though Internet technology has been widely adopted in developed countries, it is still quite new and limited in fresh produce distribution and logistics in China. The main function so far has been largely restricted to assisting acquisition and transmission of information to other supply chain partners. Generally it is only large enterprises that have their own website or lease a website to publicise or promote their products. Email, an Internet-enabled technology, is used to a greater extent for ordering and purchase of products. IC card, an electronic trade platform, has only been used for vegetable transactions in the Shandong Shouguang Vegetable Wholesale Market Co. Ltd. To be competitive in global fresh produce markets in future, it will be essential for firms in China to adopt internet technologies to increase their efficiency and decrease costs in the logistics and in chain management operations.

In summary, to accommodate both domestic and international demand for fresh produce, cold chain infrastructure, transport facilities and its management must be improved in China. Moreover, the need and demand for tightly-focused, efficient logistics services will escalate, given the expected rises in the volume of external trade and domestic distribution induced by China's WTO accession.

References

Bean, R., 2004. China, P.R: Retail and food distribution development in China. USDA FAS Gain Report No. CH4816.

Bean, R., 2005. China, P.R: Retail food sector mainland China food retail annual report 2005. USDA FAS Gain Report No. CH4838.

Carlisle, J.A. and R.C. Parker, 1989. Beyond Negotiation, Wiley, Chichester.

Chang, X., X. Dai and H. Kupper, 2003. Designing effective supply chains of fresh produce initiated by supermarkets in China: A case study on Suguo Supermarket. Int. Food and Agribus. Manag. Assoc. conference http:/www.ifama.org/conference/2003conference-/papers/Chang.pdf.

Collins, A., 2002. The determinants of retailers' margin related bargaining power: evidence from the Irish food manufacturing industry. Int. Rev. of Retail, Distrib.and Cons. Res. 12, 165–189.

Cox, A., 1997. Business success: a way of thinking about strategy, critical supply chain assets and operational best practice. Earlsgate Press, Boston, UK.

Cox, A., 1999. Power, value and supply chain management. Supply Chain Manag. 4, 167-175.

Cox, A., 2004. The art of the possible: relationship management in power regimes and supply chains. Supply Chain Man. 9, 346–356.

Cox, A., J. Sanderson and G. Watson, 2000. Power regimes. Earlsgate Press, Helpston.

Fang, X., 2001. Nong chan pin faxing zai Zhong guo. (The distribution of agricultural products in China). http://www.china-hypermarket.com/freshsupplychain/htm .

Fang, X., 2002. Shengxian gongyinlian xianzhuang yu fazhan fangxiang (The development of the fresh agriculture-food supply chain and trends). http://www.chaoshiren.com.cn-/scripts/articles/show.php?id=517.

FAO, 2005. on-line database. http://apps.fao.org/faostat/collections?version=ext&-hasbulk=0&subset=agriculture.

Han, Q., 2000. Vegetable Wholesale Market Construction in China. Agricultural Economy, 2000, 42-44.

Hingley, M.K., 2005. Power imbalance in UK agri-food supply channels: learning to live with the supermarkets. J. Market. Manag. 21, 63-88.

Hingley, M.K., 2005. Power to all our friends? Living with imbalance in supplier-retailer relationships. Ind. Market. Man. Forthcoming.

Hu, D., T. Reardon, S. Rozelle, P. Timmer and H. Wang, 2004. The emergence of supermarkets with Chinese characteristics: challenges and opportunities for China's agricultural development. Dev. Policy Rev. 22, 557-586.

(IGC) Institute of Grocery Distribution, 2005. Retailing in China.

Liu, H, and Y.P. Wang, 1999. Co-ordination of international channel relationships: four case studies in the food industry in China. J. Bus. and Ind. Market. 14, 130-150.

Liu, X., B.J. Revell and F. Zetian, 2003. Measuring the competitiveness of China's export trade in fresh produce: a strategic management approach. Poster paper presented at the 25th IAAE Congress, Durban.

Liu, X., B.J. Revell and F. Zetian, 2005. Realigning the power in the Chinese vegetable supply chain. Proc. Int. Conf. Manag. Sci. and Eng. University of Incheon, Republic of Korea Harbin Institute of Technology Press, pp. 725-731.

Ministry of Commerce Department of Market System Development, 2003. Report of the rural markets development in China. http://scjss.mofcom.gov.cn/aarticle/ap/200312/2003120-0164478.html.

National Bureau of Statistics, 2003. China statistical yearbook 2003. No. 22. China Statistics Press. pp. 617.

Peng, X. and C. Xiao, 2000. Shichang yu nongye chanyehua (Market and Agricultural Vertical Integration). Economic Management Publishing House, Beijing.

Richardson L.D., E.H. Swan and J.D. Hutton, 1995. The effects of the presence and use of channel power sources on distributor satisfaction. Int. Rev. of Retail Distrib. and Cons. Res. 5, 185-201.

Sako, M., 1992. Prices, quality and trust. Cambridge University Press, Cambridge.

Sanderson J, 1999.Passing value to customers: on the power of regulation in the industrial electricity supply chain. Supply Chain Manag.. 4:4, 199-208.

Wu, C., 2003. China, market development reports. China logistics profile 2003. USDA Gain report number: CH 3833.

Xian, V., 2003. China P.R: Retail Food Sector Report 2003. USDA FAS Gain Report No. CH3615.

Zhang X., X. Fu and J. Yang, 2001. The evolution of the Chinese vegetable supply chain. http://www.vegsys.nl/files/413897e6f7dc097ee4b4e8c5c332f77d.pdf.

Bargaining power in mango supply chains: An experimental gaming approach

G. Zúñiga-Arias, S. Meijer, R. Ruben and G.J. Hofstede

Abstract

Transactions between producers, traders, retailers and consumers in the mango supply chain are characterised by contractual arrangements concerning outlet choice, price, volume, quality and frequency. We used a gaming simulation approach to identify the intrinsic agency characteristics that typically result in specific contracts and assess the underlying differences in perceived bargaining power and that are conducive for reaching these arrangements. A stylised mango chain game is developed that permits the appraisal of different sets of delivery transactions. Agency roles are defined for all participants in the mango supply chain. Attributes of all transactions are recorded, also permitting contract breach, hold-up and repeated contracts. Bargaining power is assessed at a 10-point Likert scale and revenue distribution is measured in terms of money. The game has been played five times with different groups of mango producers, resulting in a data set of 82 transactions and 43 bargaining power positions. The game design closely mimics the negotiation conditions in the Costa Rican mango supply chain. The game results confirm the important role of trust and information exchange for reaching mutually acceptable contracts. Bargaining power is strongly related to the negotiation skills, wealth and partnerships of the negotiator. Revenue distribution is related to the bargaining power of the trade partner, risk and the length of the contract. Transparency and agency cooperation thus play important roles for improving the efficiency of mango supply chains in Costa Rica.

Keywords: bargaining, transactions, contracts, simulation gaming, mango, Costa Rica

1. Introduction

The organization of the mango supply chain in Costa Rica includes a wide number of different agents involved in numerous transactions and oriented towards multiple market outlets. Mango transactions differ in terms of volume, quality, price and delivery frequency and the produce can be sold at local market through wholesalers and/or at international markets through multinational trading companies. Relationships between producers associations, (local and international) traders, retailers and consumers are structured through a complex sequence of delivery transactions.

Contract choice in supply chains is based on a broad number of structural and behavioural factors (Williamson, 1998). Agency characteristics related to partnership, negotiation skills and wealth might influence bargaining power. In addition, the supply volume and the quality of the produce influence the price. Behavioural factors related to the trust amongst agents and the perceptions regarding contract breach also influence contract choice (Muthoo, 2002). We are particularly interested to disentangle the factors and motives underlying the engagement in different contractual arrangements. Uncertainty and asymmetric information are particularly important in shaping contract choice, and may be relevant for understanding specific business strategies. Such supply chain interdependencies can be analysed in an experimental setting.

We developed a stylised mango game simulation framework to assess the specific attributes that influence contract choice. The game includes a description of agency roles for all participants in the mango supply chain. Attributes of their transactions are recorded, permitting contract breach, hold-up and repeated contracts. Random uncertainty is introduced in production and trade to account for unknown external factors (*i.e.* climatic hazards or price risk). All agents are instructed to strive for profit (income) maximization. The game is played with 7-13 mango producers and has been repeated five times. After an introduction to the game and the gaming attributes (game board, game money, information sheets, product fiches and transaction forms), the game was played in three rounds and followed by a debriefing with the participants. The results of the game include a set of (single or repeated) transactions of mangos with specified price, volume and quality attributes, and records regarding the time involved in each transaction. In addition, all players answered a detailed questionnaire at the end of the game to assess their bargaining power perceptions.

This article provides an analytical framework and an empirical assessment of the agency characteristics and their perceptions that influence specific contractual arrangements. The analysis is conducted at two levels; first the perceived bargaining power of the participants is explained, and hereafter an appraisal is made of the overall supply chain performance. Performance of contracts is measured in terms of revenue of the participants at the end of the session, permitting also a comparison of single and repeated transactions. Finally, the outcomes are explained by relating agency characteristics with contractual attributes, using multiple regression analysis and logistic regression. This permits an understanding of underlying structural parameters and behavioural motives that give rise to particular perceived bargaining power and strategies of the participants.

An important aspect of supply chain transactions is related to the distribution of bargaining power amongst agents (Nash, 1950; Rubinstein, 1982). Contract choice is therefore influenced by the anticipation of the behaviour and response of potential trading partners. Considerations of fairness and expectations regarding trust and reciprocity are of key importance for engaging in single or repeated transactions (Berg *et al.*, 1995). Individual perceptions of bargaining power and differences in transaction costs can therefore be attributed to the place and position of agents within a network hierarchy. We therefore devote particular attention to an analysis of the agency bargaining power influencing contract choice, the transactional attributes that shape their bargaining power perceptions, and the revenues of the stakeholders.

The remainder of this article is structured as follows. First, we discuss the organization of the mango supply chain in Costa Rica, identify the major agents and analyse the key attributes underlying their differences in bargaining power. Hereafter, we outline the design of the mango game that has been designed to mimic the agency behaviour. Attention is given to the definition of the agency roles and the contractual options. This is followed by an analysis of the participant perceptions toward bargaining power paying particular attention to the underlying motives for increasing the revenue of the participants in a transaction. Finally, we discuss the usefulness of this gaming approach for a better understanding of the organization of the mango supply chain and derive some policy implications for improving the efficiency of the mango market in Costa Rica.

2. The mango supply chain in Costa Rica

The main mango varieties grown in Costa Rica and suitable for export are Tommy Atkins, Kent, Keith, Palmer and Smith (Mora, pers. comm.; Jiménez, pers. comm.; MAG, 2003). In 2003, the area planted with mango covers 8,350 ha in Costa Rica and the quantity produced reached 35,000 tons. During the last seven years (1996-2003) the surface planted increased with 7% and the quantity produced rose by 75% (Jiménez, pers. comm.). The yield increase is mainly due to technological progress (Montero and Cerdas, 2000). Costa Rica has about 1,950 mango producers of which 60% cultivate less than 5ha, 35% cultivate between 5 and 20 ha and 5% own more than 20ha (Mora, pers. comm.; SEPSA, 2001). The producers are organized in different ways; large and medium-size producers are linked to international trading companies and small producers are not organized or associated to co-operatives or producers associations.

Zuniga-Arias, Ruben and Verkerk (unpublished data) and Millet (2003) analysed data from the Ministry of Agriculture and main sources in situ for describing the structure of the mango supply chain in Costa Rica, distinguishing between production for the local and the export market. The organization of the mango supply chain is relatively simple for the export market and much more complex for the local market (see Figure 1). For the export market there is a straightforward relation with the traders of the product, but at the local market there are a wide number of different intermediaries involved. These agents play multiple roles in the chain: they buy mangoes from the producers, sell the produce to the CENADA wholesale market, or buy from CENADA to deliver to local outlets like local retailers and the wet market.

Global competition contributed to replacing spot markets by contractual production and vertical integration. More coordination and collaboration may lead to improved efficiency in production and distribution channels and to more product and market innovations (Hendrikse, 2003). Chain governance refers to the organization of transactions, while a governance structure consists of a collection of rules, institutions, constrains structuring the transactions between the various stakeholders (Hendrikse, 2003). A governance structure affects the size of the surplus that will

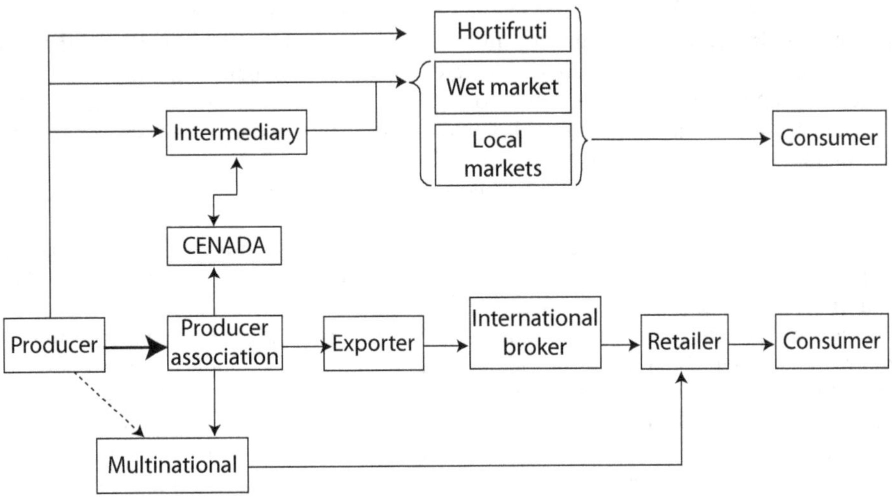

Figure 1. Mango supply chain in Costa Rica.

be generated by its effect on investment, the efficiency of bargaining, and risk-aversion (Zingales, 1998). In the case of the mango supply chain form Costa Rica, the replacement of the spot market has been done in for the export market as mention before, but for the local market spot market is still the main governance structure in the market; nevertheless, there are few example of hybrids forms of cooperation between actors such as the relational contract and the formal contract. Vertical integration is present in the export market where the multinational company controls many of the actions of the producer associations. The presence and necessity for meeting the international quality standards such as EUREPGAP, increase the presence and control of the exporters and retailers in the actions of the mango producers in Costa Rica.

3. Analyzing bargaining power in the supply chain

Muthoo (2000) defines a bargaining situation as a setting where individuals engage in mutually beneficial exchange but maintain conflicting interests over the terms of the trade. In simple words, in a bargaining situation two agents have a common interest to cooperate, but have conflicting interests over exactly how to cooperate. Muthoo (2000) explains bargaining as any process through which the agents try to reach an agreement. This process is typically time-consuming and requires the agents to make offers and counter-offers to each other. Institutionalists like Chamberlain and Kuhn (1965) consider bargaining power as the ability to secure an agreement on one's own terms. Chamberlain (in Leap and Grigsby, 1984) defines bargaining power as the cost of agreeing (or disagreeing) with the opponent. A party's bargaining power increases as the cost of disagreeing with an opponent decreases. Slichter (1940) defined bargaining power as the cost to one agent of imposing a loss upon another agent. For Lindblom (1948), bargaining power is best defined to include all the forces that enable a buyer or a seller to set or maintain a price. Following the same statement, Fossum (1982) recognizes two separate aspects of bargaining power: the power inherent to the economic positions of the parties (the environment) and the attributes and skill of the negotiator (individual characteristics).

The notion of bargaining power is rooted in power-dependency theory (Emerson, 1962; Yan and Gray, 2001), which states that one actor's bargaining power is derived from another's dependency. More specifically, Actor A's dependence on Actor B is directly proportional to A's motivational investment in goals mediated by B, and the inversely proportional to the availability of those goals to A outside of the A-B relation. Therefore an actor can develop or gain bargaining power is it is able to reduce dependency toward other actors (Cook, 1977; Bacharach and Lawler, 1984; Yan and Gray, 2001). In simulation gaming, this behaviour is related to the actions of the participants to maintain outside options open as a way to develop multiple outlet choices for the product and to reduce the problems of facing a take-or-leave offer from the buyer.

Muthoo (2000, 2002) defines seven factors that determine the bargaining power of any particular party: (1) Impatience, (2) Risk of breakdown, (3) Outside options, (4) Inside options, (5) Wealth, (6) Commitment tactics and (7) Asymmetric information. Agency bargaining power is greater if someone is more patient relative to other negotiators. The bargaining power is lower under the occurrence breakdown. Outside option will increase bargaining power only if they are sufficiently attractive. Inside option refer to the payoff that an actor obtains during the bargaining process, while the parties in the negotiation are in temporary disagreement. Commitment tactics refer increase in bargaining power occasioned by the costs to back down a decision (sustain the word). Finally, information asymmetry delivers bargaining power to the actor with more information.

For this study, we elaborated an analytical model to assess the different determinants of bargaining power. Based on the determinants of the bargaining power as defined by Muthoo (2000), we establish an experimental bargaining score and use it to explain the supply chain revenue distribution amongst the agents (see Figure 2). The dependent variable represents the revenues of the buyers and sellers and the independent variables are the perceived bargaining power of the buyer and seller, the risk attitude of buyer and seller, bounded rationality and contract choice in the exchange transactions.

Determinants of bargaining power

Muthoo (2002) stresses wealth as a determinant of bargaining power. The actor who perceives himself as wealthier will exert more bargaining power during negotiation. On the other hand poor actors will not be able to practice bargaining power during negotiation. Fossum (1982), found that the bargaining power is dependent on the personal characteristics/skills of the negotiator.

Kochan (1980) explains that the triggering forces affecting bargaining power are environmental variables, such as organizational characteristics, structural characteristics and the negotiation process. Dunning (1981) and Dicken (1986) found a relationship between bargaining power and endowment-specific factors in the exchange among international companies, the own country and the host country, specifically in the electricity sector. In the same line, Moon and Lado (2000) proposed that major sources of bargaining power are company size, technological intensity, advertising intensity, intra multinational corporations sourcing, export intensity, product diversity, ownership, industry competition and strategic importance of industry.

Lee *et al.* (1998) performed a literature review on international join ventures and found that their bargaining power increased with the perceived strategic importance of partners (Yan and Gray, 2001); with resources linkage between partners (Lecraw, 1984; Kuma and Seth, 1998); and with the availability of alternatives facing the partner firm (Yan and Gray, 2001). According to Leap and

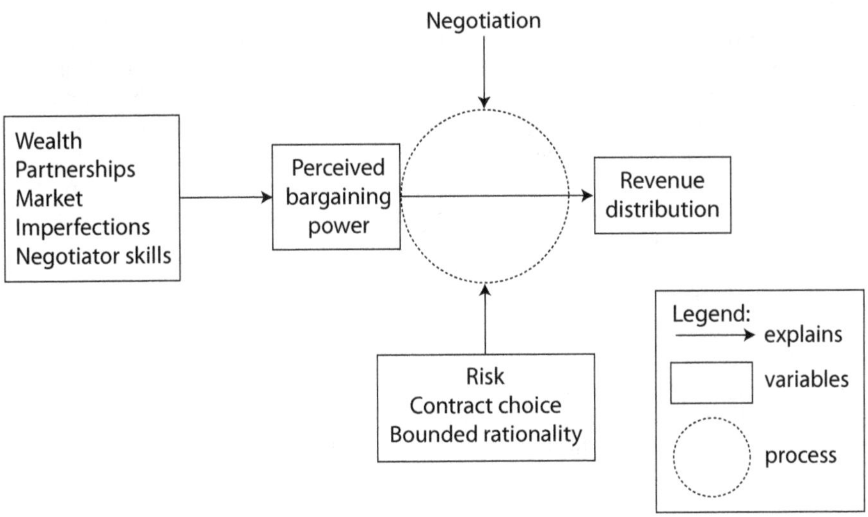

Figure 2. Analytical model.

Grigsby (1986), this points out some key factors that affect bargaining power, such as availability and control of resources, potential and enacted power and absolute, relative and total power.

Rubinstein and Wolinsky (1985) and Leap and Grigsby (1986) emphasized that the bargaining positions and hence the agreement reached in any particular meeting will be affected by the conditions prevailing in the market. These will include the chances that each of the negotiation parties have for meeting other partners in the event that the agreement in the current negotiations is delayed, including the expected length of time required to achieve any alternative transaction and the expected behaviour of alternative partners too. These are outside options' in terms of Muthoo.

Determinants of the buyer's and seller's revenue

Revenue can be explained using many characteristics. We decided to use behavioural characteristics to explained such as contract choice, risk, bounded rationality and perceived bargaining power. One characteristic of the contract is that it is a mechanism to share risk, reduce the uncertainties produced by bounded rationality and a ay to determine who has the bargaining power in a relationship.

Close relationships are a key element of Supply Chain Management, but industrial experience has shown that it is difficult to contract across inter-organizational borders. Therefore, agents need to share both risk and rewards amongst members of the supply chain (Agrell *et al.*, 2002). Contracts can be considered as mechanisms to share the risk and reduce the uncertainties of the transactions. Asymmetric information in supply chains can be used by agents to reduce their own risk and costs in detriment of the whole supply chain. Agrell *et al.* (2002) have shown the close interaction between risk-sharing, contracts and information asymmetries between actors of the chain. They found that reallocation of bargaining power may not necessarily lead to better performance and coordination in the supply chain. There is little or no evidence either that rental contracts are preferred in situations where agents are more risk-averse (Roumasset, 1993).

Contracts are usually incomplete due to bounded rationality and the fact that humans cannot forecast *ex-post* behaviour in an agency relationship. Hendrikse (2003) stressed that incomplete contract can become more complete by allocating authority to somebody. This person decides, according to his own interest, what is most desirable in the prevailing circumstances. Therefore, an important issue in organizing the firm is the allocation of control and authority. This allocation involves trade-offs in the choice of governance structure, because moving authority downstream in the chain or network entails taking power away upstream (Hendrikse and Bijman, 2002). Relational contracts or soft contracts are used to enforce formal contracts due to the relationship with reputation and trust (Hendrikse, 2003) and - depending on the cultural background of the people – also involve friendship to enforce formal contracts. The enforcement of formal contracts with relational contracts will be reflected in the reduction of transaction cost.

A major assumption is that the principal has all the bargaining power when designing a contract. This may be reasonable if there is competition between various, sufficiently homogeneous agents (Inderst, 2002). This bargaining power is represented with take-it-or-leave-it selling or buying option. A commonly accepted notion is that negotiations for the purchase of an indivisible good is futile if the reservation price of the potential buyer is bellow that of the potential seller. Therefore, bargaining over contracts in situations where the principal owns all bargaining power will have no effect in the selection of the contract type and the governance structure (Pitchford, 1998).

Argyris and Liebeskind (1999) and Stinchcombe (1985) argued that the contracting actor's ability to influence the terms and conditions of contracts is highly contingent on their present bargaining power (Buvik and Reve, 2002). High concentration of buyers in the demand side will increase the bargaining power of buyers (Heide and John, 1992).

Higher bargaining power increases the stakeholders marginal revenue and hence its provision of effort (Chemla, 2002). Suppose that two firms hold complementary patents which could only be used together. Should they agree on a way to share revenues, they can collect a flow of revenues as the technology is used; but as long as they disagree, nobody can use the patents, so potential revenues are lost. In this context, switching costs can be arbitrarily small, so the additional lawyers' fees of re-drafting an agreement would suffice (Caruana *et al.*, 2006). Taking into account that the revenue is the total income producer by one actor, one can say that the more revenue received in a specific transaction the more bargaining power actor will obtain, considering the income that a particular stakeholder gets from a negotiation as an expression of bargaining power (Dobbelaere, 2003).

Risk sharing is an important topic in contract design because it affects both the cost of risk-bearing and the motivation to behave in a certain way (Bogetoft and Olesen, 2004). According to Fafchamps (2004), if a producer is a risk taker then the effect over the revenues will be positive, therefore this study compared agricultural case studies from Madagascar, Benin and Malawi.

Muthoo (2002) stresses that asymmetric information reduces the bargaining power and Ayala (1999) pointed out that the less available information, greater the uncertainty and risk to make wrong decisions. Because asymmetric information is also part of bounded rationality we can say that the actor with lower bounded rationality (lower information problems) will have more revenues, following Fafchamps (2004) previous statement.

The negotiation process

The negotiation process in during the sessions has a time periods of ten minutes, where participants bargain about several contract attributes, and tried to share risk and practice their perceived bargaining power in the presence of bounded rationality in both sides of the agreement. The main idea is to assess the effect of those behavioural characteristics to the revenue of the participants.

Rao (2003) defines bargaining cost as *"the transaction cost involved in negotiation between and/or parties to settlement or bargain. These cost include the cost of searching, drawing up agreements, monitoring and enforcing the agreement and the opportunity cost of bargainers time"*. Williamson (1975, 1981, 1985) defined the core dimensions of a transaction: asset specific, frequency of delivery and uncertainty. These dimensions affect the selection of different governance structures between spot market and vertical integration. Due to bounded rationality and opportunistic behaviour, hybrid contract form and the vertical integration will replace the spot market when there are highly specific assets involved in the transaction.

In most bargaining situations at least one player has more information about the commodity than others, this fact may lead to control a larger share of the value added due to information asymmetry. This issue is illustrated at the markets of "lemons" *i.e.* in the second hand cars market where the seller is assumed to have more valuable information than the buyers; Muthoo (2001) defines that at least one player's reservation value is his private information. Information-sharing in groups may lead to a reduction of transaction cost, likewise it will increase trust and social cohesion and hence increase the bargaining power of the partners of the group (Weinberger and Jütting, 2001).

4. The mango chain game

We developed a gaming simulation to test the operational model in an experimental setting. The Mango Chain game is a board game build to assess the bargaining power and value added distribution amongst agents in the mango supply chain. (Sterrenburg and Zuniga-Arias, 2004; Meijer and Zuniga-Arias, 2005). The objective of the game players is profit maximization through optimal exchange arrangements with other players. The game defines specific roles for four major agents (*i.e.* producer association, multinational, independent exporter and retailer) and involves 7-13 participants. The product traded are coloured fiches representing mango in different qualities. Each session takes about 2-2.5 hours and is followed by 60 minutes for de-briefing.

Duke and Geurts (2004) emphasize that the gaming-simulation approach is relevant for strategic problem solving, because in a well-structured, transparent and effective way can produce knowledge in different type of data, insights and tacit knowledge that can be integrated by the different participants of the sessions. These games offer a safe environment to test strategies in advance. Roth (in Kagel and Roth, 1995) explains that an experiment is something that you control and in the case of economic experiments are the study of the human behaviour in a control environment, these experiments can be of three types designed to test theories, designed to test game-theoretic hypothesis and those concerned with the study of industrial organization. Experiments are useful for observing behaviour, while (computer) simulation are used to studying theory. The mango chain game has part of both, it has been constructed under the umbrella of a specific set theories and the aim is to get an understanding of the behaviour of the actors regarding bargaining, quality, uncertainty among other things.

The mango chain game is played on a game board placed on a large table, with enough room to walk around the board for all participants. The game board shows the structure of the mango chain (see Figure 3), based upon a case study of this chain in Costa Rica. (Millet, 2003). Participants receive game money, information sheets describing the roles for each player, products to be traded (divided in two quality categories), ships and boats for transport steps on the game board, transaction forms and consumer market forms.

The game leaders control special cards to apply randomized uncertainty in production and transport. Production is simulated by game leaders. There are four 'production fields' per producer association representing producers that belong to the association. Each of these field produces one ton of mango per round in a quality determined by the quality of the production field. The amount of mango per field has bandwidth of 50%: by picking a coloured fiche from a dark box the producer organization has ¼ chance for 50% less production, ½ chance for normal production and ¼ chance for 50% more production. In a similar vein, consumptive demand is simulated by game leaders. Retailers inform the game leaders how much they want to sell to the consumer market. The game leader then determines prices per unit based upon the total supply. Consumers have unlimited consumption possibility, but the price decreases when supply becomes larger. Separate prices are maintained for the different mango qualities. The domestic market buys all left-over products from producer associations, multinational and independent exporters at a fixed low price regardless of the quality.

During transportation of the mangoes there is a chance that quality decreases. Quality levels present in the game are: extra good, normal and perished. Perished goods are destroyed immediately at no costs, thus never occur on the board. For both the trucks and the boats a dice is thrown. All trucks experience the same good or bad luck at once, as do the boats. There is a chance of $1/6^{th}$ that the quality will decrease one step. This models real-world uncertainty in transportation due to storage, delays or rough handling. In a similar vein, goods that are left over after a round decrease one step in quality, as they get older.

The mango chain game defines four specific agents with their respective roles:

- *Producers associations* (PA) maintain as main goal to earn as much money as possible with their mango yields during each round from the deliveries of the associated producers. By selecting a coloured fiche from a dark box they are informed about the volume of production from the producers. They can sell to the multinational company, to independent exporter or directly to retailers. The associations have their own trucks to transport the mangoes to the local market outlets or to the harbour. Producer associations can influence the quality level of their supply by investing a fixed amount of money per production field. After an investment, all yields from this field will be of high quality.

- There is only one *multinational company* (MC) in the game, representing a dominant actor in the supply chain of mango. This is the only role that can be played by more than one person. Multi-person multinationals can be at several places at the same time to make simultaneous arrangements and gather more information. The initial operating capital of the multinational company is large. The goal of the multinational is to optimize revenues, similar to other traders. The multinational buys from producer organizations and sells to retailers. The company owns its own boats to transport mangoes to foreign clients.

- *Independent exporters* (IE) represent an alternative for the multinational. They have the same goal, but are considerably smaller which becomes clear from the single person role and a more limited operational capital. They buy from producer association and sell to retailers. Each independent exporter has a boat to transport mangoes to foreign retailers. One more boat is available from the game leaders for free when an independent exporter has more than one client per round.

- *Retailers* (R) have exclusive access to the consumer market. They buy from the multinational, independent exporters or directly from producer associations. The first two deliver the mangoes right on the doorstep of the retailer. The producers associations need to rent a ship at high costs from the game leaders to pick up goods in the Costa Rican harbour. The goal perceived by the retailers is similarly defined as income optimization.

The game sessions described in this paper had Costa Rican mango producers as participants. For each session the researchers visited a local producer association. It had been negotiated in advance that the local leader of the association arranged that enough participants of his association would show up. The association arranged a room to play too. Most of the participants owned a small to medium sized mango yard. Few were employee of a medium sized mango yard. Role assignment to participants in a session was random. The participant selection made that social relationships between people in the real-world were present in the gaming simulation. If social relationships like friendships influence bargaining power, the selection setup facilitated that this can appear in the game too.

During the design of the mango chain game the decision had been to play the game with mango producers only. For the first study of bargaining power the perspective of local producers interested the researchers most. A second reason was that in the Costa Rica case arranging combined sessions would be practically impossible to organise. Having only producers who know each other makes that retailer, independent exporter and multinational roles are played from the perspective of producers. The participants enact the roles as they perceive them in real-world. Real-world authority and pure market-like transactions between producer associations and multinationals are absent.

The sessions took 2 to 2,5 hours of actual gaming time per session. During this period 3 to 4 rounds could be played. While playing the game, participants go back and forth between phases of contract negotiation and contract fulfilment. During the contract negotiation phase, people walk around the room and there is no central leader for the process. During the contract fulfilment phase all participants are standing around the (large) game board and a game leader conducts the group through the activities that need to be fulfilled.

Contracts are signed on pre-structured forms with a variables duration, price, quantity, quality and allocation of risk. Contract choice is determined by the duration of the contract. (1 round, 3 rounds, everlasting). Risk allocation is about the two types of risk and uncertainty present in the game; (1) variability in the supply from the simulated producers, and (2) quality loss with a chance of 1/6 in each transport stage.

Measurement of concepts

Wealth: Producer associations have an asymmetric distribution of the quality of their associated production fields. This ranges from all high quality fields to all low quality fields. This is measured in percentage of high quality and medium quality fields. The amount of money is different for distinct roles of the participants. Thus, wealth differences are visible. Furthermore wealth is a self-perception of the real-life situation of the participant that he brings into the game. This is measured with the questionnaire on a ten-point Likert scale.

Partnership: The participants know each other before the game as they belong to the same real-world producer association. Some will be friends and some will like each other. These relationships they take with them into the game. Attitude towards the importance of good partnerships was measured in the questionnaire on a 10-point Likert scale.

Market imperfections: Asymmetric information about the price structure of the consumer market. Retailers know more than the other participants. Contracts are closed in face-to-face deals. Finding a business partner can be hard for some participants. Market access costs are modelled in the transportation costs. Producer associations have trucks to move products to multinationals or independent exporters. Multinationals and independent exporters have boats to transport materials to retailers. If retailers want to bypass the multinationals or independent exporters they need to rent a boat from the game leaders or from a multinational or independent exporter. This costs money. There are also imperfections in terms of information. On one side the access to information is different; the multinational has all the information at the start of the game but producer associations only have part of that information. On the other side the capability of understanding and using that information is different between participants. The perceived importance of market imperfections for bargaining power is measured in the questionnaire on a 10-point Likert scale.

Negotiator skills: The participants are mango growers, not professional traders. Their real-world negotiation skills are taken into the game negotiations too. The perceived importance of negotiation skills in bargaining in the game is measured in the questionnaire on a 10-point Likert scale.

Perceived bargaining power: Within the game, the seven determinants of bargaining power of Muthoo (2000) are present as described in Table 1. The perceived bargaining power was measured after the game in the questionnaire and reflected the perceived bargaining power within the game.

Risk: Is the self-perceived risk and uncertainty in the supply quantity, in the quality after transport and in consumer market prices, as has been described before. In a contract the risk for non-delivery of the agreed products is decided upon. Measurement of risk is via the questionnaire. (See section 5 for an elaborate description)

Contract choice: The length of the contract as depicted on the contract form (one, three or infinite rounds). This is measured per contract and attributed to both buyer and seller.

Bounded rationality: Decision-making under partial information. Participants stick to their business partner sometimes without checking other options in the market. It is a statistically significant grouping of all items about this aspect that could be depicted under question 13 of the questionnaire. (Appendix 1)

Negotiation: Activity to agree on a contract after finding a possible business partner. Not measured.

Revenue distribution: Distribution of all money earned in the chain (being the difference in consumer prices and production prices (0) over the participants. Per player this is the result of the financial benefits of contract fulfilment. Measured in money at the end of the game.

Table 1. Appearance of the determinants of bargaining power in the game.

Determinants of bargaining power (Muthoo, 2000)	
Inner options	Keep products for a round, taking a decrease of quality for granted.
Outer options	Other suppliers or buyers. For PA: domestic market.
Impatience	Time pressure of the negotiation phase: waiting too long makes that you cannot get other agreements anymore.
Risk of breakdown	When time is over and no agreement has been reached goods will degrade in quality. For PA's this risk is smaller as they have the domestic market with a fixed (low) price for their surplus.
Commitment tactics	For MN with 2 or 3 employees only. (In the game we saw people coming up with fantasy stories about their 7 hungry children and an ill wife that need to be fed, but this is not in the game structure itself).
Asymmetric information	Retailers know the price structure of the consumer market. Because it is a board game all participants can see the supply and the risks along the game. Contracts from previous rounds are visible too, indicating if suppliers will be able to fulfil their contracts.
Wealth	Differences in quality of production between producer associations and as differences in the initial amount of money the roles in the games have.

5. Materials and methods

The analysis of bargaining power and revenue distribution is based on two data sets collected during and after the game sessions. During the game, transaction forms are used that register the contract choice and the different exchange attributes (price, volume and quality). After the game, all participants filled in a questionnaire to enable the debriefing, with information about risk attitudes, access to information, perceptions of market access, and determinants of bargaining power. (See Appendix 1) The transaction data set is constituted of 82 records (contracts) for a specific seller and buyer, while there are 43 questionnaires available from the debriefings.

Figure 3. shows the distribution of the revenues of the participants of the sessions, besides the margins gotten from the different quality and commercialization chains they used. Producers get better prices when they do business with the multinational. Retailers who chose to buy directly from the producer associations are the ones getting ore revenue from trade.

Table 2 shows that the preferred contract type was the one round contract. Nevertheless, participants kept signing short terms contracts with the same persons. Hence the short-term contract becomes a long-term agreement with the possibility of dissolution or renegotiation. The behaviour of the participants in the different games is similar. In round zero they tried both contract types and then they keep signing one round contracts.

Following Muthoo (2002), all variables are measured at a Likert scale ranging from 0 (low) to 10 (high). Descriptive statistics are reported in Table 3. We can notice multinationals have the largest bargaining power, possess more negotiation skills and have highest wealth. Producers have relatively low wealth, but have the second-highest bargaining power. This is most probably due to their hold-up capacity. Multinationals and exporters are strongly dependent on partnership relationships to guarantee continuous sourcing. All agents are affected by market imperfections.

In addition to the agency attributes, all participants in the game were asked to answers some questions regarding their risk perception. We used two methods for risk assessment: (1) a direct approach where participants were asked to define their risk attitude (risk averse, risk neutral,

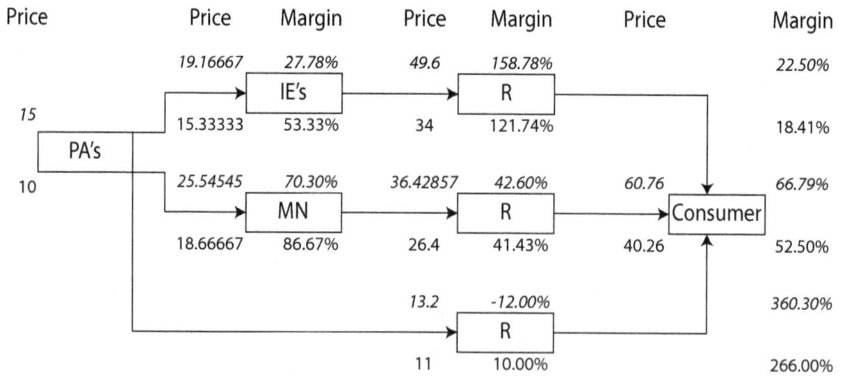

Figure 3. Revenue distribution in the different chain configurations. Italic numbers mean superior quality and regular numbers mean normal quality.

Table 2. Type of agreement signed between participants per round and location.

% within round			Round				
Location			0	1	2	3	Total
Paquera	Contract	One Round	75.0%	50.0%	100.0%	100.0%	76.9%
		Three Round	25.0%	50.0%			23.1%
	Total		100.0%	100.0%	100.0%	100.0%	100.0%
Jicaral 1	Contract	One Round	80.0%	100.0%	100.0%	100.0%	92.9%
		Three Round	20.0%				7.1%
	Total		100.0%	100.0%	100.0%	100.0%	100.0%
Jicaral 2	Contract	One Round	85.7%	75.0%	100.0%	75.0%	84.2%
		Three Round	14.3%	25.0%		25.0%	25.8%
	Total		100.0%	100.0%	100.0%	100.0%	100.0%
Abangares 1	Contract	One Round	100.0%	100.0%			100.0%
	Total		100.0%	100.0%			100.0%
Abangares 2	Contract	One Round	83.3%	75.0%	100.0%	83.3%	85.7%
		Three Round	16.7%	25.0%		16.7%	14.3%
	Total		100.0%	100.0%	100.0%	100.0%	100.0%

Table 3. Descriptives of the agency attributes.

	Producer Associations		Multinationals		Independent exporters		Retailers	
	Avg.	S.D.	Avg.	S.D.	Avg.	S.D.	Avg.	S.D.
Bargaining Power	7.32	2.60	8.17	1.17	6.57	1.27	5.91	1.64
Wealth	5.36	2.95	7.81	1.92	5.69	3.14	5.73	2.28
Negotiation Skills	7.07	2.02	9.17	0.75	7.41	2.16	6.73	2.32
Good Partnership	7.86	2.50	9.00	0.89	8.06	1.50	6.73	2.76
Market Imperfections	8.43	1.80	8.73	0.94	8.13	0.94	8.38	1.38

Note: Values measured with Likert scale (from 0 to 10)

risk-taker) before the simulation game; and (2) an economic experiment for risk measurement where the participants were asked to select amongst different lottery choices. Cross-tabulation and Chi-square test were performed to identify significant association between these variables (Table 4). After verifying the independence of both risk indicators, a principal component was performed in order to construct an index of risk perceptions.

Another aspect of the game refers to differences in information regarding markets and product characteristics for each of the agents. We constructed a proxy variable for the 'bounded rationality' faced by participants during the sessions. All agents were asked to identify the different types of problems they faced for meeting an agreement, and these answers were classified into problems related to information and other problems (Table 4). Finally, a ratio was created of the relationship between problems. This was used as independent variable in subsequent regressions.

Table 4. Relationship between risk perception and the risk experiment[1].

	Producer associations		Multinationals		Independent exporters		Retailers	
	Avg.	S.D.	Avg.	S.D.	Avg.	S.D.	Avg.	S.D.
Risk attitude[2]	2.0	2.27	2.2	0.49	2.3	0.38	2.5	0.33
Risk attitude[3]	1.36	0.13	1.60	0.25	1.75	0.25	1.38	0.18

[1] Pearson Chi-Square is significant, therefore a relation exist between these variables
[2] Scale form 1 to 3, where 3 is risk taker
[3] Scale from 1 to 2, where 2 is risk taker

After the construction of the variables, an average of these different variables has been performed in relation to the specific role in the simulation. These averages were added to the transaction data set. In the transaction form we have among other attributes the contract choice and the buyer and seller. Thus, this buyer or seller roles are the specific roles of the participants, and the key variable to join the data sets.

The data analysis for operationalising the conceptual model includes two stages. Multiple regression models are estimated for the determinants of (perceived) bargaining power related to the agency attributes. Hereafter, logistic regressions techniques to disentangle the differences in bargaining power between buyers and sellers, relying on the attributes of the contractual partners as major explanatory variables. Finally, the index of bargaining power is used as an independent variable for explaining the outcome of the game as reflected in the revenue distribution in the mango supply chain.

6. Results

The data analysis is performed in two steps. In the first step, the agency characteristics and attributes that influence the perceived bargaining power (PBP) of the game participants are determined. Results are reported in Table 5. We observe that variables reflecting negotiation skills, partnership and wealth are all significant and exhibit a positive effect on the perceived bargaining power of the participants. These variables together explain 56 percent of the variance. Bargaining power proved to be independent from market imperfections.

The second step of the analysis relates the main factors from the conceptual framework that could explain the distribution of revenue between buyers and sellers. To be able to perform this analysis, we have constructed two different models, one for the buyer's revenue and another for the seller's revenue. The perceived bargaining power obtained in the first stage is disaggregated for sellers and buyers depending on their contractual arrangements. The information problems related to uncertainties faced by the players during the game simulation are also specified specifically for sellers and buyers. The risk perception of buyers and sellers has been treated in a similar way. The variable contract choice is a binomial choice, since the participants were only able to sign one round or three round contracts on the contract form. (The 'infinite' option has never been used.)

Table 5. Determinants of perceived bargaining power.

	Coefficient	Significance
Constant	-1.968	0.847
Market Imperfections	2.958	0.811
Negotiation Skills	0.325	0.017**
Partnership	0.375	0.001***
Wealth	0.194	0.055*

Adjusted $R^2 = 0.558$
* sig. at 10%, ** sig. At 5%, *** sig. at 1%

The determinants of the buyer's revenue are reported in Table 6 and explain 47% of the variance. Factors negatively influencing the buyer's revenue are the perceived bargaining power of the other agents and their perceived risk. However, the agents perceived bounded rationality nor the contract choice have a significant impact on the revenue distribution. Consequently, with a lower bargaining power and a higher risk aversion of the selling agent, the buyer may expect to receive higher revenues.

The determinants of the seller's revenue are reported in Table 7 and explain 35% of the variance. The perceived bargaining power of the seller, his risk attitude and the selected contractual arrangement negatively influence the seller's revenue. The bounded rationality of the buyer is not significant. The main difference with the factors influencing the buyer's revenue is that the contract arrangement is having a significant impact on the seller's revenue. Since there were only two types of contracts (one round and three rounds), the negative sign of the coefficient means that the long-term contract decrease the revenues of the seller.

For both traders the perceived bargaining power that the trade partner has of himself negatively influences the revenues. This means that powerful traders do get more revenues at the expense of the less powerful trade partner. This is supported by Muthoo (2002).

For both traders the risk perception that the trade partner has negatively influences the revenues. This means that the risk present in the game makes risk averse buyers willing to pay less for the

Table 6. Determinant of the buyers revenue (N=82).

	Coefficient	Significance
Perceived Bargaining Power Seller (PBPS)	-0.775	0.048**
Perceived Risk Seller (RS)	-1.321	0.001***
Perceived Bounded Rationality (PBRS)	-0.670	0.380
Contract choice	-0.317	0.108

Adj. $R^2 = 0.466$
* sig. at 10%, ** sig. at 5%, ***sig. at 1%

Table 7. Determinants of the seller's revenue (N=82).

	Coefficient	Significance
Perceived Bargaining Power Buyer (PBPB)	-0.379	0.039**
Perceived Risk Buyer (RB)	-2.025	0.027**
Perceived Bounded Rationality (PBRB)	-1.057	0.236
Contract choice	-0.264	0.043**

Adj. R2 = 0.352
* sig. at 10%, ** sig. at 5%, sig. at 1%

mangos if they have to take some of the risk. Risk averse sellers ask more to be willing to take the risk. Again this is in line with Muthoo (2002).

The contract choice is only significant for the seller's revenue. This means that long-term contracts are negative for the revenues of the seller but not for the buyer.

7. Discussion and conclusion

This article presents an experimental and innovative approach for analysing human behaviour in the supply chain, using simulation gaming as a tool that can be used in research regarding bargaining power and value added distribution. This provides an important extension to simulation gaming that has been used before mainly for training purposes and policy making. (Duke and Geurts, 2004) The main advantage of reliance on simulation games is that these can be effectively used to identify less tangible and unobservable issues involved in transactions such as quality market segmentation, trust-enhancing agreements, and efficient chain configurations among others. Nevertheless, playing more sessions will be required to gather reliable information and to reach adequate regression estimates.

In the analysis, buyer characteristics explain seller's revenue and vice versa. As expected, with a lower perceived bargaining power of the buyer (seller), the revenue of the seller (buyer) tends to increase. Therefore, stakeholders with more bargaining power will take advantage of the other actors. Higher risk aversion of both the buyers and/or the sellers will lead to more revenue for the other actors derived from the exchange relationship. In the same vein, long-term contracts in the buyer-seller relationship will lead to lower revenues for sellers.

The latter result is surprising, since contract choice is only significant for the sellers' revenue. Sellers are well aware of the type of market in which they are engaged, mainly based on short-terms contracts, which gives them the opportunity to remain flexible towards the changes in demand and supply that they do not control. Otherwise, sellers are trying to establish long-term relationships, but could rely on repeated short-term contracts with the same partner. Mainly these contracts will rely on trust or friendship.

An important incentive to take into account in the transaction relationship between buyer and seller is that the sellers relying on short-term contract with the buyer ask not only for price, delivery time and quality attributes but also require certain risk-sharing possibilities to be able to reduce the risk averseness of the seller in detriment of the buyers' revenue. It is rational to believe that buyers do not want to change the status quo and are therefore not interested in reducing their revenues. Sellers must increase their bargaining power for improving their wealth, searching for good partnerships and increasing their negotiation abilities. In the case of Costa Rica this might be achieved by following training in negotiation skills deliver by the National Learning Institute (INA), while the search of partners might be done in conjunction with International Trade Promoter (PROCOMER). Moreover, the government has operational and institutional options to improve the bargaining power of sellers, while on the other hand sellers might also improve their bargaining power by cooperation. The latter option was observed during the game sessions where producers associations that negotiations as a group were more successful compared to producer associations who worked alone.

As multinationals in the game have the highest perceived bargaining power (Table 3), independent exporters score lower and retailers less; the game was realistic in creating these differences in comparison with the study of Millet (2003). The producer association in the game have a relatively high bargaining power compared to the personal information the participants gave us during the game sessions. This might be due to the modelling that left out the larger growers, thus giving the producer associations all of the supply market.

Finally, wealth and information have a clear impact on bargaining power. Both rely on intangible assets and tangible assets. In terms of intangible assets, the government might increase its presence by promoting the training of the producers with organization devoted to capacity building. In terms of the tangible assets the government could establish credit facilities for small and medium-size producers that might be conditional on membership of a producer organization. This could increase the producers' bargaining power because - as was observed in the game - access to credit, resources and cooperation were key factor in the outcome of the negotiation process.

Acknowledgements

The authors wish to acknowledge the contributions by Sietse Sterrenburg for his assistance in the design of the mango game and the execution of the field trials.

References

Agrell, P., R. Lindroth and A. Norrman, 2002: Risk, information and incentives in telecom supply chains. International Journal of Production Economics 90, 1-16.

Argyris, N. and J. Liebeskind, 1999. Contractual commitments, bargaining power, and governance inseparability: Incorporating history into transaction cost theory. Academy of Management Review 24, 49-63.

Ayala, J., 1999. Instituciones y Economía. Una introducción al neoinstitucionalismo económico. Fondo de Cultura Económica, México.

Bacharach, S. and E. Lawler, 1984. Bargaining: Power, tactics, and outcomes. Jossey-Bass, San Francisco, C.A.

Berg, J., J. Dickhaut and K. McCabe, 1995. Trust, reciprocity and social history.Games and Economic Behavior 10, 122-142.

Bogetoft, P. and H. Olesen, 2004. Design of production contracts. Lessons from theory and agriculture. Copenhagen Business School Press. Denmark.

Buvik, A. and T. Reve, 2002. Inter-firm governance and structural power in industrial relationships: the moderating effect of bargaining power on the contractual safeguarding of specific assets. Scandinavian Journal of Management 18, 261-284.

Caruana, G., L. Einav and D. Quint, 2006. Multilateral bargaining with concession costs. Journal of Economic Theory (in press).

Chamberlain, N.W. and J.W. Kuhn, 1965. Collective bargaining. 2nd ed. New York: McGraw-Hill.

Chemla, G., 2005. Hold-up, stakeholders and takeover threats. Journal of Financial Intermediation 14, 376–397.

Cook, K., 1977. Exchange and power in networks of international relationships. Social. Q. 18, 62-82.

Dicken, P., 1986. Global shift, industrial change in a turbulent world. Harper & Row, London.

Dobbelaere, S., 2003. Estimation of price-cost margins and union bargaining power for Belgian manufacturing. International Journal of Industrial Organization 22, 1381-1398.

Duke, R and J. Geurts, 2004. Policy games for strategic management. Pathways into the unknown . Dutch University Press, Amsterdam, The Netherlands.

Dunning, J.H., 1981. International production and the multinational enterprise. George llen & Unwin, London.

Emerson, R., 1962. Power-dependence relationships. American Sociological Review 27, 31-41.

Fafchamps, M., 2004. Market institutions in sub-Saharan Africa: theory and evidence. The MIT press, Cambridge, Massachusetts, London, England.

Fossum, J.A., 1982. Labor relations development, structure, process. Business Publications: Plano, Texas.

Heide, J. and G. John, 1992. Do norms matter in marketing relationships? Journal of Marketing 56, 32-44.

Hendrikse, G. and W. Bijman, 2002. On the emergence of growers' associations: self-selection versus market power 29, 255-269.

Hendrikse, G., 2003.Governance in chain and networks: A research agenda. ERIM Report series Research in Management.

Inderst, R., 2002. Contract design and bargaining power. Economics Letters 74, 171-176.

Kagel, J. and A. Roth, 1995. The handbook of experimental economics. Introduction by Albin Roth. Princeton University Press, Princeton, New Jersey.

Kochan, T., 1980. Collective bargaining and industrial relations; from theory to policy and practice. Homewood, III.: Richard D. Irwin.

Kumar, S. and A. Seth, 1998. The design of coordination and control mechanisms for managing joint venture-parent relationships. Strategic Management Journal 19: 579-599.

Leap, T. and D. Grigsby, 1986. A conceptualization of bargaining power. Industrial and Labor Relations Review 39, 202-213.

Lecraw, D., 1984. Bargaining power ownership, and profitability of transnational corporations in developing countries. Journal of international Business Studies 27, 877-903.

Lee, J., W. Chen and Ch. Kao, 1998. Bargaining power and the trade-off between the ownership and control of international joint ventures in China. Journal of International Management 4, 353-385.

Lindbloom, Ch., 1948. Bargaining power in price and wage determination. Quarterly Journal of Economics 62, 396-417.

Ministerio de Agricultura y Ganadería MAG., 2003. Censo Agrícola de la Región Pacífico Central. San José, MAG.

Meijer, S. and G. Zuniga-Arias, 2005. Learning experiences with the Mango Chain Game. In: Proceedings of the 9th international workshop on experimental learning in industrial management, May 2005, Espoo, Finland.

Millet, D., 2003. The producers organizations in the Costa Rican mango chain: Governance structure and performance. MSc. Thesis Management Studies Group and Development Economics Group, Wageningen University.

Montero, M. and M. Cerdas, 2000. Manejo poscosecha del mango para el mercado fresco. Primera edición. San José, Costa Rica. Centro de Investigaciones Agronómicas, Laboratorio de Tecnología Poscosecha.

Moon, Ch. and A. Lado, 2000. MNC-host government bargaining power relationship: A critique and extension within the resource-based view. Journal of Management 26, 85–117.

Muthoo, A., 2000. A Non-Technical Introduction to Bargaining Theory. http://privatewww.essex.ac.uk/`muthoo. Revision September 2004.

Muthoo, A., 2001. The Economics of bargaining. http://privatewww.essex.ac.uk/`muthoo. Revision September 2004.

Muthoo, A., 2002. Bargaining theory with applications. Cambridge University Press.

Nash, J., 1950. The bargaining problem. Econometrica 18, 155-162.

Pitchford, R., 1998. Moral hazard and limited liability: The real effect of contract bargaining. Economic Letters 61, 251-259.

Rao, P.K., 2003. The economics of transaction cost. Theory, methods and applications. Palgrave Macmillan. London

Roumasset, J., 1995. "The nature of the agricultural firm," Journal of Economic Behavior & Organization, vol. 26(2), pages 161-177.

Rubinstein, A., 1982. Perfect equilibrium in a bargaining model. Econometrica 50, 97-109.

Rubinstein, A. and A. Wolinsky, 1985. Equilibrium in a market with sequential bargaining. Econometrica 53.

SEPSA, 2001. Boletín Estadístico, Sector Agrícola Costarricense. Ministerio de Agricultura y Ganadería.

Slichter, S., 1940. Impact of social security legislation upon mobility and enterprise. American Economic Review 30, Supplement, 44-77.

Sterrenburg, S. and G. Zuniga-Arias, 2004. The Costa Rica chain game. In: Experimental learning in chain and networks. G.J. Hofstede *et al.* (eds.) Proceedings of the 8th international workshop on experimental learning in chain and networks, 24-27 May 2004, Wageningen, The Netherlands.

Stinchcombe, A., 1985. Contracts as hierarchical documents. In: A.L. Stinchcombe and C. Heimer (Eds.), Organization theory and project management. Bergen: Norwegian University Press.

Weinberger, K. and J. Jütting, 2000. Women's participation in local organizations: conditions and constrains. Wold Development 29, 1391-1404.

Williamson O., 1975. Markets and hierarchies: analysis of antitrust implications. New York: Free Press.

Williamson, O., 1981. The economics of organization: The transaction cost approach. American Journal of Sociology 87, 548-577.

Williamson, O., 1985. The economic institutions of capitalism. The Free Press. New York.

Williamson, O., 1998. Transaction Cost Economics: How it works; where it is headed. The Economist 146, 23-58.

Yan, A. and B. Gray, 2001. Negotiating control and achieving performance in international joint ventures: A conceptual model. Journal of international Management 7, 295-315.

Zingales, L., 1998. Corporate governance. In: J. Eatwell (ed.), The new palgrave, Macmillan, London.

Appendix 1. Simulation systematization

Date: ________________ Place: ____

1) Role in the simulation: 1. Producer Assoc. 2. Multinational 3. Independent exporter 4. Retailer

2) The simulation is: 1. Very easy to understand and play 2. Easy to understand and play 3. Neither easy nor hard 4. Hard to understand and play 5. Very hard to understand and play

3) The simulation is: 1. Really close to reality 2. Neither similar nor different to reality 3. It is not similar to reality

4) How much profit did you make? ________ 5) How much loss did you have? ________

6) Whom did you do business with? ____

7) Why? 1. They where closer 2. I know them for very long time 3. I trust them 4. They offer the best options 5. I had not many options 6. They irradiate trust

8) During the simulation , your buying and selling expectations. Never changed, always changed. Select a number between 1 and 10 where 1 is never changed and 10 always changed

Maintain expectations 1 2 3 4 5 6 7 8 9 10

9) Why do you think this happened? ____

____

10) On a scale of 1 to 10 where 1 is low bargaining power and 10 is high bargaining power. How do you consider your bargaining power?

1 2 3 4 5 6 7 8 9 10

11) Please mark with a X the number you feel that represents you best on the different items.

I consider my self to be … in the simulation

Very patient	0 1 2 3 4 5 6 7 8 9 10
A risk taker	0 1 2 3 4 5 6 7 8 9 10
To have a lot of buyers	0 1 2 3 4 5 6 7 8 9 10
To take time to make important decisions	0 1 2 3 4 5 6 7 8 9 10
I am a wealthy participant in the simulation	0 1 2 3 4 5 6 7 8 9 10
I can negotiate prices	0 1 2 3 4 5 6 7 8 9 10
I am not afraid of breakdown negotiations	0 1 2 3 4 5 6 7 8 9 10
Main business partner is a good partner	0 1 2 3 4 5 6 7 8 9 10
I hold on to my word	0 1 2 3 4 5 6 7 8 9 10
I maintain the business expectations	0 1 2 3 4 5 6 7 8 9 10

12) Write a number between 0 and 10 to describe the perceived bargaining power of the participants of the simulation

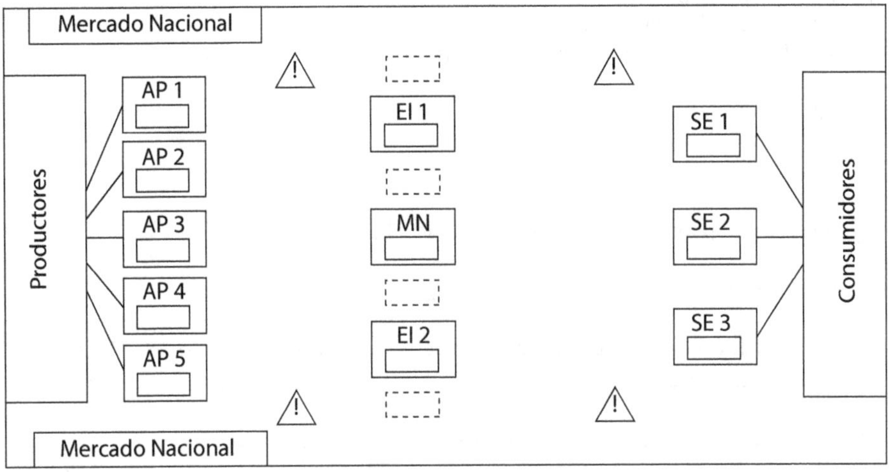

13) Mark with an X, which were the problems you faced in the simulation?

1 ☐ Few information to make decisions
2 ☐ Disadvantage to learn the norms and rules of the game
3 ☐ Problems of produce quality in relation to other participants
4 ☐ Few opportunities to make business with other participants
5 ☐ Not knowing anybody beforehand in the simulation
6 ☐ Supply problems
7 ☐ Problems with not meeting the quality requirements of the buyers
8 ☐ Problems when negotiating with other participants
9 ☐ The buyers always look for paying the least
10 ☐ There was a lot of uncertainty in the simulation
11 ☐ I could not look for new partners
12 ☐ I attached myself to a long-term relationship and I had no benefits out of it
13 ☐ I always traded in the spot market because I could not sign any contract

14) Mark with a X the number. In the simulation who do you consider has the most bargaining power?

The buyer	0 1 2 3 4 5 6 7 8 9 10
The one with information	0 1 2 3 4 5 6 7 8 9 10
The one with a better economic situation	0 1 2 3 4 5 6 7 8 9 10
The one with many buyers	0 1 2 3 4 5 6 7 8 9 10
The strategic business partner	0 1 2 3 4 5 6 7 8 9 10
The participant with control over activities of other participants	0 1 2 3 4 5 6 7 8 9 10
The business partner with more dependency	0 1 2 3 4 5 6 7 8 9 10
The seller that must sell his perishable product	0 1 2 3 4 5 6 7 8 9 10
The one who contributes most financially	0 1 2 3 4 5 6 7 8 9 10

15) You consider your self to be 1. Risk taker, 2. Neutral to risk, 3. Risk Adverse

16) Which option do you pick?
Option A: To get $30, 100% certainty
Option B: To get $40 with 80% chance, and to get $0 with a chance of 20%

17) How old are you? _________

18) Sex: Masc. Fem.

19) Do you know the people in the simulation beforehand? 1. Yes 2. No

20) Formal Education?_______________________________

21) Main activity: 1. Student, 2. Lecturer, 3. Entrepreneur, 4. Public sector, 5. Mango Producer 6. Other: ____

22) Farm size? ____________ 23) Family size? _________

24) Monthly income? 1. 1.000 a 50.000 colones, 2. 50.001 a 100.000 colones, 3. 100.001 a 250.000 colones, 4. 250.001 a 500.000 colones 5. More than 500.000 colones

¡MUCHAS GRACIAS!

Appendix 2. The Mango Supply Chain sessions

Rol	Session 1 ASOFRUL	Session 2 ASOFRUPA	Session 3 ASOFRUL	Session 4 COONAPROSAL1	Session 5 COONAPROSAL 2
AP1	AP's **cooperate**. If producers work by them self then the buyers get better deals and abuse from the AP's.	At the begining they did not undestood what to do. The PA's set low prices an were **not cooperating**, he did a 3 round contract because he was testing partners. The main problem was that the other PA's were selling cheaper than him. The activity is too fast and he had not much information.	At the begining they played separdly then they did **cooperation** and bargaing as a group. the union was to mantain volume and look for better prices and long term relationships, they also thought in doing business directly to the Retailers but because of risk and uncertainties they prefered to make it small and play conservative.	PA's **cooperate** and tried to sell to the retailers. They tried to control prices, volumes and selling options.	It only worked with one buyer, because of fear to loose he sold the produce a low price. c He think he should better sell produce to the Retailers. He was not capable (have enough resources) to invest in quality
AP2	Idem	The business is to sell, he was afraid of losing with the 3 round contract,	Idem AP1	Idem	Secure 1/2 of produce in a 3 round contract and for the rest look for the best price
AP3	Idem	Direct sells to the Retailers, between the IE and MN there were always agreements, and the producers could not overcome that. I sold at the price they ofered	I signed a 3 round contract with a Retailer, and everything I produce I send it to the retailer, at the end I had to sell to buyers that only need small portions therefore I had to be fasterto sell.	Idem	Secure red quality with only one buyer during the game, the yellow produce he sold it any where. He promote 3 round contracts, but never improve quality
AP4	Idem		En el inicio firmo un contrato de 1R, esperando que llegara un comprador pagando mejor, pero siempre llego el mismo. Siempre el comprador llega a ofrecer. No queria arriesgar mucho	Idem	He had 3 buyers, and sold a low price, never thought in cooperation, plaed by him self and never exploited been the player with more red produce.
AP5	Idem				

Appendix 2. Continued

Rol	Session 1 ASOFRUL	Session 2 ASOFRUPA	Session 3 ASOFRUL	Session 4 COONAPROSAL1	Session 5 COONAPROSAL 2
EI1	Objective to sell first and then try to buy a fix price	I bought cheap and to protect my self and I sold expencive, I did not want to pay much for the produce	I tried to make a 3 round contract with AP1 and AP2, but i could not sign it with the retailer, because I wanted to check prices and volumes behavor through the rounds to get the better price possible.	He did a cooperation with MN and his rol in the **cooperation** was to make business with the local market and MN business with the export market to overcome the coooperation of the PA's and control commercializartion	Only bought red produce but low quantities, because the Retailers where not paying much. The retailers where very agresive with the prices, allways looked for red porduce, because it was sefer and had better prices
EI2	Dedicated to buy red quality to secure prices and volumes		I looked for the person I knew better to make business with him.		
MN	Pay in advance to build trust. Paid to incentive to quality, to eliminate other buyers. He said 3 round business are too risky	First I learnt to commercializate, then I bought cheap produce, then I check the total amount in the market to be able to bargaing prices and buy at low price. Because of the production rislk I did not sign a 3 round agreement.	He looked first for 1 round contracts then to promote trst he moved to 3 round contracts	**Cooperation** with the EI, he did the export market and somehow they verticalized the chain.	Allways did 1 round contracts but with the same person (long term relationship). The stretegy was to llook for a VI with the Retailer he tryed to buy one but he could not. he uses 1 round contracts to get to know the person
SE1	They where not very aggressive and had no experience (information), to go directly to the PA's	I tried to buy directly from the producers, to buy at low price, 3 round contracts too risky	He tried to look fast for partnersand want to assure the vendor. With the buyer he has confidence, trust , firnedship, he looked up for informationof the other players, to use it in his own benefit, he continued to sign 1 round contracts because there was no need to take more risk.	He bought olnly yellow produce, but he could not sell it to the consumer	He did not fully understood what to do, the MN envolved him, he learnt buy trial nad error. the negociation process was manage by the multinational

Appendix 2. Continued

Rol	Session 1 ASOFRUL	Session 2 ASOFRUPA	Session 3 ASOFRUL	Session 4 COONAPROSAL1	Session 5 COONAPROSAL 2
SE2		I tried to buy from the producers at a price lower than the production cost. Only 1 round contracts because prices and volume might fall and then 3 round contract becomes too risky	He did not understood what to do in the game, lack of information and communication, signed a 3 round contract and had a bad outcome, after that he signed 1 round contracts	Dod business with MN, signed an exclusivity contract, but then broke the agreement to do business with EI due to price	He looked for something safe with the MN, but he felt that was not a good business, therofere broke the relationship and did business directly to the PA's the business was better off, MN put pressure on the PA's and the rel;ationship broke, he had to do business againg with the MN but the conditions changed.
SE3					He did a 3 round contract and secure 1/2 of the production od PA's, but the PA's had no fix buyer. He needed more time to know the business well. To have both quality produce make commercialization easier
General OBS	Retailers where low aggressive but they thyed to cooperate	They did not saw the possibility to increase quality, time was much pressure, PA behave like the lossing link in the chain, they waited for the leader to make the first move, they know how to cooperate but they did not do it in the session	If PA's cooperate they control the price, volume and direct relations, but this because information problems they just sell to the one offering the best bid. The cooperation is feasible but there are things to consider such as size, produce, type of people involeved, etc. they never tryed to upgrade quality.	In real life the intermediaries tryed to lower our prices therefore we try to skip them. MN and EI cooperate to control the commercialization market. In real life SE are the ones with power. They said that yellow produce is easier to sell but that red product gives them more profits, they said that the client never asked for quality. Because they did not knew each other they did not signed long term contracts and only 1 round contracts.	The real market does not pay for quality, they only ask for requisites, in the real life to skip the intermediaries is just for security but we will not get better prices. the strategy should be to look for long term relationships, based in trust, lower the intermediation margind and manage volumes.

Part 6. Performance

Modelling the recall costs of consumption milk in relation to time

A.G.J. Velthuis, C. van Erve and M.P.M. Meuwissen

Abstract

The relation between the direct recall costs of one batch of consumption milk and time was investigated. To do this, we developed a Monte Carlo simulation model of a farm to consumer supply chain associated with one batch of consumption milk of 150,000 kilograms. The stochastic variables of the model were the length of stay of the milk in each stage of the supply chain, which were described by a minimum, maximum and most likely value. The direct recall costs of recalling a batch of 150,000 kilograms are less than €100,000 in the first 16 hours after the milk is collected from the farms, but increase rapidly to more than €200,000 mostly due to the costly media announcement that has to be done when possibly the milk has reached the consumer. With this study we showed that a modelling approach can be very useful to study the relation between recall costs and time.

Keywords: recall, costs, time, consumption milk

1. Introduction

Due to various food scandals in the past consumers are worried about the safety of their food. An important aspect of a food safety problem to evolve into a scandal is the situation that the source of contamination is not detected in a reasonable time period and that the contaminated food cannot be recalled in an early stage (Meuwissen, 2003b). A scandal has a negative influence on the consumer confidence and needs to be prevented. To maintain consumer confidence it is therefore important that the food industry minimizes the time lag between the announcement of a food safety problem and the moment at which the source of the contamination is traced and all concerned products are tracked and recalled.

A recall is a request by the manufacturer of a defect product to return the product. It is the most extreme action a company can take since the losses can be big (Skees *et al.*, 2001). In addition to the direct costs of a recall, the losses may include loss of customer confidence and the company's good name, although these are hard to quantify (Kramer *et al.*, 2005).

There are different ways to reduce possible recall costs. First, costs can be reduced by shortening the period between the moment of contamination and the moment at which the recall can be executed (Teratanavat and Hooker, 2004). This can be achieved by installing a fast and good monitoring system, which is a continuous process of sampling and testing the product, the ingredients or the environment to control and guarantee the quality and safety of the food. Furthermore, an effective traceability system can also reduce the time lag and can decrease the damage in case of calamity (Van Dorp, 2004). A traceability system provides a set of data about the location of food and food ingredients of each batch along the production chain (Meuwissen, 2003a). Another way to reduce the recall costs is to reduce the size of the batch to be recalled (Dupuy *et al.*, 2005). A batch is a definite quantity of product produced under conditions that are considered uniform.

Intuitively, everybody will agree that time is an important factor in reducing the recall costs. However, the relation between time and recall costs has to our knowledge never been quantified. Testing the relationship in practise is impossible, since a lot of recalls are needed to provide enough data. Therefore, a model might be a good alternative. This paper aims to demonstrate a modelling approach to investigate the relation between the time at which the recall of consumption milk takes place and the recall costs. We choose to model the consumption milk supply chain, since consumption milk is a simple product that exists only of one ingredient (*i.e.* raw milk) and is therefore an easy product to model.

This paper describes a simulation model that models the milk stream through the milk supply chain in relation to time. Knowing the distribution of one batch of milk over the supply chain, it is possible to calculate the direct recall costs in case that a contamination of the batch has been detected at some point in the chain. With this model we were able to investigate how the direct recall costs of consumption milk depend on the speed of tracking and tracing and the speed of monitoring on possible contaminations. The objective of this study was to study the relation between the recall moment of consumption milk and the recall size, the direct recall costs and the distribution along the milk supply chain.

2. Materials and methods

2.1 Model description

The aim of the simulation model is to predict the amount of milk in each stage of the milk supply chain at a certain moment in time. The simulation model is a Monte Carlo model and considers the flow of consumption milk through a farm-to-consumers supply chain (Figure 1). The input values of the model, which were validated by two experts working at two Dutch dairy co-operations, are given in Table 1. The central unit in the model is the silo at the milk processing plant. We assume that this plant mainly produces consumption milk. For the baseline scenario we assumed that the content of the silo at this plant equals the batch size, which is 150,000 kilograms of milk. Implicitly it is assumed that the silo is cleaned before being refilled. Given the batch size, the number of truck loads can be calculated by assuming that the ratio of small (10 tons), medium (20 tons) and big trucks (30 tons) that fill the silo equals 3/3/2.

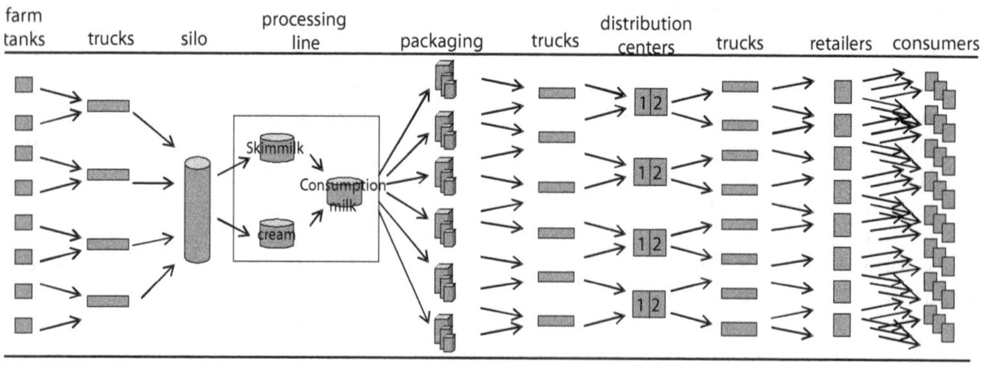

Figure 1. The milk supply chain.

Table 1. The number of units, the amount of milk per unit, the sub-batch modelled and the descriptive values that describe the distribution of the length of stay of milk at each specific stage of the milk supply chain.

Variable	Number	Kg/unit	Sub-batch (kg)	Length of stay (in hours)		
				Most likely[1]	Minimum	Maximum
Truck2	8	18,750[2]	18,750[2]	3.0	1.0	6.0
Silo	1	150,000	100	10.0	1.0	24.0
Processing lines	1	150,000	100	1.5	1.0	2.0
Packaging stage	1	150,000	100	1.5	1.0	2.0
Truck	15	10,000	10,000	3.0	1.0	5.0
Distribution centre 1	20	7,500	7,500	8.0	1.0	12.0
Distribution centre 2	20	7,500	7,500	4.0	1.0	12.0
Truck	150	1,000	1,000	3.0	1.0	5.0
Retail	1,260	119	119	12.0	2.0	72.0
Consumer	50,000	3	3	12.0	0.5	72.0

[1] We assumed a triangular distribution
[2] Three trucks of 10,000 kg, three of 20,000 kg and two of 30,000 kg

The milk flows from the silo through the processing line, where the milk is separated, standardized, pasteurized and homogenized. At the end of the processing line the milk is packed in different sizes of cartons varying from ½, 1, 1½, and 2 litre sizes. For simplicity we assume that all the raw milk will be processed to consumption milk and ignore the fact that a part of the cream is used in other dairy products.

After being packed, the milk is transported by 10 trucks to 20 different distribution centres, which is based on the assumption that the milk company delivers milk to two Dutch retail organizations (*i.e.* Schuitema and Laurus). The model distinguishes two sites of the distribution centre, *i.e.* the storage site (distribution centre 1) and the pick-up location (distribution centre 2). This is done because in case of a recall a media announcement needs to be done when a part of the contaminated batch is located at or has passed the pick-up location of one of the distribution centres. Hundred-and-fifty trucks distribute the milk from the distribution centres over 1260 supermarkets. This number is based on the fact that an average distribution centre delivers to 63 supermarkets (www.schuitema.nl or www.laurus.nl). Finally the milk is bought by 50,000 households in sub batches of 3 kilograms and will be consumed finally.

The stochastic variables of the model are the length of stay of each sub-batch of milk in each stage of the chain. The time line of the model starts at the moment that the milk leaves the farm (t=0) and for simplicity we assume that each truck leaves the first farm at t=0. The input values to describe the distributions of each stage of the supply chain are given in Table 1. Per sub-batch a time value is drawn from the corresponding distribution. For example if the first number drawn from the triangular distribution that describes the silo is 1.52, it means that a sub batch of 100 kilogram leaves the silo 1.52 hours after the milk has been collected at the farm.

The simulation model predicts the amount of milk in each stage of the milk supply chain at the moment a recall is executed. This moment is called the 'recall moment' and is defined as the period between t=0 (*i.e.* the moment at which the milk leaves the first farms) and the moment at which the contaminated milk will be recalled from the different stages of the supply chain. For each recall moment 1000 recalls of consumption milk are simulated and the distribution of the amount of milk in each stage of the supply chain is determined. This distribution is used to calculate the recall costs per stage and in total. This Monte Carlo simulation model is built in Microsoft Excel 2003 with the add-in programme @Risk 4.5 (Palisade-Corporation, 2004).

2.2 Calculation of the recall costs

Direct recall costs of a recall include costs for media announcements, transportation, warehousing costs, extra labour and destruction cost (Meuwissen *et al.*, 2005). Indirect recall costs can include the losses due to lost market share, costs made for brand rehabilitation and crisis response expenses. Brand rehabilitation might be necessary to re-establish the reputation and market share of the affected product by investing in advertising campaigns, special promotions, and consumer education. Crisis response expenses include fees and expenses of outside consultants retained exclusively for the function of responding to the product contamination and recall. In this paper only the direct recall costs are quantified.

The direct recall costs are calculated following the partial budgeting approach. A partial budgeting is based on the principle that a small change in the organization of a business, or in this case the supply chain, will have one or more of the following effects: eliminate or reduce some costs, eliminate or reduce some returns, cause additional costs to be incurred and cause additional returns to be received. The net effect will then be the sum of positive financial impact minus the sum of negative financial impact (Dijkhuizen and Morris, 1997).

Table 2 gives a description of the items in each component of the partial budget and the calculated costs due to this item in Euros per kilogram of recalled milk. All figures are verified with two experts working at two Dutch dairy cooperations.

The additional costs made to recall milk located at all stages but the consumer stage include transport costs to the destructor, destruction costs and extra cleaning costs of equipment that has been in touch with the contaminated batch. The transport costs at the truck stage are lower, since no extra truck is needed to transport the milk. The destruction costs after the packaging stage are higher, since not only the milk but also the packaging has to be destroyed. The additional costs at the consumer stage are the costs made to refund the purchase price of the milk. Consumers are asked to throw the milk away and to send the barcode and batch number by mail to the milk company. The milk company checks the send information and will refund the purchase price to each consumer (Anonymous, 2005). The costs include therefore the postal stamp (€0.39) and addition labour at the milk company to refund the purchase price of €0.72 per litre to the consumer. We assumed that the extra labour is 5 minutes per request which covers 3 litres of milk for €8 per hour. Furthermore, we assume that all consumers of not-consumed milk will send the information to the milk company.

The costs of making a media announcement were assumed to be €10,000 for each advertisement in a National paper multiplied by five National papers. Plus €15,000 extra costs of advertising in other papers and magazines. The total costs for a media announcement were therefore €75,000. The costs for a media announcement were treated as fixed costs in the model and were added to the total costs if at least one litre of the contaminated batch is located or passed the distribution centre 2 stage.

Table 2. Partial budgets of recalling milk from different stages of the milk supply chain.

	Components of the partial budget[1]					
Location of batch	Additional costs	€ / kg	Reduced returns	€ / kg	Reduced costs	€ / kg
Truck	Transport	0.01	Selling price	0.61	Manufacturing costs	0.29
	Destruction	0.10				
	Cleaning	0.01				
Silo	Transport	0.05	Selling price	0.61	Manufacturing costs	0.27
	Destruction	0.10				
	Cleaning	0.01				
Processing	Transport	0.05	Selling price	0.61	Manufacturing costs	0.15
	Destruction	0.10				
	Cleaning	0.01				
Packaging	Transport	0.05	Selling price	0.61	Manufacturing costs	0.07
	Destruction	0.10				
	Cleaning	0.01				
Distribution centre 1	Transport	0.05	Selling price	0.67	Manufacturing costs	0.08
	Destruction	0.13				
	Cleaning	0.01				
Distribution centre 2	Transport	0.05	Selling price	0.67	Manufacturing costs	0.03
	Destruction	0.13				
	Cleaning	0.01				
	Media announcement[2]	-				
Retail	Transport	0.05	Selling price	0.67	Manufacturing costs	0.03
	Destruction	0.13				
	Cleaning	0.01				
	Media announcement	-				
Consumer	Media announcement	-	-	-	-	-
	Cost for refunding the purchase price[3]	1.07				

[1] There are no additional returns when recalling a batch of milk.

[2] The costs of a media announcement (€ 75,000) were treated as fixed costs in the model and were added to the total costs if at least one litre of the contaminated batch is located or passed the distribution centre 2 stage.

[3] This includes the postal stamp (€0.39) and addition labour at the milk company to refund the purchase price of €0.72 per litre to all the consumers (i.e. 5 minutes per consumer to refund 3 litres of milk for €8 per hour).

The milk that has been destroyed cannot be sold anymore and therefore the selling price will be missed. These are the reduced returns of the Partial Budget. The selling price equals either the product value at the end of the packaging stage (*i.e.* when the milk is leaving the production plant) for all the stages before this stage or the product value at the end of the retailer stage (*i.e.* when the milk is sold) for the stages after the packaging stage.

The manufacturing costs are the costs made (variable and fixed) at a specific stage of the chain before the milk is going to the next stage of the chain. For the stages truck, silo, processing and packaging the manufacturing costs equal the product value at the end of the transport 1 stage minus the product value at the former stage. For the stages distribution centre 1, 2 and the retail the manufacturing costs equal the product value at the end of the retail stage minus the product value at the former stage. The product values at the end of each stage are €0.31 (farm tank), €0.32 (truck), €0.34 (silo), €0.46 (processing), €0.54 (packaging), €0.59 (transport 1), €0.61 (distribution centre 1), €0.66 (distribution centre 2), €0.67 (transport 2), €0.69 (retail) and €0.72 (consumer), respectively.

We assume that there are no additional returns when recalling a batch of milk. The destroyed milk will not be re-used for other purposes, so the salvage value is zero. And we assume that there is no recall insurance in place, so that possible insurance payments are zero too. The recall costs for the milk located at the stage of the milk collection truck are only calculated for one average truck, since the size of one truck is the (sub-) batch size in this specific stage of the chain.

3. Results and discussion

Figure 2 shows the estimated direct recall costs per stage of the consumption milk supply chain in relation to the recall moment. For example, if the recall moment is at 12 hours, the recall costs are €0, €53,000, €9000, €8000, €13,000, €0, €0 and €0, at the stages truck, silo, processing, packaging, distribution centre 1, distribution centre 2, retail and consumer, respectively.

For each stage of the supply chain the direct recall costs increases and decreases thereafter with the recall moment. This is logic, since the recall costs are lineally related to the amount of milk in each stage and this amount increases and decreases with time for each single stage. At consumer level the amount of milk decreases too since the milk is consumed.

The recall costs are low at truck level. For example, the recall costs are less than €10,000 if the contaminated milk is detected within the first hour. This is because the amount of milk to be destructed is less than the batch size of 150,000 kilograms, since the contamination is detected in the sub-batch of milk that is still in a truck. So, only the milk of a single truck should be destroyed. Another scenario of a very early recall moment is that the contamination is detected in the silo, while not all trucks were unloaded yet. In this case, only the milk in the silo has to be destroyed and not the milk in the truck stage. At this moment, the Dutch dairy industry takes samples at the moment the milk is pumped into the truck and with help of a quick test (before the milk is pumped into the silo) the test results are know. This test checks the presence of the most important antibiotics. So this strategy agrees very much with the results of this study.

The maximum recall costs at a specific stage can be found at the retail stage (*i.e.* around 48 hours these costs are on average €103,900). This is probably an underestimation of the costs at this stage, since only the destruction of the contaminated batch of milk at the retail stage is considered in the calculations. In practice, however, often milk of other batches are send back to the producer with the reason that it takes too much time for the retailer to check all the batch numbers of the packages. Thus, it would be a good strategy for the milk company to set up a monitoring and a

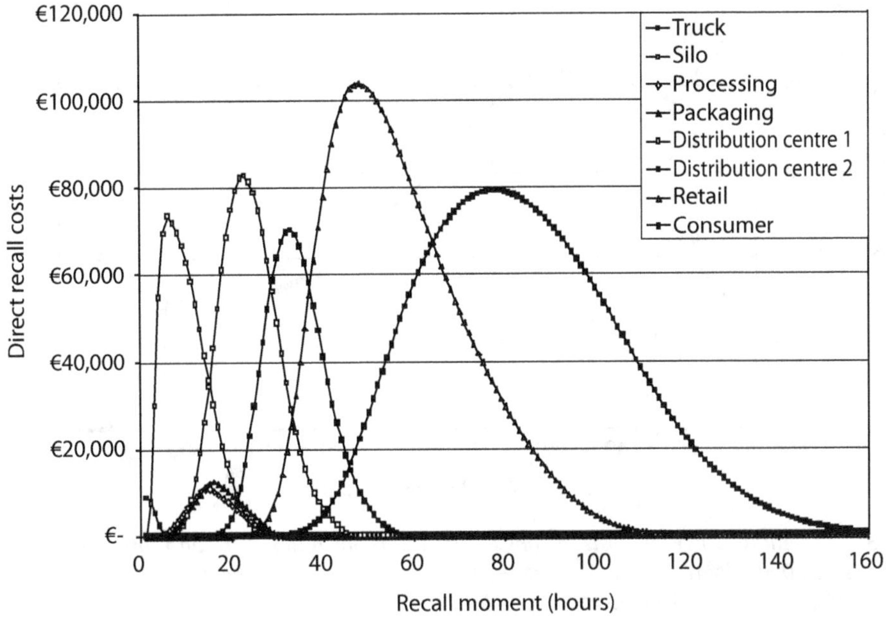

Figure 2. The direct recall costs made in each stage of the consumption milk supply chain for different recall moments. The batch size is 150,000 kilograms. The costs for a media announcement are excluded.

traceability system that enables to recall the milk in case of a contamination before it is located at retail level.

If the milk company is able to recall within 10 hours after collecting the milk from the farm around 95% of the direct recall costs are located in stages belonging to the milk company (*i.e.* truck, silo, processing and packaging). This implies that the costs are for a great part predictable and the unpredictable costs made at the retail level are low. If the recall moment is later than 24 hours more than 95% of the direct recall costs are located at retailer or consumer level (*i.e.* distribution centres, retail and consumer), which is undesirable.

Figure 3 shows the total direct recall costs in relation to the recall moment. The total recall costs are high if the recall moment is between 16 and 160 hours. This is because a costly media announcement needs to be done when at least one litre of the contaminated batch passed the pick-up location of the distribution centre (*i.e.* stage distribution centre 2). The best strategy for a milk company is to organise the monitoring and traceability systems so that a possible recall can take place within the first 16 hours.

On average, the variability in total recall costs is low. The difference between 5% and 95% percentiles within the first 16 hours is around €2300. This difference is within the period between 24 and 60 hours around €2100 and €6850 within the period between 60 and 130 hours. However, the variability is relatively high in two short periods; within the period between 16 and 24 hours the difference between the two percentiles is €78,000 and within the period between 170 and 185 hours it is €75,000. This is because the moment at which the first milk has reached the consumer

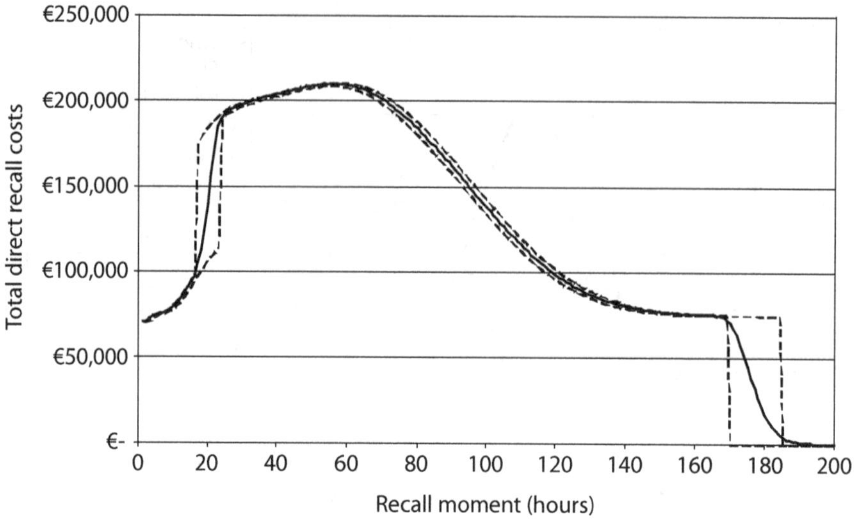

Figure 3. The total direct recall costs of consumption milk for different recall moments. The batch size equals 150,000 kilograms. The line represents the mean, whereas the dotted line represents the 5% and 95% percentiles of 1000 iterations.

and the moment at which the last litre of milk has been consumed vary and these two moments determine when a costly media announcement has to be done.

The total direct recall costs will be zero if the recall moment is later than 185 hours, when all the milk has been consumed. Hypothetically, it might be a good strategy to wait as long as possible before initiating the recall. However, the indirect recall costs (*e.g.* a reduction in market share) can be extremely high when waiting so long. Furthermore, the intangible indirect costs can also be high. For example, the risk of illness of consumers due to consumption of the contaminated milk, the risk of decreasing consumers confidence in the concerned food, the company's name or in the whole sector are unwanted consequences of a recall that are probably related to a later recall moment.

4. Sensitivity analysis

Table 3 contains the results of a sensitivity analysis of isolated changes in 12 time variables of the model; *i.e.* a reduction in the most likely or maximum value of six different stages of the supply chain. It can be concluded that the total direct recall costs are not very sensitive towards a reduction in the most likely or maximum value of the length of stay of milk in six different stages of the milk supply chain. However, a small reduction in recall costs is observed at a recall moment of 96 and 120 hours, especially when the maximum value of the retail or consumer stage is decreased.

5. Conclusions

This paper aimed to demonstrate a modelling approach to investigate the relation between the time at which the recall of consumption milk takes place and the recall costs. The developed

Table 3. Sensitivity analysis for isolated changes in 12 time variables of the model.

			Total direct recall costs at recall moment				
Scenario description			24 hours	48 hours	72 hours	96 hours	120 hours
Base case		-	€ 190,154	€ 208,179	€ 198,136	€ 147,523	€ 98,587
Most likely length of stay at stage:	New value	Default					
Truck	1.5	(3.0)	€ 189,879	€ 208,685	€ 197,330	€ 146,326	€ 97,845
Silo	5.0	(10.0)	€ 190,880	€ 208,660	€ 195,380	€ 143,520	€ 96,442
Distribution centre 1	4.0	(8.0)	€ 189,838	€ 208,598	€ 196,003	€ 144,236	€ 96,710
Distribution centre 2	2.0	(4.0)	€ 187,676	€ 208,303	€ 197,081	€ 145,901	€ 97,720
Retail	6.0	(12.0)	€ 189,191	€ 208,988	€ 194,033	€ 142,577	€ 96,430
Consumer	6.0	(12.0)	€ 188,439	€ 207,353	€ 192,899	€ 142,417	€ 96,578
Max length of stay at stage:	New value	Default					
Truck	4.0	(6.0)	€ 190,188	€ 208,315	€ 197,144	€ 145,822	€ 97,661
Silo	18.0	(24.0)	€ 192,353	€ 209,157	€ 195,273	€ 142,400	€ 95,292
Distribution centre 1	10.0	(12.0)	€ 189,395	€ 208,274	€ 197,176	€ 145,991	€ 97,604
Distribution centre 2	8.0	(12.0)	€ 189,176	€ 208,658	€ 196,266	€ 144,369	€ 96,614
Retail	48.0	(72.0)	€ 187,698	€ 209,448	€ 192,692	€ 125,589	€ 84,418
Consumer	48.0	(72.0)	€ 188,437	€ 207,505	€ 186,510	€ 123,744	€ 84,284

simulation model predicts the amount of milk in each stage of the milk- supply chain at a certain moment in time and calculates corresponding direct recall costs. From the modelling results it can be concluded that the direct recall costs are less than €100,000 in the first 16 hours after the milk being collected from the farms, but the costs increase rapidly to more than €200,000 with an increasing recall moment later than 16 hours. This is due to the costly media announcement that has to be done when the milk has reached the consumer.

The dairy supply chain has been modelled before to answer other food safety research questions. Wein and Liu (2005) build a deterministic model that considers the flow of a Californian milk supply chain to investigate a bioterror attack on the food supply chain (*i.e.* botulinum toxin in milk). In the corresponding appendix of this study the amount of milk that needs to be recalled in case of an attack was calculated. For the calculation, it was assumed that the milk stream over the chain was in a steady state, which means that the amount of milk in each stage will not change in time. Implicitly it was assumed that no batches can be distinguished within this milk stream, which is the worse case scenario. The amount of milk to be recalled in this calculation is probably an overestimation; although in case of a bioterror attack it might be a good strategy to recall all the milk present.

We choose for a simulation approach to include stochasticity, since the distribution of one batch of milk over the chain over time is not fixed. And in case of calculating the direct recall costs this variation is important to incorporate.

Now we analysed the direct recall costs, it would be very interesting to relate also the indirect recall costs to the recall moment. This will be a challenge, since a lot of assumptions on consumer behaviour are necessary to be able to say anything about this. Anyhow, the relevance of this study depends on the question to what extend the indirect recall costs exceed the direct recall costs and at which moment. Until now this question stays unanswered.

Another addition on this study is to investigate the relation between batch size and direct recall costs. It might be an effective strategy for a dairy company to reduce the batch size to prevent extreme recall costs in the future.

With this study we showed that a modelling approach can be very useful to study the relation between recall costs and time. A model provides useful information to the supply chain manager (*i.e.* often the milk company) about the costs of a possible recall in relation to the recall moment. This information can be helpful when thinking about a strategy how to monitor the quality of the milk and to detect a contamination. The design of a monitoring scheme depends on a lot of variables, technical ones but also economic ones (*i.e.* the costs of a monitoring system but also the costs of possible recalls prevented).

References

Anonymous 2005. Melk uit handel. Ede post (Ede).

Dijkhuizen, A.A. and R.S. Morris, 1997. Animal Health Economics, 1 Edition. Post graduate Foundation in Veterinary Science, University of Sydney, Sydney, 306 pp.

Dupuy, C., V. Botta-Genoulaz and A. Guinet, 2005. Batch dispersion model to optimise traceability in food industry. Journal of Food Engineering 70, 333-339.

Kramer, M.N., D. Coto and J.D. Weidner, 2005. The science of recalls. Meat Science 71, 158-163.

Meuwissen, M.P.M., N.I. Valeeva, A.G.J. Velthuis and R.B.M. Huirne, 2005. The insurability of product recall in food supply chains, In: J. Wijnands (ed.) Quantifying the Supply Chain. Kluwer Academic, Dordrecht, pp. 145-157.

Meuwissen, M.P.M., A.G.J. Velthuis, H. Hogeveen and R.B.M. Huirne, 2003a. Technical and economic considerations about traceability and certification in livestock production chains, In: A.G.J. Velthuis, L.J. Unnevehr, H. Hogeveen and R.B.M. Huirne (eds.) New approaches to food-safety economics. Kluwer Academic Publishers, Dordrecht, pp. 49-62.

Meuwissen, M.P.M., A.G.J. Velthuis, H. Hogeveen and R.B.M. Huirne, 2003b, Traceability and certification in meat supply chains. Journal of Agribusiness 21, 167-181.

Palisade-Corporation 2004. @RISK (Newfield, Palisade Corporation).

Skees, J.R., A. Botts and K.A. Zeuli, 2001. The potential for recall insurance to improve food safety. International Food and Agribusiness Management Review 4, 99-111.

Teratanavat, R. and N.H. Hooker, 2004. Understanding the characteristics of US meat and poultry recalls: 1994-2002. Food Control 15, 359-367.

Van Dorp, C.A., 2004. Reference-data modelling for tracking and tracing. Wageningen University, Wageningen.

Wein, L.M. and Y. Liu, 2005. Analyzing a bioterror attack on the food supply chain: The case of botulinum toxin in milk. Proceedings of the National Academy of Sciences of the United States of America 102, 9984-9989.

Alliances between wholesalers and retailers in the fresh vegetable industry

J.H.M. Wijnands, W.B. Koot and S. Rosenkranz

Abstract

This paper focuses on the governance structure and possibilities of improving the profit by collaboration between wholesalers and retailers in the fresh vegetable chain. The framework of Porter's five forces model describes the actual governance characteristics of the wholesalers and retailers in the fresh vegetable industry. Transaction Costs Economics classify the governance structure as a spot market. The bargaining power of the wholesalers should be regarded as smaller compared to the retailers. Collaboration offers opportunities of cost control in the chain, such as the ability to exploit market value and cost savings. A micro-economic model shows that in particular reduction of waste levels is beneficial for the supply chain as well as for the consumers. The collaboration profit is distributed unequally between retailer and wholesaler. Based on spot market assumptions, in some cases a chain actor experiences a negative impact on the profit of collaboration, whereas the total chain profit increases. Redistribution should be bargained in order to have an incentive for each chain actor. This paper shows that even if chain collaboration increases the total chain profit, achievement is doubtful due to the unequal distribution of the additional profit among the chain participants.

Keywords: wholesaler, supermarket, collaboration, fresh vegetables chain

1. Introduction and goal

Chain and network research advocates cooperation as a new way of doing business. This view abandons the notion of business as a sequential chain of transaction based activities (Verhallen *et al.*, 2004). The benefits in supply chain management are generally highlighted from a governance perspective. Little attention got the distribution of rewards among chain actors related to their activities contributing to the chain objectives (Van der Vorst, 2004). Den Ouden *et al.*(1997), as one of the exceptions, show that the additional cost incurred were distributed unequally over the pork chain actors. No reference is found concerning the fresh vegetable chain, which is the topic of this paper. Actors in the Dutch greenhouse industry are aiming at a higher value added of fresh products and at higher shares of the consumers' expenditures (LTO, 2004). The profit margins are small: on average 1.5% for the wholesaler and 3% for the retailer (HBD, 2005a, b). The governance structure of the industry determines the possibilities of gaining higher value added. The first research question will therefore be: 'What is the governance structure of the greenhouse vegetables supply chain?'. This clarifies the possibilities of price setting. Next to price setting the cost structure and returns of the actors, determine the possibilities of cost savings. The second question deals with opportunities of collaboration in the supply chain. The next questions are 'How is the profit of each actor determined?' and 'What is the impact of cost savings by collaboration on the economic performance of each chain actor separately?'. This paper shows that even if chain collaboration increases the total chain profit, achievement is doubtful due to the unequal distribution of the additional profit among the chain participants.

Section 2 outlines the research framework. Section 3 deals first with market structure of the greenhouse vegetable industry and second with the governance structure. Section 4 deals with the cost savings options and Section 5 with the performance structure of the actors based on a microeconomic model. The results and analyses are subject of Section 6. Section 7 addresses the discussion and Section 8 the conclusions.

2. Research framework

Well-known and frequently used business management and microeconomics theories are combined as research approach in this paper. The most appropriate theory will be used to address one of the aforementioned research questions. Porter's five forces model is used to describe the industry context of the fresh vegetable industry (Porter, 1980). The model of Porter deals especially with the bargaining power of the different actors. Transaction Cost Economics is used to analyse the governance structure of the vegetable chain (Williamson, 1979). This approach describes also the nature of competition between wholesalers and supermarkets in the fresh vegetable industry. This analysis provides information which strategies the players (could) follow. The collaboration and cost savings option are derived from a literature search. The specification of the profit of each actor will be based on the microeconomic theory and will depend on the governance structure of the chain (see *e.g.* Kreps, 1990). The microeconomic model of wholesalers and retailers and the interdependencies will be mathematical formulated. Actual data are used to calculate the impact on profit of some cost savings options.

3. Governance structure

3.1 Introduction

Product and production characteristics distinguish fresh food chains from industrial product chains as well as from non-perishable agricultural chains. Managing these chains is complex because of the characteristics of the product. Perishability, variability in price, quantity and quality, seasonality, bulkiness, and geographic specialization all complicate efficient organization of perishable supply chains. The often large number of small and medium-sized producers further complicates matters since this entails high transaction costs, increases variability of production, and the geographic spread increases distribution costs (Kohls and Uhl, 2002). Perishability prevents stock taking: the salvage value of the product decreases to zero within a few days. The fact that production, trade and consumption are spread over the world makes it one of the few truly global supply chains. These characteristics require advanced organization and management to result in efficient and sustainable global chains.

Table 1 characterize the Dutch fresh vegetable chain. This chain is strongly international since the Netherlands is one of the leading world exporters and importers of fresh vegetables, fruit and flowers. (Wijnands *et al.*, 2004). This paper will focus on retailers and wholesalers.

First the industry context of the retailer and wholesaler will be described based on Porter's five force model and second the governance structure on Transaction Costs Economics.

Table 1. Characteristics of the Dutch greenhouse vegetable chain.

Supply chain level	Number	Main asset	Ratio investments/ annual turnover	Operational decision on:	Flexibility of decisions
Retail chains	Tens	Store brand Outlets	Middle	Shelf space profitability	Monthly
Wholesalers	Hundreds	Logistics (Service brand)	Low	Cargo load Points of delivery	Daily
Producers	Thousands	Production assets	High	Crop variety	Yearly

3.2 Retailers

Vegetable retailers are the large supermarkets, the greengroceries, the week markets and out of home food suppliers or foodservice organizations (for example aiming at hospitals or company restaurants). The most important players are the supermarkets, they account for 74% of the total consumption value of fresh vegetables in the Netherlands (HBD, 2005a). In Germany the supermarkets (service-, discount- and small supermarkets) distributed over 80% of the fresh vegetables in 2003 (Behr *et al.*, 2004). In this research, the supermarket will be regarded as the only retailer and therefore only the relation between the wholesaler and supermarket will be investigated.

Threat of entry

Economies of scale are an important factor in the case of entry barriers. Retailers are characterized by an increasing scale and decreasing numbers of firms. The economies of scale involve declining average costs of joint logistics, joint buying and joint marketing of the supermarket companies. To achieve economies of scale regional concentration is necessary, especially with respect to the logistic costs. Economies of scale in the supermarket industry and the capital requirements may prevent new entrants entering the market, but it will not deter entry of established supermarket companies in a specific region. An example of such competition is the entry of the established company Wal-Mart on the European market. Economies of scale enable *e.g.* the fierce price competition between the Dutch supermarkets chains. The price competition reduces the threat of new entrants.

Second, product differentiation can be a barrier to entry. Possible entrants have to compete with the loyalties that established firms have created by their store format and brand. Increasing differentiation of the international operating supermarket chains is a major development (Colla, 2004). Thus, a consumer can choose a supermarket that fits best with his own preferences, because in general he can choose between different retailers.

Switching costs are a third entry barrier. Switching costs are transaction costs *e.g.* agreeing new terms of delivery, temporary supply shortage due to new partner seeking, quality change or disturbance of the relationship. Quality of the fresh vegetables is very important with respect to the consumer's favourable supermarket. The supermarkets are focused on keeping the quality constant at a high level. Switching between suppliers can jeopardize this focus, and therefore it creates an entry barrier. The greenhouse vegetables industry applies several standards (*e.g.* EurepGAP, BRC) and quality control (in the Netherlands by Kwalitieits Controle Bureau (KCB)). Within the industry, quality standards are transparent and maintained and therefore switching costs are a small entry barrier.

Threat of substitutes

Other supermarket formats or other goods that attract consumers can be seen as a threat of substitutes in the retail market. Fresh vegetables are in the 'destination category'. Being in this category means that the consumer is willing to switch store for this product, so it is very important to align the policy with the consumers' preferences. Fresh vegetables are in one of the few categories with almost all own and no brands or labels at all (Abbenhuis, 2005a, b; Fearne and Hughes, 1999). In this case, retailers can exert relatively large influence on positioning fresh vegetables and the bargaining power of wholesalers is relatively low. It is possible that consumers substitute a store format for another. The consumer can also substitute products: different varieties, alternative forms or other products. For fresh vegetables, the most important substitutes are made up by the different varieties: *e.g.* tomatoes on the vine, cherry tomatoes, beef tomatoes or round tomatoes.

Bargaining power of buyers

With respect to the supermarkets, the consumers are the buyer group. Consumers have no bargaining power with respect to market share: the amount an individual buys is not significant for the supermarket. The supermarkets try to satisfy the preferences of a large part of all consumers. Research showed that consumers are mainly focused on products that are fair priced, *i.e.* good price/quality ratio (CGEY, 2002). So, the consumers are price sensitive, but they also take into account the quality (especially with fresh vegetables). The consumer is not exerting its bargaining power directly, but store switching by consumers is the driving force behind the competition in the supermarket industry.

Bargaining power suppliers

Suppliers with bargaining power can threaten to reduce the quality or to raise the price. With respect to the supermarkets, the supplying wholesale companies are the supplier group. The wholesale market is less concentrated compared to the retail market, which restricts their bargaining power (Section 3.3). Therefore, the wholesalers are not able to exert considerable influence on prices, quality and terms of supply. The supermarkets with a market share of 74% are by far the most important consumer channel for the fresh vegetables (HBD, 2005b). Because the supermarkets are such important buyers, the wholesalers are cautious in exerting their bargaining power. If the wholesaler's product is differentiated or characterized by switching costs, it provides him some bargaining power.

Degree of rivalry between existing competitors

Rivalry is the situation in which a firm sees the opportunity or feels the pressure to improve its position. The forecast is a few very large supermarket chains (Rabobank, 2001). A small number of large supermarket chains on the market could reduce competition theoretically, because they can exert some level of monopoly power. However economic theory also predicts a situation in which a small number of large players compete with each other; in the Bertrand model these companies do compete with prices (Cabral, 2000). In conclusion, rivalry among supermarkets exists despite of the high concentration and high market share.

3.3 Wholesalers

Threat of entry

Fresh vegetable wholesaling can be performed at a small scale (*i.c.* truckload). First economies of scale can be realized by a geographical concentration of the distribution area. This means efficient logistics. Second, due to the size, economies of scale are possible *e.g.* in overhead costs, like order picking and planning or in ICT-use. However, the economies of scale of the wholesalers are not insuperable entry barriers. Differentiation can be of more importance. By differentiating their product or services, they will be able to enforce deals with the supermarkets due to their special quality, new variety or special service. New entrants that provide a better price/quality ratio, variety or service than the incumbent will gain the supermarket orders. With respect to the wholesalers, hardly any entry barriers exist. New entrants can relatively easy cooperate with producers and become competitive.

Threat of substitutes

The wholesalers provide the retailer with a very large range of varieties of products from different production regions. The differentiation in the range of products between wholesalers is very low. This means that the retailer can easily find another wholesaler with an almost identical range of products within comparable quality standards.

Bargaining power of buyers

Supermarkets have a certain level of bargaining power which they use to claim prices and conditions that are not always suitable to their suppliers (Traue, 2005). However, a certain level of costs must be covered; otherwise, the suppliers will end the collaboration. Fresh vegetables generally are exchangeable according to the view of the buyers. Fresh vegetables are characterized by hardly any brand recognition (Abbenhuis, 2005a, b; Fearne and Huhghes, 1999) and therefore it is easier for a buyer to switch from one supplier to another.

Bargaining power suppliers

With respect to the wholesalers, the producers have to be seen as the suppliers. Producer has relatively little bargaining power. Compared to wholesalers, producers operate on a small scale. Individual bargaining power is still small, despite the development of wholesale companies that increased the power of producers. The bargaining power of producers is even threatened by the logistic possibility of international sourcing: substitutes are plentiful available from several parts of the world.

Degree of rivalry between existing competitors

The total turnover of the wholesale companies is increasing, but instead of attracting more companies to the market, the trend is that over time the wholesale market is characterized by a declining number of firms active on the wholesaling of vegetables and fruit. The degree of rivalry in the wholesale market is relatively high. This results in a market that is characterized by competition. The number of firms and their market share is of course not a guarantee that they operate competitively, but it is at least stimulating it.

In the Netherlands, over 1000 wholesalers operate on the market for fresh vegetables. This is a large number compared to the number of retailers. The largest wholesaler in the Netherlands is The Greenery, with a turnover of approximately €1,600 million, followed by a few with a turnover between €150 and €400 million (Wijnands *et al.*, 2004). The majority of the exporting wholesalers have a turnover below €50 million(HBAG, 2004).

3.4 Governance structure

Transaction Cost Economics provides a framework for determining the governance structure (Williamson, 1979). The degree of collaboration companies exhibit is determined primarily by asset specificity and by the uncertainty and their influence on the governance structure of the market (Bijman, 2002; Nooteboom, 2004; Williamson, 1979). Asset specificity follows from the fact that the investments will lose value if the asset is used in another transaction. In the case of wholesalers and supermarkets this asset specificity can for example be made up by investments in a specific variety which is associated with a specific supermarket label. The market in which the supermarkets and wholesalers are operating is characterized by a low level of asset specificity. If wholesalers are collaborating with supermarkets on the fresh vegetables market almost no investments lose significant value if they are used in other transactions. Branding is not very common for fresh vegetables. Masten also concludes in a literature study that agriculture transactions are characterised by low asset specificity (Masten, 2000). Collaboration is also affected by uncertainty the supermarkets and wholesalers experience. Uncertainty deals with *e.g.* switching, prices, quantity or quality. The high frequency (daily) of transactions is decreasing this uncertainty. The wholesalers can predict the behaviour of the buyers. From the previous can be concluded that the fresh vegetables industry is generally characterised by low levels of uncertainty and asset specificity. Therefore based on the classification of Transaction Cost Economics, in the fresh vegetables industry the spot market is the main governance structure (Bijman, 2002; Douma and Schreuder, 1992).

Gereffi *et al.* (2005) use in their Global Commodity Chains theory three slightly different variables, to determine the governance structure of the market (see Table 2). The first variable is the complexity of transactions; this refers to the complexity of the required information and knowledge of a transaction. Fresh vegetables transactions require some specific knowledge and information, especially to guarantee the quality and to control the perishable characteristics of the products, which lead to some complexity. The complexity in the market could increase if supermarkets demand for product differentiation, specified packaging, varieties, labelling or just-in-time delivery. The complexity of transactions can generally be characterised as low and only in some very specific cases as high. The second variable is the ability to codify transactions. Codification is the extent to which information and knowledge can be classified and transferred easily to third parties. Codification helps to stimulate the efficiency of the transmission of the required information, because transaction specific investments are becoming redundant. In the fresh vegetable industry, codification is commonly used. Not only international retail codifications like the BRC, HACCP and EurepGAP, but also industry standards like grading, packaging or payment agreements are of importance. The capabilities in the supply-base are the third variable. This variable refers to the capabilities of wholesalers to meet the requirements of the supermarkets. The capabilities of the wholesalers are considered as high (Abbenhuijs,

Table 2. Key determinants of Governance Structure in a supply chain.

Governance type	Complexity of transactions	Ability to codify transactions	Capabilities in the supply-base	Degree of explicit coordination and power asymmetry
Market	Low	High	High	Low
Modular	High	High	High	↑
Relational	High	Low	High	│
Captive	High	High	Low	↓
Hierarchy	High	Low	Low	High

Source: Gereffi et al., 2005

2005a, b). Wholesalers and their supplying producers are capable of adapting their supply to the demand specifications that are set by the supermarket. Also from Gereffi *et al.* (2005) it follows that the main governance structure is spot market or collaboration with the modular governance structure. These two governance types will be the focus in the next sections.

In conclusion the governance structure of the retail and wholesale in the fresh vegetable industry can be classified as almost perfect competition according to Transaction Cost Economics and the Global Commodity Chain approach. The segment where the supermarkets operate might have some characteristics of an oligopolistic market. Only a few large players affect the market largely. The competition between these firms is assumed to be substantial, because 'Bertrand' price setting competition is characterizing the market (Tirole, 1988). The supermarkets active on the market possess relatively more bargaining power than wholesalers.

4. Collaboration options and benefits

Collaboration between wholesalers and supermarkets is in fact a modular governance structure. This means; "*...some arrangements between buyer and seller, entered into freely, to facilitate a mutually satisfying exchange over time, which leaves the operation and control of the two businesses substantially independent*" (Hughes, 1994). If the preferences of the supermarkets are converging to collaboration and value creation, the sales potential of the wholesalers is decreasing, as they stay focussed on the non-collaborative spot market. The demand-driven-supply-chain forces the wholesalers to comply with the preferences of the supermarket. The partnerships must offer mutual benefits to both actors of the partnership. Modular partnership is, in contrast to integration, characterized by the lack of equity sharing and the absence of drastic obligations like a management take-over. Throughout this paper, this type of partnerships will be used as the main form of collaboration.

The successfulness of collaboration will be determined by the following factors (Duffy and Fearne, 2004; Fearne and Hughes, 1999; Zanquetto-Filho *et al.*, 2003):

1. Strategic orientation. Growth strategies of retailers are based on location, size, competitiveness of product range and price, and increasingly on differentiation, with an own label for fresh produce. Wholesalers have to respond to the demand of retailers. Wholesalers that are willing

to enter such a partnership receive a growth of their sales potential by investing in customer specific preferences.

2. Ability to exploit (*i.e.* add value) market information. Sharing product information between supermarkets and wholesalers can result in an increase of the value added to a certain market. The value added will result from the ability to achieve strategic objectives; setting competitive prices or reductions of wastes are easier to achieve if knowledge about costs and market data is shared.

3. Organizational structure and business culture. The organizational structure and business culture are hard to measure and are closely linked to the strategic visions of the partners. It aims at meeting the preferences of the customer at every level of the business. As follows from the fresh vegetables, industry wholesalers and supermarkets differ in their approach to partnerships. Supermarkets have more power and therefore they are inclined to change between wholesalers on a regular basis. This causes uncertainty for the wholesalers and makes it difficult to realize long-term relationships with the supermarkets.

4. Cost control. Cost control can be used to realize costs reductions by indicating which costs can be cut. Examples of cost reductions from collaboration are a decrease in waste, handling and logistics costs. Also using the final packaging can reduce costs. Wholesalers who are implementing good cost control are more suitable for operating in partnerships.

5. Innovation. Innovation is the strategic plan that stimulates long-term competitive advantage. This results from the fact that the fresh vegetables market is characterised by over-supply. Over-supply increases the need for innovation to stay ahead of competitors by developing new products, varieties or services. Supermarkets are therefore especially interested in wholesalers that are representing innovative firms; because these companies could bring partnership long-term advantages such as strengthen the retailers' brand.

This paper is focussed on two source of cost control for which collaboration is conditionally:

- Waste reduction. Partnership between wholesaler and retailer stimulates positively the communication. Improved communications could result in lower stock levels and a better fit between the supply and demand. A higher responsiveness will reduce the waste level at wholesale level, at retail level or at both levels at the same time.

- Decreasing handling and transportation costs. Improved communications can result in realizing cost saving options, such as less logistic, packaging or handling costs. These costs are included in the variable costs as described in Tables 3 and 4 of Section 5.

The benefits of cost control can either improve directly the profit margin or reduce the consumer price. In the first case, it is assumed that the collaboration firms benefit from the first mover advantage and can set their prices above the marginal costs. However, their competitors will take over this innovative competitive advantage on a very short notice. In the second case, the increased demand by lower prices should contribute to better profits. Secondly, the two sources of cost control that will be taken into account require successfully applying the aforementioned factors that determine a successful collaboration. Otherwise the cost control can not be managed effectively.

5. Performance structure

5.1 Theoretical framework of the performance functions

Collaboration between retailers and wholesalers will be evaluated by cost and profit functions. The general formulations of these functions are taken from Industrial Organization textbooks (Tirole, 1988). For simplicity, two actors represent the industry: one retailer and one wholesaler. Both represent the large number of firms as is shown in section 3. In the equilibrium, the consumer quantity and price are standardized to one unit.

The dependency of quantity on prices is represented by an elasticity coefficient (equation (1)). The profit functions depend on the quantity, price and costs (equations (2) and (5)). The variable costs depend on the quantity and are divided in purchase costs of products and other variable costs: mainly providing services (Van Dalen, 1992). The decreasing share of fixed costs realizes economies of scale if larger quantities are handled. Collaboration costs, the fourth group, increase due to monitoring: the product specification should fully comply with the agreements. The perishable nature of fresh vegetable is taken into account in the loss parameters (equation (4) and (7)). Collaboration aiming at reducing waste levels influences these parameters

Consumer demand		$Q_r = 1 + \varepsilon(P_r - 1)$	(1)
Retail level	Profit	$\pi_r = P_rQ_r - C_r$	(2)
	Costs	$C_r = P_wQ_w + acv_rQ_r^2 + bcv_rQ_r + cf_r + cc_r$	(3)
	Where	$Q_r = (1 - \alpha)Q_w$	(4)
Wholesaler level	Profit	$\pi_w = P_wQ_w - C_w$	(5)
	Costs	$C_w = P_pQ_p + acv_wQ_w^2 + bcv_wQ_w + cf_w + cc_w$	(6)
	Where	$Q_w = (1 - \beta)Q_p$	(7)

Symbols	acv, bcv	Variable cost parameters
	cc	(Fixed) cost of collaboration
	C	Costs
	cf	Fixed costs
	P	Price of quantity sold
	Q	Quantity sold
	π	Profit
Parameters	α	Product losses at retail level
	β	Product losses at wholesale level
	ε	Price elasticity (negative number)
Indices	w	Wholesaler
	r	Retailer
	p	Producer

5.2 Empirical framework of the performance functions

Consumer demand

In equation (1) the quantity demanded by consumers depends on the price of the product. For groceries with low stockpiling propensities the price elasticity in the introductory and growth stage are stronger than in the mature and decline stage. The price elasticities are -4.10 respectively -2.62 (Bijmolt *et al.*, 2005). As fresh vegetables cannot be classified in the introductory and growth stage, the elasticity of -2.62 is chosen.

Wholesaler's and retailer's margins

The empirical data on profits and costs in this section represent the spot-market, the actual governance as shown in section 3. Table 3 summarizes the costs for the wholesalers. It shows the limited size of the profit margins achievable in the fresh vegetables industry for the wholesalers. The purchase costs, approximately 84%, reduce the possibilities of a high profit margin largely. The operating expenses that are involved in this market are taking an important part of the gross profits that remains, approximately around 14-15%. This leaves a profit margin of roughly 1.5% for the wholesalers. From this structure follows the importance of transport and packaging, with a share of 9% in the net sales. An important reason for supermarkets and wholesalers to participate in a partnership is to reduce the costs. Transport and packaging play an important role in the cost structure and they might be reduced by collaboration of the two partners.

The cost structure and profit margin of retailers is different between all groceries and fresh products such as greenhouse vegetables (Table 4). The perishable nature of fresh produce results in higher costs for transportation, packaging, handling and storage (*e.g.* cold store). On all groceries, the gross profit is 24%: for fresh products 35% (HBD, 2005a). It is assumed that costs and net profit for fresh products have the same distribution as for all groceries.

It can be concluded that supermarkets and wholesalers are both receiving relative small profit margins. The margin of the supermarkets however is substantially higher than the margin of the wholesaler. Under the assumption of perfect competition – marginal cost are equal price-, the theoretical cost functions are calibrated with these data. It assumed that the figures of Table 3 and 4 capture the maximum profit.

Table 3. Profit-and-loss account for an average fruit and vegetable wholesaler.

	% of net sales
Purchase costs	84.0
Fixed costs (personnel, housing, capital)	5.5
Variable costs (transport, handling & packaging)	9.0
Profit Margin	1.5
Net Sales	100

Source: Greenery, 2003; Van Os, 2000

Table 4. Profit-and-loss account of the supermarket industry for all products and for fresh vegetables for 2003.

	All products % of net sales	Fresh vegetables % of net sales
Purchase costs	76	65
Fixed costs (personnel, housing, capital)	15	22
Variable costs (transport & packaging)	7	10
Profit Margin	2	3
Net Sales	100	100

Source: HBD, 2005a

Waste levels

Relatively high waste levels characterize the fresh vegetable market. Nowadays, the food chain from producer to consumer accounts for 32-39% waste. The loss at retail level is estimated at 2 till 6% and at wholesale level at 2 till 10% (Pegge *et al.*, 2004). This paper takes 4% for the retail and 6% for the wholesale level as starting point. This results in an α of approximately 0.04 and in a β of 0.06. The wholesaler must buy approximately 111 units of product from the producer. After waste deduction 104 units will remain. The supermarket needs these 104 units to sell eventually 100 units to the consumer.

Collaboration costs

Wholesaler and retailer experience costs that arise from the quality and safety standards they should maintain. Collaborating with a supermarket increases the quality level the wholesaler has to guarantee, because they collaborate on providing a specialized and high quality product. The retailer has costs to maintain the high quality and to improve attractiveness of the products. The costs are in majority labour costs: inspection costs during acquisition and quality monitoring during the sales process. The costs of control involved account for approximately less than 0.1% of net sales (Collinson, 2001; Valeeva, 2005). Therefore, the collaboration costs are at each level estimated at 0.1% of the total sales.

6. Results

6.1 Reduction of wastes

As is stated before collaboration can reduce the waste levels. The impact on the profit depends on the possibility of price setting and on the actor who realizes the waste reduction. The micro-economic theory has been followed if monopoly price setting is not possible: marginal costs are equal to marginal revenues. This is indicated as spot market. In the second case a monopoly price setting is assumed: the prices remain at the actual level. This assumption can only be realized if other wholesalers and retailers do not copy the approach of waste reductions. So the assumption will than be that at each level more actors are involved.

Figure 1 summarizes the impact on profits of a waste level reduction. The left hand side presents the results of waste reduction at retailer level. The retailers will gain and the wholesalers have a lower profit. Due to the lower wastes, the retailer buys a smaller quantity of products from the wholesaler. The higher consumer demand does not compensate for the lower demand at wholesale level. Even with monopoly prices, the profit of the wholesaler will decrease due to collaboration costs. Without a redistribution of the higher chain profit, the wholesaler has no incentive to collaborate. Waste level reduction in the spot market at wholesale level first decrease the profit due to the collaboration costs (right hand side Figure 1). After a waste reduction from 6% to 5.5%, the chain profit increases and again retailers benefit most. The profit of the retailers is already higher than the actual level at a waste level of 5.5%. At wholesale level, the profit increases if the waste level is below 4.75%. The profit increase of the wholesaler is many times higher if he can set a monopoly price at the actual level. The collaboration costs of the retailer result than in a profit loss. The analysis shows in both cases that waste reduction of more than 0.5% points is beneficial for the chain profit even with spot marker prices. These exercises show that collaboration will increase the profit of the chain as a whole, but some actors in the chain have a negative incentive to collaborate. Without a redistribution of the additional chain profit, such beneficial cooperation will not occur (Section 6.3).

6.2 Reduction of variable costs

A reduction in the service part of the variable costs affects not always positively the chain profit. Figures 2 shows the results based on spot market prices of a reduction in the variable costs of the retailer (left hand) respectively of the wholesaler (right hand). The actor who reduces the variable costs will be confronted with decreasing profits as the variable costs decrease. As the marginal revenue is equal to marginal costs, the spot market prices will decrease. The marginal costs depend on the variable costs. The marginal costs will therefore decrease with decreasing variable costs. The higher consumer quantities sold compensates insufficiently for the lower

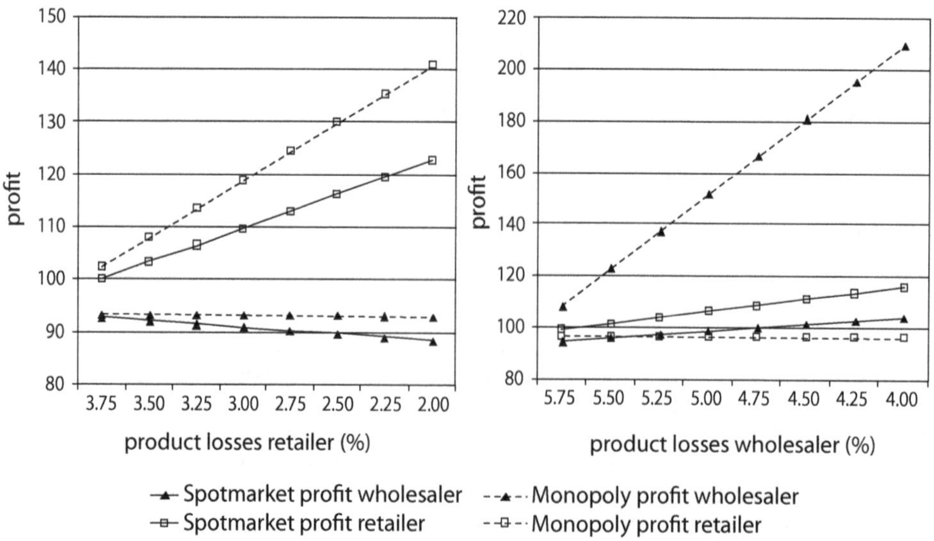

Figure 1. Impact on profits (actual level 100%) of lower waste levels due to collaboration.

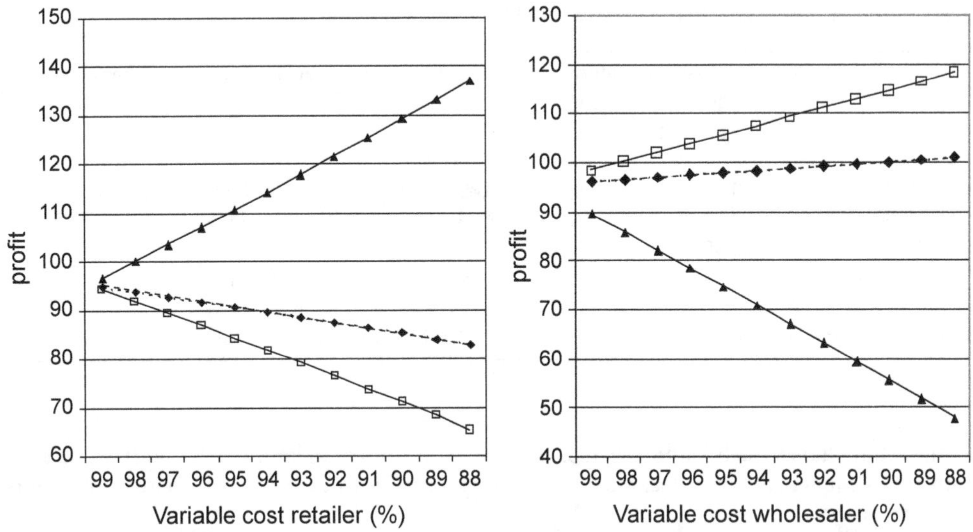

Figure 2. Impact on profits (actual level is 100%) of lower variable costs due to collaboration.

margins. Only if the variable costs of the wholesaler decrease more than 10% the net–result will be positive. A reduction of variable costs is therefore in most cases only beneficial if the two collaboration actors can set monopoly prices and redistribute the additional profit.

6.3 Redistribution of profits.

Collaboration must offer mutual benefits and these benefits should only be achievable by collaboration. As shown in section 6.1 and 6.2 benefits are not distributed equally between chain partners. In some cases, one actor has no incentive to collaborate at all. Therefore, the profit of each actor should be larger than without collaboration. This means also that the joint profit should be higher than the sum of profits in the spot market without collaboration. The redistribution of the joint profit results in the next conditions:

$$\pi_r \leq \lambda\pi^c \quad (8)$$

$$\pi_w \leq (1 - \lambda)\pi^c \quad (9)$$

$$\pi^c \geq \pi_r + \pi_w \quad (10)$$

Symbols
joint profit of collaboration
redistribution coefficient of the joint profit between retailer and wholesaler.

To illustrate the redistribution of the profits two case will be analyzed: product losses at retail level of 3 and 2% instead of the actual level of 4%. Table 5 summarizes the profits of each actor separately and of the chain in total. At the minimum level of the profit of the retailer remains at the actual level and all the additional profit is directed to the wholesaler. At the maximum level

Table 5. Profit levels at spot market prices and lower and upper value.

	Actual (calibrated as % of consumer expenditures)	3% losses at retail level	2% losses at retail level
Profit wholesaler	0.98	0.89	0.86
Profit retailer	3.00	3.29	3.68
Profit chain	3.98	4.18	4.54
Profit chain in % actual level	100	105	114
Lower and upper value	$\lambda = 0.75$	$0.72 \leq \lambda \leq 0.77$	$0.66 \leq \lambda \leq 0.78$

of λ the retailer takes the full additional profit. If stays at the actual level - 0.75 - both actors earn relatively an equal profit increase. The lower and upper value of determine the bargaining limits of the two actors.

7. Discussion

Besides the benefits of lower costs and higher quantities demanded some extra positive benefits from collaboration exist that are hard to measure. Section 4 summarizes some effects and key success factors. An example is the destination effect, which refers to the positive output effect on other products that the retailer experiences due to the increased traffic. The consumers are willing to change store (destination) for higher quality of fresh vegetables. Therefore, the supermarket attracts more consumers resulting in higher sales in all categories. Another positive side effect of collaboration is the quality increase of the product that will be realized by means of improving the knowledge exchange. A quality effect would be especially interesting for service and quality orientated supermarkets. Whereas the discount orientated supermarkets, that are representing an increasing part of the market, are more interested in the price decline that could be realized by collaboration. Of course, one should bear in mind some comments on these positive outcomes. Collaboration could for example fail due to frictions between both managements. Moreover it can be questioned how long the first mover advantage of collaboration will last.

The fact that the actual market situation is assumed a spot-market can be discussed. Due to transaction costs switching between wholesalers by retailers or between retailers by wholesalers induce costs. Mutual, non-contractual agreements exist between the actors on the market, which stimulates a longer-term relationship. However, wholesalers are substitutable from the supermarket's point of view and therefore a spot-market is assumed. Although changing supplier may induce some switching costs to both actors, switching is relatively easy and at low costs in the actual market.

The results of section 6.1 and 6.2 are based on the assumption of a redistribution of the increased chain profit. In reality, an actor whose profit increases most might be reluctant towards such a drastic type of collaboration. Contracts are needed to guarantee that the other actor will get its part of the joint profit. The lower and upper value of the redistribution coefficient, which is determined theoretically, needs further research on the negotiating process between the wholesaler and retailer.

The increased demand due to the price decline is in economic theory explained by the substitution and income effect. The substitution effect means that the product compared to other products becomes relatively cheaper and therefore the consumer substitutes others and buys more of the product. The income effect refers to the change in consumption caused by the fact that the consumer is able to buy more of your product with the same amount of money. This is not taken into account explicitly.

8. Conclusion

The fives forces model of Porter indicates that

- a high level of competition and rivalry between actors exist;
- switching costs are relatively low;
- substitutes are available;
- wholesaler have a low level of bargaining power compared to the retailers.

The theories of Transaction Costs Economics as well as Global Commodity Chain classify the greenhouse fresh vegetable industry governance as spot market. This research quantifies the benefits of waste reduction and cost saving achievable by collaboration in the fresh vegetables market. Reduction of the waste levels of the product purchased is even beneficial at long run spot market prices. A reduction of variable costs dealing with the logistics and services are only beneficial with monopoly price setting. With spot-market price setting it will be hard to increase the profit at industry level with lower other variable costs: mainly the consumers will benefit from it. The conventional wisdom that collaboration increases the profit, due to cost reductions is not always obvious. In the case the chain profit can increases, some actors can experience a profit decrease whereas the others experience all benefits. To make collaboration beneficial for all partners, the higher profit needs to be redistributed. Further research will be necessary whether a redistribution of profits between wholesalers and retailer will be achievable.

References

Abbenhuijs, R., 2005a. Kees de Wit, CEO The Greenery: AGF-sector is sneller volwassen dan we ons realiseren. Vakblad AGF, 24-27.

Abbenhuis, R., 2005b. Retailer en producent samen naar de consument. Vakblad AGF, 20-21.

Behr, H.C., S. Illert, C. Rampold, U. Sturm and M. Weiss, 2004. Marktbilanz Gemüse 2004. Zentrale Martk- und Preisberichtstelle GmbH, Bonn.

Bijman, W.J.J., 2002. Essays on Agricultural Co-operatives: governance structure in fruit and vegetable chains. Erasmus University, Rotterdam.

Bijmolt, T.H.A., H.J.van. Heerde and R.G.M. Pieters, 2005. New Emperical Generalizations on the Determinants of Price Elasticity. Journal of Marketing Research 42, 141-156.

Cabral, L.M.B., 2000. Industrial Organization. The MIT Press, Cambridge.

CGEY, 2002. Searching for the global Consumer: A European Study of changing lifestyles and shopping behaviour. Cap Gemini Ernst & Young, Utrecht.

Colla, E., 2004. The outlook for European Grocery Retailing: competition and format development. International Food and Agribusiness Management Review 14, 47-69.

Collinson, C., 2001. The business Costs of Ethical Supply chain Management: Kenya Flower Industry Case Study. Natural Resource Institute, Kent.

Den Ouden, M., R.B.M. Huirne, A.A. Dijkhuizen and P.van Beek, 1997. Economic optimization of pork production-marketing chains. II. Modelling outcome. Livestock-Production-Science 48, 39-50.

Douma, S. and H. Schreuder, 1992. Economic approaches to organizations. Prentice Hall, New York.

Duffy, R. and A. Fearne, 2004. Buyer-supplier relationshops: an investigation of moderating factors on the development of partnership characteristics and performance. International Food and Agribusiness Management Review 7.

Fearne, A. and D. Hughes, 1999. Success factors in the fresh produce supply chain: Insights from the UK. Supply Chain Management. 4.

Gereffi, G., J. Humphrey and T. Sturgeon, 2005. The governance of Global value chains. Review of international Political Economy 12, 78-104.

Greenery, 2003. Jaarverslag. The Greenery, Barendrecht.

HBAG, 2004. Exporthandel: aantal bedrijven per omzetklasse.

HBD, 2005a. Branches in detail. Supermarkten. Hoofdbedrijfsschap Detailhandel, Den Haag.

HBD, 2005b. Branches in detail 2005; Groentezaken. Hoofdbedrijfsschap Detailhandel, Den Haag.

Hughes, D., 1994. Breaking with tradition: Partnerships and alliances in the European food industry. Wye college Press, 218 pp.

Kohls, R.L. and J.N. Uhl, 2002. Marketing of agricultural products. Prentice Hall, New Jersey.

Kreps, D.M., 1990. A course in Microeconomic theory. Harvester Wheatsheaf, New York, 850 pp.

LTO, 2004. Schoon, Veilig en Gezond. Afzetkansen voor glasgroenten. LTO, The Hague.

Masten, S.E., 2000. Transaction-cost Economics and the Organization of Agricultural Transactions. In: M.R. Baye (ed.), Advances in Applied Microeconomics: Industrial Organization, Vol. 9. Elsevier, New York, pp. 173-195.

Nooteboom, B., 2004. Governance and competence: How can they be combined? Cambridge Journal of Economics 28, 505-525.

Pegge, S.M., I. van Dijke and A.J.M. Timmermans., 2004. Coalitievorming beperken derving van versproducten in de agroketen: enquête resultaten. Agrotechnology & Food Innovations (285), Wageningen.

Porter, M.E., 1980. Competitive strategy: techniques for analyzing industries and competitiors. The Free Press, New York.

Rabobank, 2001. De smaak van samenwerking.

Ontwikkelingen in de voedingstuinbouw: van solisme naar partnership in de ketens. Rabobank, Utrecht.

Tirole, J., 1988. The theory of industrial organisation. MIT-press, Cambridge.

Traue, T., 2005. Thema: Kozentration im Lebensmittelhandel; Medaille mit zwei Seiten. Minderer Tageblatt.

Valeeva, N.I., 2005. Cost-effectiveness of improving food safety in the dairy production chain, Wageningen University, Wageningen.

Van Dalen, J., 1992. Quantitative studies in wholesaling. Erasmus University Rotterdam, Rotterdam, 293 pp.

Van der Vorst, J.G.A.J., 2004. Supply chain management: Theory and Practice. In: T. Camps, *et al.* (eds.), The emerging world of chains and networks. Bridging theory and practice. Reed Business Information, 's Gravenhage, pp. 105-128.

Van Os, E.B.J., 2000. Structuur van de groothandel in groenten. LEI, Den Haag.

Verhallen, T., V. Wiegerinck, C. Gakeer and T. Poiesz, 2004. Demand driven chains and Networks. In: T. Camps, *et al.* (eds.), The emerging world of chains and networks. Bridging theory and practice. Reed Business Information, 's Gravenhage, pp. 129-146.

Wijnands, J.H.M., S.D.C. Deneux, R.A.F.van Paassen, E.H. Poot and S.C.van Woerden, 2004. Internationalisatie en concurrentiekracht van de Nederlandse vruchtgroentesector. 5. LEI, Den Haag.

Williamson, O.E., 1979. Transaction costs economics: The governance of contractual relations. Journal of Law and Economics 22, 233-262.

Zanquetto-Filho, H., A. Fearne and N.D. Pizzolato, 2003. The measurement of benefits from enablers for supply chain partnerships in the UK fresh produce industry. Chain and Network Science 3, 59-74.

Part 7. Quality, safety and sustainability

Hybrid forms in the supply of safe food

G. Martino and C. Perugini

Abstract

The paper analyses some crucial organizational issues arising in the supply of safe food. Within the theoretical framework provided by Transaction Costs Economics, behavioural uncertainty and asset specificity are taken into account in order to analyse the influence of the characteristics of food on the choice of the structure of governance. Sound theoretical reasons lead to the contention that hybrid forms are efficiently chosen by agents to tackle the questions posed by food safety. This idea is supported by empirical evidence from one of the most important food distribution companies in Italy. The study supports previous empirical evidence and thus confirms the validity of the approach in studying complex relationships along the supply chain.

Keywords: governance choice, hybrid forms, centralization

1. Introduction

Many natural and behavioural factors underlie the accidents and flaws which influence the degree of safety of the products supplied to final consumers, and they make it necessary for suppliers to organize their relationships with the aim of coping with the consequent threats. The possibility of reacting in a coordinated manner to exogenous shocks and the flexibility of adaptation play an important role in the organization of the supply of safe food. Conversely, agents invest growing amounts of resources in order to improve the safety standards of the goods supplied. Human capital, brands, labels and other adopted solutions generate different types of asset specificity, the consequences of which must be tackled with the tools provided by theory. The very nature of the questions posed by food safety, *i.e.* the complex links between information flows, asset specificity, uncertainty, and organizational solutions, suggests that the study of the relationship between governance forms and levels of food safety supplied could find appropriate answers inside the analytical framework of the Transaction Costs Economics (Williamson, 1996).

Although with significantly diversified approaches, researchers have largely recognized the advantages of this theoretical view in analysing quality and safety issues in the agro-food system (Menard and Valceschini, 2005) and in particular have emphasized the prominent role of informative flows in supporting exchange activities (Holleran *et al.*, 1999; Hobbs, 2003). Both public approaches to regulation (Henson and Caswell, 1999) and international trade agreements (Marette *et al.*, 2003; Hobbs, 2003) tend to modify the institutional environment, thus influencing the choice of governance form. Hobbs (2003, 2004) has shed light on the role of information issues and on the influence of search and monitoring costs and organizational choices, and has highlighted the emergence of transaction costs associated with regulatory devices. Inside the analytical structure of Transaction Costs Economics (TCE), this paper investigates the relationship between the food safety strategy and the choice of the governance forms. It also contributes to the organizational approach of studying food safety issues (Grunert, 2005). Theoretical reasons as well as empirical evidence lead us to the contention that hybrid organizations (Williamson, 1991; Ménard 1996, 1997, 2004) represent effective solutions because of the opportunity they provide for improving safety standards and, consequently, favouring firms' adaptation and the fulfilment of consumers' needs.

The paper is organized as follows. Section 2 discusses the fundamental aspects of food safety and places them within the general framework of TCE. The study emphasizes the analytical role of the hybrid structures of governance, thus Section 3 is devoted to briefly discussing the systematic conceptual bases for their understanding provided by Ménard (1996, 1997, 2004). Section 4 examines possible organizational answers to the questions posed by food safety and provides the theoretical background that supports the choice of hybrid forms of organization. Section 5 provides empirical support for the main idea of the paper. The case study concerns the strategies and the supply chain relationships stimulated by a large Italian food distribution company. Section 6 supplies some final remarks.

2. The theoretical bases of the study: a Transaction Costs Economics approach to food safety

Food safety organizational issues are centred around three critical areas: (i) distribution of information, mainly related to final and intermediate transaction due to the features of intermediate and final food products (Hobbs, 2004); (ii) coordination needs, since the degree of safety depends upon the activities of all the supply chain's agents; (iii) governance options in relation to strategic positioning (Raynaud *et al.*, 2004). The TCE approach seems to provide an appropriate theoretical basis for understanding the emergence of governance structures and the development of institutional devices, making these solutions credible in guaranteeing adequate enforcement (Ménard and Valceschini, 2005).

2.1 Distribution of information and uncertainty

According to a widely used classification (Tirole, 1989; Nelson, 1974; Darbi and Karni, 1973), the safety characteristics of a food product can be classified as corresponding to the *experience*, and prevalently, *credence* classes (Hobbs, 2004). So, the safety features crucially contribute toward determining the distribution of information between sellers and buyers of both raw materials and food products along supply chains, and this asymmetric distribution of information lies at the heart of many problematic aspects of food supply (Antle, 1999). Many forms of public regulation and private systems (quality systems, mandatory or voluntary certification systems, HACCP, minimum standards, labelling, traceability systems, *etc.*) have been examined, focusing on their effects on exchange flows, firms' strategy and consumption behaviour (Henson and Caswell, 1999; Sporleder and Goldsmith, 2001; Segerson, 1999). Hennessy (1996) examines the distribution of information and the organization of the agri-industrial food chain and shows, perfectly consistent with the neo-institutional theoretical framework (Coase, 1995; Williamson, 1996), that when it is difficult to identify the quality levels of raw materials or when specific monitoring systems are available, the processor is stimulated to integrate up-hill production stages (agricultural firms). Information asymmetry can feed opportunistic behaviours that accentuate contractual hazards (Williamson, 1993, 1996), and this could imply a reduction of the efficiency of the exchange (Milgrom and Roberts, 1994) as well as the supply of safety itself. There is a general causal nexus which could be summarized as follows. Credence and experience characteristics determine information asymmetry, which in turn increases the risk of opportunism. In other words, a causal nexus from characteristics of food toward behavioural uncertainty does exist. Thus, it is plausible to argue that a nexus also exists between the attributes of food products and raw materials and the distribution of information among agents, so that institutional solutions to information questions reflect the nature of the attributes of goods (Martino and Perugini, 2005).

Inside this environment, the institution of measures targeted at reducing contractual hazards becomes urgent and supports the activation of strategies explicitly devoted towards contrasting the insurgence of those behaviours (Wathne and Heide, 2000).

2.2 Safety characteristics and coordination issues

The safety level of food products depends on the behaviour of all the agents involved in the food chain. This is because the source of accidents - chemical, physical, microbiological, *etc.* - may occur at any stage of the system and because remedies and precautions have to be systematically implemented at any stage of the supply chain. They may also fail due to technological flaws or human errors. So a general coordination problem does exist: safe food supply involves a collective base of economic agents who pursue the same interests with respect to this specific target. The consequence is that coordination plays a crucial role with respect to safety strategies (Martino and Perugini, 2005). Firms' choices must be able to support collective-based coordination. They need to establish effective relationships along the supply chain in order to monitor the quality of the input or to deliver their product. This entails not only the investigation of economic dependency among actors, but also the identification of the type and structure of incentives promoting the employment of available resources towards achieving the targeted safety standard.

2.3 Positioning strategies and asset specificity

The questions raised by the features of food products and their transactions and coordination must be taken into account by the firm in shaping an effective positioning strategy in the food market. It is a delicate and difficult managerial challenge to be able to generate new areas of activity and to activate specific investments. For example, the safety ensuring systems (HACCP, voluntary or compulsory certification, private brands, *etc.*) contribute to reducing search and monitoring costs (Hobbs, 2003, 2004; Golan *et al.*, 2001; Holleran *et al.*, 1999), but at the same time they differ in many aspects and require investments to increase asset specificity. HACCP systems, for example, tend to be customized in order to meet buyers' standards (Ropkins and Beck, 2003) and this implies a more intensive specificity of the assets. Similarly, voluntary or mandatory certification (Segerson, 1999), a private brand (Hanf and Kühl, 2005) and Protected Designation of Origin (Raynaud *et al.* 2004, 2002) also increase the specificity of the assets. According to the TCE framework (Williamson, 1991; Ménard, 2004), as the specificity increases the economic dependence between parties becomes more influential on the choice of the governance structure.

Raynaud *et al.* (2002, 2004) propose an effective theoretical structure by enlarging the scope of Transaction Costs Economics. The formulation and implementation of a quality strategy inside a coordination setting entails a reciprocal connection between *positioning strategy* and *choice of governance form*. Raynaud *et al.* (2004) suggest that, on one hand, the efficiency of the connection between quality and organizational strategies depends greatly on the governance form of up-hill transactions (intermediate transaction). On the other hand, quality control can be better ensured through a relevant level of integration. The theoretical position of Raynaud *et al.* (2004) assumes that final quality attributes (including safety) depend on the behaviour of many (or all) of the agents distributed along the food chain. This assumption is shared by Hobbs (2004) and may be interpreted in the systemic perspective suggested by Williamson (1996).

The thesis of Raynaud *et al.* (2004) provides a theoretical base to the connection between the positioning strategy on the final market – which is in turn dependent on consumers' expectations – and the choice of the governance structure of intermediate transactions. This connection also provides a theoretical explanation of the need for vertical links along the supply chain. Examining how this connection develops and which variables influence them is a task of economic analysis.

3. The hybrid forms of governance

As noted in the introduction, this paper aims to support the idea that, due to the joint effects of asset specificity and uncertainty, hybrid forms appear to be the more efficient structures of governance in the field of food safety. This contention will be discussed in section. 4. In this section, mainly following Ménard (2004), we will briefly describe the basic features of these forms of governance.

The economic nature of hybrid forms has been recently analysed by Ménard (1996, 1997, 2004). Because of several factors, hybrid forms can tackle the crucial issue of *interfirm networks* - that is, how to reduce costs due to negotiations (necessary to avoid or reduce opportunistic behaviours between transaction partners) compared with the costs of administering a wide set of resources within a single firm (Ménard, 2004: 363). Hybrid forms are characterized by a mix of cooperation and competition as regards the mechanism designed to reconcile legal autonomy and interdependence (Ménard, 2004). Hybrid forms are distributed across the spectrum of the various forms of connection between autonomous entities.

Three fundamental regularities among hybrid forms emerge from empirical evidence: the tendency toward aggregation of resources (pooling), the contractual basis (contracting), and the tendency toward competition (competing). The tendency toward pooling implies three main effects: (i) the choice of potential partners is crucial, since common investments give rise to potential opportunistic behaviours; (ii) common planning systems must be provided; (iii) implementation of a proper information system among participants becomes strategically essential. Pooling requires continuity of the relationship which, in turn, means necessary cooperation and coordination. This is the core of the first problem characterizing hybrid forms: how can low-cost coordination among agents be carried out without giving up the advantages of decentralized decision-making? Contracting characterizes many hybrid forms, since the contract signed by participants is frequently a simple framework within which more or less ample room is allowed to everyone. The second typical problem is then the need to identify the best mechanisms for optimizing alignment of transactions to the governance structure by monitoring the organization and the search for solutions to emerging questions. Competition influences hybrid forms, because both partners and organizations may compete among themselves. Moreover, if asset specificity is moderate, participants may be encouraged to move into alternative and competing organizations, rendering the starting form unstable. So the third typical problem concerns the design of a mechanism allowing joint decisions in order to discipline partners, resolve conflicts and prevent free riding. These regularities clarify to what extent hybrid forms are rooted in both cooperation and competition. Whether this mix functions or not depends on whether the mechanisms devised are able to reconcile legal autonomy and interdependence.

Economic theory identifies the main factors determining the choice of a specific hybrid form and of its internal features in the mutual dependence of the investments and in uncertainty. Mutual dependence is indeed crucial in order to understand contractual hazards, whereas the importance of uncertainty lies in explaining coordination issues.

Agents can choose individual but complementary investments or common investments in order to carry out specific activities. Both strategies may face serious obstacles due to asset specificity, if the possibility of employing these investments elsewhere is weak. As regards coordination aims and the intensity of interdependence, hybrids tend to develop coordination mechanisms which, on one hand, challenge competition policies and, on the other, contribute to characterizing the same organizations (Ménard, 2004). Investing in mutual dependence is justified by the fact that the parties derive gains from these decisions. In each case, partners' selection processes (*ex ante*) and governance measures adopted (*ex post*) in order to preserve quasi-rents and stabilize the organization must be examined. In general, more specific mutual investments will lead to more opportunistic behaviour and more robust monitoring measures (Ménard, 2004). Moreover, the more interest there is in achieving quasi-rents, the more centralized coordination will tend to be (Ménard, 2004). Uncertainty is connected with the realization of investments: the more relevant it isand the more it is a consequence of investments, the more centralized coordination will be.

4. Hybrid forms and the supply of safe foods

To supply safe foods, firms must deal with several factors. It is imperative to select suppliers able to provide the degree of quality needed. Resources have to be invested in order to specify the characteristics of foods delivered along the supply chains and monitoring activities are necessary in order to ensure the characteristic requested. Moreover, the organization of the supply of safe food requires flexible and coordinated reactions. The discrete alignment principle (Williamson, 1996) suggests that agents choose the governance structure best able to minimize transaction costs. The framework supplied by Williamson (1993) also suggests that the distribution of the organizational solutions (private or public) depends on both the institutional environment (Henson and Caswell, 1999) and choices made by agents.

The information problem also influences regulatory approaches: these approaches contribute towards defining the institutional environment (Williamson, 1991) in which agents make their choices about governance structures (other relevant dimensions being national and supranational legislations, principles posed by technology, *etc.*). Hobbs (2003) highlights how, in the case of food safety, market failures emerge because of: (i) negative externalities (social costs due to diseases, *etc.*), and (ii) information asymmetry between sellers and purchasers about experience and credence attributes of goods (leading to information, monitoring and search costs).

Examining evidence in the field of safety strategies, Ménard and Valceschini (2005) have introduced analytical reasons for the diffusion of hybrid forms in the agri-food sector. Increasing empirical evidence confirms the importance of these forms in the adaptation of supply chains to food safety requirements (Réviron and Chappuis, 2005). This study contributes to the comprehension of these issues by focusing on the links between information asymmetry, opportunism and asset specificity. Private transactions are greatly influenced by asymmetric information distribution, which increases monitoring and verification costs (ante-regulation transaction costs). The more frequent and large-scale "accidents" are, the more these exogenous shocks stimulate agents and public agencies to devise reaction strategies. The main results of this process are the intensification of the structure of the institutional environment and the emergence of new (post-regulation) transaction costs, which, in turn, influence organizational adaptation. Thus the choice of governance form is due to the strategy developed by agents in order to face information asymmetry, as well as to the rules and constraints imposed by institutional innovation. Organizational adaptation is therefore a causal sequence as represented in Figure 1 (Martino and Perugini, 2005). Examples of joint investments and of pooling resources are derived from the literature and are not directly examined here.

Market exchanges guarantee a proper safety level, by means of opportunism limitation, only when adopted strategies do not entail large specific investments. Vice versa, as investment specificity grows, mutual dependency among agents (Ménard, 2004) and uncertainty (Ménard, 2004; Williamson, 1991) render hybrid forms more efficient than hierarchical and market structures.

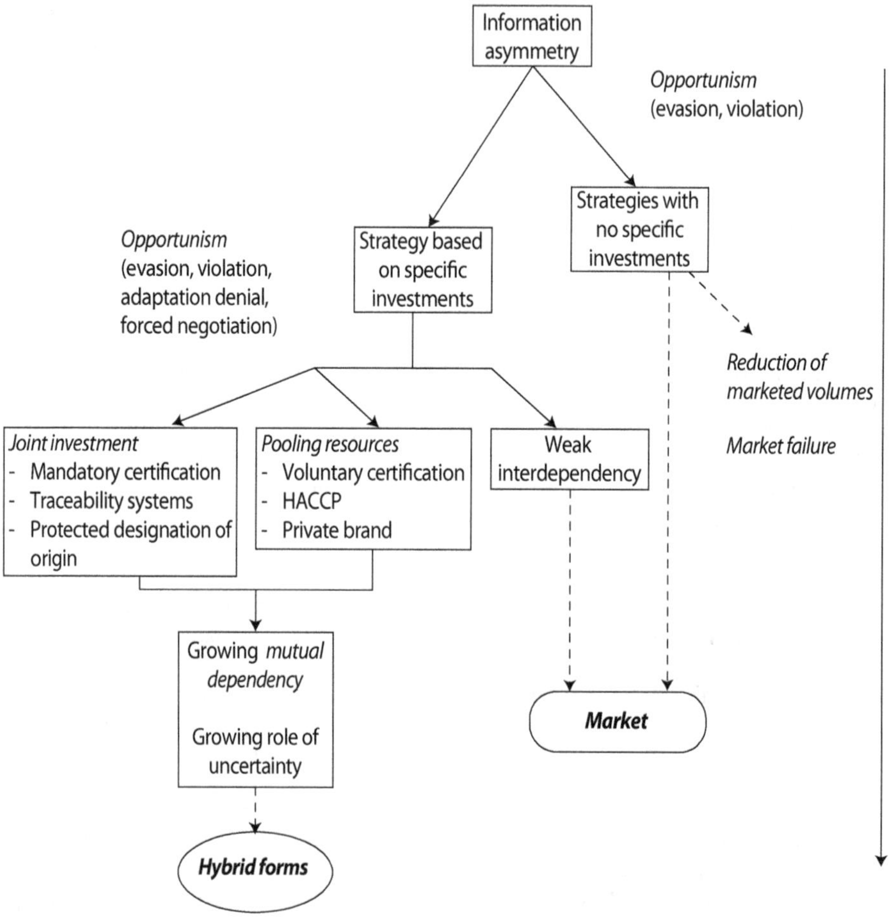

Figure 1. Information, asset specificity, and choice of governance form.

5. Empirical evidence

The research questions addressed here pertain to the provision of quality and safety approach (Grunert, 2005) and concern the forms of governance chosen by agents explicitly adopting food safety strategies. In particular, through an explanatory agri-food case study, the paper attempts to derive evidence about the form chosen (market, hybrid, hierarchy) and its features. First some evidence from literature is pointed out in order to provide an empirical basis for the idea that there is a nexus between food safety strategy and the choice of hybrid forms. Secondly, the case study aims at providing a more detailed framework, pointing out the role of both uncertainty and asset specificity in determining control and centralization.

5.1 Evidence from literature

Ménard and Valceschini (2005) have pointed out the diffusion of hybrid forms in the agri-food sector, suggesting that their emergence can be thought of as the result of three main forces: the evolution on the supply side, the evolution on the demand side and the redefinition of quality control emphasizing the control both over process and products. Food safety requirements largely contribute to the demand side evolution, thus food safety issues and the related solving strategies contribute to the diffusion of hybrid forms in the agri-food sector. A complex public and private institutional framework supports the organizational choice in this field.

Fulponi (2006) looks at the major retailers in OECD countries and points out that ensuring safety is a basic business requirement, whereas organizational choices are mainly based on the desire to protect a company's reputation from the potential damage of a safety accident. Réviron and Chappuis (2005), analysing some Swiss supply chains, identify a link between the "risk of an accident" and the "risk of damaging a firm's reputation" and show that asset specificity and the link between these risks induce the agents to choose hybrid organizational forms.

Mazè (2002) clearly illustrates the relationships between quality strategies and the organizational choice. This analytical view addresses the problems of credible guarantees in terms of organizational choice and emphasizes the role of hybrid structures. This choice provides a basis from which to deal with safety issues, namely allowing retailers to face crises caused by flaws and accidents (Sans *et al.*, 2005). On the other hand, recent empirical analysis in Spain shows that in the beef market retailers tend to establish long-term relationships with breeders, whereas food safety concerns for both retailers and consumers appear to be the main forces leading recent developments (Sans *et al.*, 2005). Further evidence has been provided by Martino and Perugini (2005), who examined a small sample of organic farms that share a common interest in food safety issues. The researchers found that hybrid forms are the more frequently chosen governance structure.

5.2 The Ancc-Coop case study

The current research concerned the Ancc-Coop, an Italian national association of consumer cooperatives that is made up of 155 large, medium and small distribution companies. These cooperatives represent about 6 million consumers. This case study is important for two main reasons (Yin, 1994). Ancc-Coop is a well established economic entity that is tightly structured and diffused over a very large geographic area. It delivers food and non-food products to millions of consumers. Furthermore, in the mid 1980s, Ancc-Coop started to elaborate a food safety strategy to cope with a growing number of accidents and flaws occurring in the food sector.

To some extent, the current Ancc-Coop quality and safety strategy and the Coop brand may stem from the food safety strategies and experiences of the past two decades (Zucchi, 2005 and interviews).

5.3 Methodology and goal of the case study

In accordance with the usual approach in the agri-food sector (Sterns *et al.*, 1998), a set of research questions was first specified. The crucial question of the research was how agri-food companies manage their supply chain relationships in order to enhance the degree of safety of the products delivered. Once the food safety strategy is identified, this issue can be addressed in terms of choice of governance structure and within the general analytical framework elaborated by Raynaud *et al.* (2004), *i.e.* the relationships between marketing choices and safety and quality strategy.

Theoretical propositions were then composed and structured in order to define a prediction pattern. The analytical generalization method was applied in which a previously developed theory was used as a template for analysing the empirical evidence from the case study (Yin, 1994) and in which results were generalized into theory.

The basic theory from which the predicted pattern of non-equivalent variables (Yin, 1994) was derived is the framework of TCE and of the economics of hybrid forms (Ménard, 2004). Firstly, the theory predicts that, in order to enhance the degree of safety of food delivered to consumers, agents efficiently chosoe hybrid organizations (Ménard, Valceschini, 2005; Réviron and Chappuis, 2005; Martino and Perugini, 2005). Secondly, the predicted pattern deals with the mutual dependence issues and the role of uncertainty (Ménard, 2004). In hybrid forms agents aim at linking their investments and then bear a certain degree of dependence and the related risk of opportunism (Williamson, 1991, 1993; Ménard, 2004). According to this analytical framework, the following nexus appear to be relevant: a) specificity → risk of opportunism → closer form of control; b) specificity → interest to quasi-rent appropriation → centralization of coordinating. Furthermore, uncertainty is particularly relevant in the field of food safety, since it may reflect contractual hazards able to intensify the trend toward centralization (Ménard, 2004). Therefore the predicted pattern includes: a) general characteristics of the hybrid forms chosen; b) specificity nexus in terms of forms of control and centralization; c) role of uncertainty in input selection and in the transformation processes. The study aims thus to show that the empirical evidence is consistent with the predicted pattern. To some extent, comparing the predicted and the empirical patterns would provide a control for the theory proposed.

The economics of hybrid form (Ménard, 2004) provide the analytical framework to link empirical data to theoretical propositions. The criteria for interpreting the findings were derived from the usual business study approach and from the general proposition linking marketing policies and quality strategies (Raynaud *et al.*, 2004).

The case study design includes multiple units of analysis and aims at examining the relations along the supply chains as they are shaped by the Ancc-Coop strategies. The paper focuses primarily on the Coop Italia consortium of cooperatives, which is an actor in a very wide and complex system. The Coop Centro Italia sub-unit has also been considered (focused interview) to allow for some large-scale operational aspects of the policies and strategies examined. Neither national nor local suppliers were taken into consideration. Various methods of data collection were used in the case study. The sources of the data were a) documents: internal reports, administrative documents, formal studies, newspaper articles; b) archival records, organizational records:

budgets; c) interviews: focused repeated interviews. To enhance the validity of the information, data triangulation among different sources was performed. The following sections summarize the interview reports (Coop Italia, Coop Centro Italia) and the other collected data.

5.4 The Coop national system: an overview

Ancc-Coop is a group of companies that share a wide set of activities, such as (i) purchasing food and non-food products, (ii) marketing, (iii) management training and (iv) projecting and engineering of distribution and logistics plants. The companies share economic values and social objectives and views that could be gathered under the general goal of emphasizing and fostering a sense of economic solidarity. The group is made up of 155 consumer cooperatives (local companies) that are legally autonomous entities associated with the Lega Coop (a general association of Italian cooperative firms) and to Ancc-Coop (National Association of Consumer Cooperatives).

The case study focuses on the whole Coop system (Figure 2) by briefly illustrating its structure, examining the food safety activities implemented, and analysing the related governance choices. Each one of the three districts represented in Figure 2 groups companies at a given territorial level (eastern; western and north-western regions of Italy). The remaining organizations (Coop Italia, Business School and Inres) perform general functions that involve and benefit all the companies of the different districts. A basic description of the functions performed by the various companies is given in Table 1.

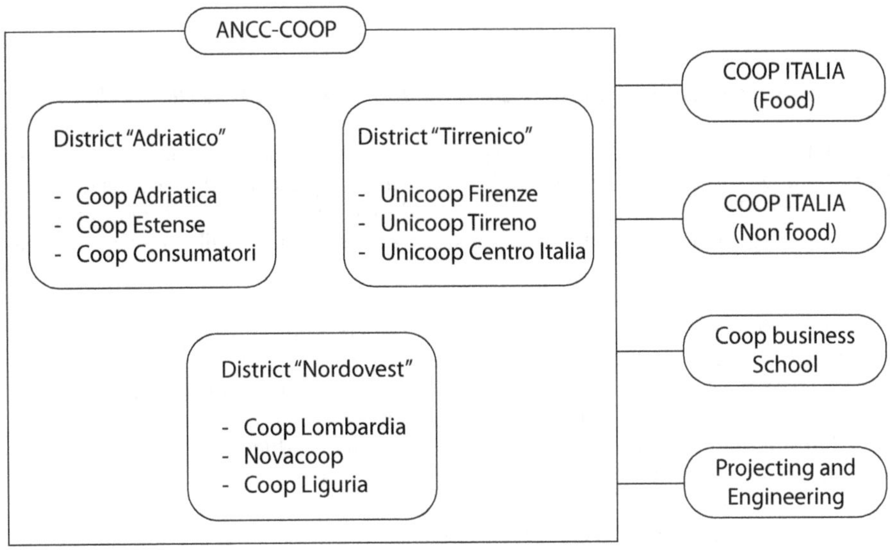

Figure 2. The Coop national system – Companies and associations.

Table 1. Coop system.

National level	Horizontal level policies
relations with Public Administration to define common strategies to define common activities	a) Consumer oriented Marketing Advising Communication b) Supplier oriented Food Safety Quality Collecting and elaborating information Logistics Knowledge creation
Territorial level policies	Single company level
relations with regional and local public administrations to specify local common strategies to specify local common activities following horizontal/national guidelines	to manage product distribution to select local suppliers to collaborate with local suppliers to collaborate in monitoring quality and food safety, to manage customer relations

Source: Ancc-Coop (2005)

5.5 Food safety and quality strategies: shaping and structuring the supply chains

In the mid 1980s Coop started to invest systematically in the field of food safety. Flaws and accidents in the agri-food sector encouraged Coop to intensify and structure the related activities. Coop Italia is largely involved in policy-making and managing horizontal functions at national level in this strategic field. The wide variety of suppliers (large and small firms) also necessitates specific efforts.

The food safety strategy concerns mainly branded products. The Coop brand is a reputation asset of the group (Raynaud *et al.*, 2004) that covers a very large number of products. Coop Italia tends to establish a persistent and durable relationship with suppliers in order to reach the expected and desired standards of safety and quality. The safety strategy was conceived for the Coop branded products, but a similar approach also covers the remaining products (Zucchi, 2005). This safety strategy relies on the following activities:

a) *Selection* of suppliers and of agricultural farmers.
b) *Contract design:* design and elaboration of purchasing contracts.
c) *Regulation*: definition of safety standards (more restrictive than legal standards); definition of production rules and tasks (protocols); standardization of analysis methods.
d) *Monitoring*: detailed inspections and controls, mainly on behalf of Coop Italia.
e) *Research:* research activity in the field of food safety in collaboration with academic research institutes.

National level suppliers (which have to be distinguished from a "basic" local supplier) have to pass through three control steps:

a) inspection at the production stage aimed at verifying compliance with hygienic standards and quality protocols;
b) sampling of products in order to verify compliance;
c) analysis (chemical, microbiological, *etc.*).

The Coop brand products are produced by suppliers directly selected and managed by Coop Italia. The whole group represents 315 processors and 132,000 farmers (Ancc-Coop, 2005), implying thousands of inspections and analysis (see Zucchi, 2005 for more details). The "closeness" between suppliers and distribution companies can be observed in the cattle, poultry and pig sectors (Ancc-Coop, 2005). Partnership is encouraged between companies and suppliers both in specifying technological solutions and in developing efficient control systems.

Broadly speaking, Coop Italia has modified its role in recent years with regards to the associated cooperatives, evolving from a simple purchaser to the coordinator of many different actors. Its current coordinating role includes establishing protocols, supervision and monitoring (Zucchi, 2005). This is especially clear in the critical meat sector (Mora and Menozzi, 2005), but it is also at the core of the organizational dimensions of other supply chains (milk; tomato; fruit and vegetables; olive oil; organic food). A traceability system has been developed that extends beyond the standard legal duties and establishes relationships with the remaining actors in the supply chains (Mora and Menozzi, 2005). The ISO 9001 certification adopted sends a positive message to consumers and contributes to enforcing the contractual relationship between Coop and its suppliers. The food safety strategy of Coop Italia basically relies on an integrated set of tools and this may be due both to the ethical values shared by the companies associated in Ancc-Coop and to the perception of more consequential uncertainty (Raynaud *et al.*, 2004).

5.6 Food safety and quality strategies: analytical remarks

From the information gathered, two relevant types of transactions appear to be crucial: the transactions between national suppliers and Coop Italia (T_{NS}) and the transactions between local suppliers and local Coop companies (T_{LS}). Coop Italia influences T_{LS} through various tools: a) defining purchasing policies and strategies; b) defining safety and quality policies and strategies; c) specifying production protocols; d) specifying the system of inspection and controls. But Coop Italia also influences the T_{NS} as a horizontal actor. To examine the link between organizational choices and food safety strategy, the analysis can only focus on T_{NS}. The horizontal level policies show the specific approach to the centralization of the coordination activity. This centralization is due above all to the need to endow the brand Coop with a common base including: marketing activities; logistics planning activities; purchasing activities; safety and quality activities. The food strategy adopted aims at enhancing the credibility of the quality signal, involving an increase of the governance costs of intermediate transactions (Raynaud *et al.*, 2004).

The Coop Italia approach aims at gaining efficiency by widening the scale of each activity. But it is also a powerful tool to establish common principles in order to support and systematically enhance the reputation needed for the brand Coop. *Ex ante* activities supporting the brand and the related strategies of food safety and quality are concerned with the selection of potential suppliers. Horizontal activities thus reduce search costs due to: a) the enlarged scale efficiency in search and selection activities; and b) the possibility of supporting the selection process by widening the number of potential suppliers and the consequent possibility of substituting them. *Ex post* activities are mainly related to monitoring. The common levels of purchasing and safety-quality policies are also an effective means to connect these policies to the related activities. Due to the food safety and quality policies and activities, the purchasing activity cannot be considered as a simple "buying" activity. The transaction between the supplier and Coop Italia implies that the goods exchanged satisfy specific standards. This may increase the monitoring cost of the whole activity. So, the horizontal structure of the food safety policies and activities also reflects a specific characteristic of the governance structure chosen. Actually, the form of governance of the T_{NS} reduces the monitoring costs by relying on the possibility of performing control at the lower cost granted by the horizontal activities. This approach requires the supplier to ensure that: a) the control activities are performed by Coop and by itself; and b) its processes meet standards and productive protocols. This in turn means that the supplier is constrained in its decision making process, even though it remains a legally autonomous entity.

Reducing search and monitoring costs is at the basis of the horizontal structure of several activities. This entails a lower level of transaction costs than an alternative, decentralized system of searching and monitoring. On the other hand, economizing by centralizing basically allows Coop Italia to provide the group a way to manage the distribution of quasi-rents. Both search and monitoring reduce the probability of free-riding (Raynaud *et al.*, 2004). Namely, monitoring activities provide the basic information shared between supplier and Coop Italia. This reduction of asymmetry of information can be seen in theoretical terms as minimization of contractual hazards due to *ex post* opportunism. Broadly speaking, one may suggests that the *ex ante* and the *ex post* activities establish a framework aimed at supporting the credibility of the quality signal (Figure 3).

This system of activities depicted in Figure 3 allows Coop Italia to stay in touch with suppliers. This is the basis of the definition of partnerships. It contributes to the governance of the supply chain by: (i) ruling the competition among suppliers by selection and inspections; (ii) providing safeguards to contractual relationships: the selection process reduces the risk of bad quality; the threat of a suppliers' exit is not critical, due to large demand supported by Ancc-Coop, whereas a penalty may be imposed to suppliers; (iii) contributing to the identification of common plans of investments and establishing communication channels. The hybrid nature of the governance form is indicated firstly by the associations between cooperatives and, according to Raynaud *et al.* (2002, 2004), also by the fact that:

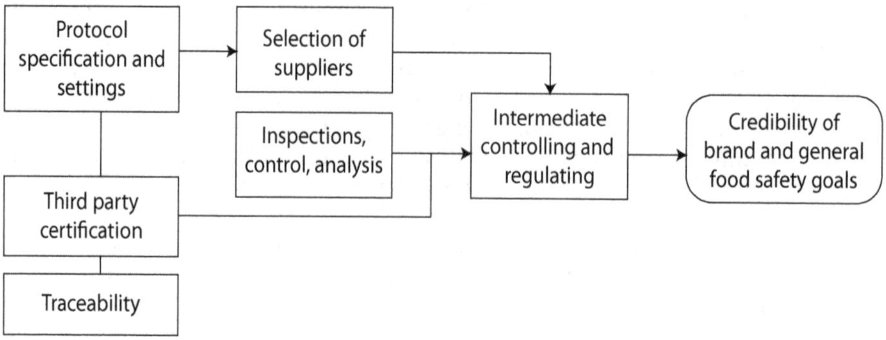

Figure 3. Credibility and food safety strategy.

a) there are no impersonal relations between cooperative companies and their suppliers;
b) there are *ex ante* restrictions on partner choice;
c) relations are formalized (written contracts);
d) duration may be variable, even if it is usually long.

Table 2 summarizes the main evidence emerging from the case study organized according to the analytical elements of the predicted pattern.

Table 2. Values observed in the case study.

Analytical elements	Values observed in the case study		
Investment strategy	Complementary investments: a) Coop: Logistics plants, laboratories; research activities; selection of suppliers; definition of production protocols; monitoring activities. b) Suppliers: Dedicated organizational framework of production processes; personnel training; monitoring activities; logistics plants; genetic resources.		
Degree of specificity	Types of specificity	Coop	Suppliers
	Physical assets	High (logistics plants, laboratories)	High (plants)
	Human capital	High (selected suppliers, trained personnel)	High (trained personnel)
	Dedicated resources	Low	High (genetic resources, internal organizational framework)
	Site specificity	Low-Medium (local suppliers)	Medium-High (plants; absence of alternative buyers)
	Brand	High (Brand Coop)	-
	Time	Medium-Low (Resources invested for seasonal patterns of consumption)	High-Medium (fruit and vegetables, milk)
Risk of opportunism	Vertical free riding, lowered by: a) *ex ante* selection process; b) production and processing protocols; c) centralized monitoring activities		
Main contractual hazards	Bilateral dependence Problems of measuring		
Implemented form of control	Inspections, laboratories		
Degree of centralization of coordinating	Highly centralized due to: a) centralization of policy-making; b) centralization of *ex ante* activities of suppliers selection; c) definition of production protocols; d) centralization of *ex post* monitoring activities.		

6. Final remarks

This paper addresses some key questions posed by food safety issues to suppliers, using the analytical framework of Transaction Costs Economics. The first part of the paper (sections 2 to 4) pointed out relevant issues related to distribution of information among agents along supply chains, to the need for coordination and to the link between positioning strategy and governance choice. The choice of governance form appears to be a crucial decision-making problem for the firms in this field. This discussion also suggests that: a) agents have to deal with various organizational choices, which are quite complex because the safety-ensuring systems often require specific investments; and b) food characteristics and the need for flexible adaptations require specific organizational structures. The study proposes that, in the supply of safe food, hybrid forms of governance are efficiently chosen by agents and are widespread.

In the second part of the paper we provide empirical support for our contention by means of a case study of an important Italian distribution company (Ancc-Coop). This enhances the empirical evidence supplied elsewhere (Martino and Perugini, 2005). Although more empirical evidence is necessary to soundly support the proposed analytical view, the results so far achieved are consistent with relevant literature on the topic. The case study also suggests that the "large-scale" food safety strategy, supporting the supply of food to millions of consumers, strongly relies on complex ordering. Actually, according to Ménard and Valceschini (2005), the private micro-institutions illustrated appear to both be embedded in public macro-institutions and to emerge as a response to endogenous needs.

Acknowledgement

Research funded by Italian Ministry of Education, University and Research-Cofin 2003: Responsabilità del produttore ed investimenti nell'offerta di alimenti non dannosi per la salute del consumatore: aspetti giuridici, economici e finanziari.

References

Ancc-Coop, 2005. Rapporto Sociale del Sistema Coop Nazionale 2004. Coop Italia, Casalecchio di Reno, Italy.

Antle, M.J., 1999. Benefits and costs of food safety regulation. Food Policy 24, 595-623.

Coase, R.H., 1995. Mercati, imprese, diritti. Il Mulino, Bologna, Italy.

Darby, M.and E. Karni, 1973. Free competition and the optimal amount of fraud. J. of Law and Econ. 16, 67-88.

Fulponi, L., 2006. Private voluntary standards in the food system. The perspective of major food retailers in OECD countries. Food Policy 31, 1-13.

Golan, E., F. Kuchler, L. Mitchell, C. Green and A. Jessup, 2001. Economics of food labelling. Jour. of Cons. Pol. 24, 117-184.

Grunert, K.G., 2005. Food quality and safety: consumer perceptions and demand. Eur. Rev. of Agr. Econ. 32, 369-391.

Hanf, J.H. and R. Kuhl, 2005. Branding and its consequences for German agribusiness. Agribusiness 21, 177-189.

Henessy, D.A., 1996. Information asymmetry as a reason for food industry vertical integration. Am. Jour. of Agr. Econ. 78, 1034-1043.

Henson, S. and J.Caswell, 1999. Food safety regulation: an overview of contemporary issues. Food Policy 24, 589-603.

Holleran, E., M.E. Bredhal and L. Zaibet, 1999. Private incentive for adopting food safety and quality assurance. Food Policy 24, 669-683.

Hobbs, J.E., 2003. Information, incentives and institutions in the agri-food sector. Can. Jour. of Agr. Econ. 51, 413-429.

Hobbs, J.E., 2004. Information asymmetry and the role of traceability systems. Agribusiness 20, 397-415.

Marette, S., JC. Bureau, B.Coestier and E. Gozlan, 2003. Regulating food safety in European Union. In: A.G.J. Velthuis, L.J. Unnevehr, H. Hogeveen and R.B.M. Huirne (eds.), New approaches to food-safety economics. Kluwer Academic Publishers, pp. 99-114.

Mazè, A. 2002. Retailers' branding strategies. Contract design, organisational change and learning. Chain and Net Sc. 2, 33-44

Martino, G. and C. Perugini, 2005. Food safety and governance choice. Quaderni del Dipartimento di Economia, Statistica e Finanza, Università degli Studi di Perugia.

Ménard, C., 1996. On clusters, hybrids and other strange forms: the case of the French poultry industry. Jour. of Inst. and Theor. Econ. 152, 154-183.

Ménard, C.,1997. Le pilotage des forms organisationnelles hybrids. Rev. Econ. 48, 741-750.

Ménard, C., 2004. The economics of hybrid forms. Jour. of Inst. and Theor. Econ. 160, 345-376.

Mènard, C. and E. Valceschini, 2005. New institutions for governing the agri-food industry, European Review of Agricultural Economics 32, 421-440.

Milgrom, P. and J.Roberts, 1994.Economia, organizzazione e management. Il Mulino, Bologna, Italy.

Mora, C. and D. Menozzi, 2005. Vertical Contractual Relations in the Italian Beef Supply Chain. Agribusiness 21, 213-235.

Nelson, P., 1974. Advertising as information. Jour. of Pol. Econ. 81, 729-754.

Raynaud, E., L.Sauvée and E. Valceschini, 2002. Governance of the agri-food chains as a vector of credibility for quality signalization in Europe. 10th EAAE Congress, "Exploring diversity in the European agri-food system", August 28-31, Zaragoza, Spain.

Raynaud, E., L.Sauvée and E. Valceschini, 2004. Fit between branding strategies and governance of transactions. 8th Annual Conference of the New Institutional Economics "Institutions and economics and political behavior", Tucson, Arizona, September 30 - October 4.

Réviron, S. and J.M. Chappuis, 2005.Effects of the Swiss retailers' strategy on the governance structure of the fresh food products supply chains. Agribusiness 21, 237-252.

Ropkins, K.and A.J. Beck, 2003. Using HACCP to control organic chemical hazards in food wholesale, distribution, storage and retail, Trends in Food Sc. and Tech. 14, 374-389.

Sans, P., G. de Fontguyon and J. Briz, 2005. Meat safety as a tool of differentiation for retailer. Spanish and French examples of meat supply chains brands. Int. Jour. of Retail & Distr. Manag. 33, 618-635.

Segerson, K., 1999. Mandatory versus voluntary approaches to food safety. Agribusiness 15, 53-70.

Sporleder, T.L. and P.D. Goldsmith, 2001. Alternative firm strategies for signaling quality in the food system. Can. Jour. of Agr. Econ. 49, 591-604.

Sterns, J.A., D.B. Schweikhardt and C.H. Peterson, 1998. Using case studies as an approach for conducting agribusiness research. Int. Food and Agrib. Man. Rev. 1, 311-327.

Tirole, J., 1989. Teoria dell'organizzazione industriale. Hoepli, Milan, Italy.

Yin, R.K., 1994. Case study research. Design and methods. Thousand Oaks, London and New Delhi, Sage Publ.

Wathne, K.H. and J.B. Heide, 2000. Opportunism in interfirm relationships: forms, outcomes, and solutions. Jour. of Mark. 64, 36-51.

Williamson, O.E., 1985. Le istituzioni economiche del capitalismo. Franco Angeli, Milan, Italy.

Williamson, O.E., 1991. Comparative Economic Organization: The analysis of discrete structural alternatives. Admin. Sc. Quart. 36, 259-296.

Williamson, O.E., 1993. Opportunism and its critics. Manag. and Dec. Econ. 14, 97-107.

Williamson, O.E., 1996. The mechanism of governance. New York, Oxford University Press.

Zucchi, M., 2005. Qualità e garanzia dei prodotti: l'impegno di Coop. Coop Italia, Casalecchio di Reno, Italy, January, 1-14 (Report delivered to the authors)

Quality assurance and vertical alliances: Case studies in the UK and Brazilian beef chains

M.D. de Barcellos, G.C. Ferreira and L.M. Vieira

Abstract

Consumer demands for food safety and quality imposed several changes in the relationship between beef chain members. As a consequence, consumers' demand created a favourable environment for the development of co-ordinated actions towards the achievement of better quality and safety assurance. These co-ordinated actions are related to the development and management of quality assurance schemes and vertical alliances. Both vertical alliances and quality assurance schemes are strategies that have been allowing beef chain agents to interact in a co-operative way, increasing their ability to satisfy consumers' demands for food safety and quality standards. This paper is part of a broader study on strategic alliances, but here we focus on the producer point-of-view of the advantages and disadvantages of joining a vertical alliance and a quality assurance scheme in the UK and Brazil. The results indicate that producers in both locations have a similar perception of the advantages and disadvantages of these alliances. The success factors in the vertical alliances involve commitment, trust and payment of premium prices to producers, according to producers in the UK and Brazil. Even though these two initiatives come from different context and time of length, findings suggest that vertical alliances and quality assurance schemes effectively contribute to increase beef chain co-ordination and performance, improving the relationship existent between the agents and providing added value and safe products to end consumers.

Keywords: quality assurance schemes, vertical alliances, beef chain

1. Introduction

The literature on alliances between companies is growing in reference to the agri-food chain analysis. One possible reason is that these alliances, also known as strategic alliances (Perosa, 1999; Bankuti and Machado Filho, 1999; Rocha *et al.*, 2001) have been increasing in number, not only in Brazil, but also in other countries. In general, these authors have characterised the structure of the alliances and their objectives, and whilst the advantages have been approached in a discrete way, the disadvantages have largely been ignored.

In this sense, this paper uses the example of two specific strategic alliances established between beef producers, abattoirs and retailers, to analyse the perception of the beef producer on both the advantages and disadvantages of being in an alliance. In both examples, producers had established a partnership for the production of a differentiated beef product.

We are going to present the core concepts of this research by firstly analysing the issue of quality assurance schemes and secondly discussing the role of alliances in supply chains in a broader perspective. Next, we discuss the establishment of these alliances as a form of increasing the competitiveness of beef in an agri-food chain. In this section the method and the procedures used in this research are described. The fourth section presents the results and discussion of the cases and, finally the main managerial implications and conclusions are synthesized.

2. Research framework

In this section, we present the conceptual basis of the research, regarding quality assurance schemes (QAS), and strategic alliances in supply chains and agri-food chains.

2.1 Quality assurance schemes

The CODEX Alimentarius defines quality assurance as *"all those planned and systematic actions necessary to provide adequate confidence that a product or service will satisfy given requirements for quality"*. A quality assurance scheme (QAS) is defined as the *"organizational structure, procedures, processes and resources needed to implement quality assurance"* (Sterns *et al.*, 2001: 6). There is a diversity in the system used to assure quality, ranging from basic good manufacturing practices (GMP) to specialty labels (*e.g.* fair trade or organic).

QAS can be created with different aims. Spriggs *et al.* (2000) compared a British and a Canadian beef QAS and whilst the former was developed to improve food safety and recover consumers' confidence, the latter focused on improving production methods. The comparison of these two QAS also showed how the different drivers can lead to different levels of success of the schemes. Loader and Hobbs (1999) described how the different activities within a company (operations management, information technology, marketing and distribution) were used in the formulation of a QAS. Furthermore, these authors identified a supply chain strategy, where the close monitoring of activities of upstream suppliers reduced the need for analysis and monitoring of products when they reached the firm. They also provided evidence that identified shifts towards greater co-ordination in Canada and the US, but not as strongly as in the UK examples (Loader and Hobbs, 1999).

Supply chains establish standards at all levels when joining or creating a QAS. An example of standards established within a supply chain is described by Northen (2000) based on a British beef QAS. It is summarised in Table 1:

This QAS is an example of a set of private voluntary standards adopted in the supply chain. It was developed after the BSE crisis in the UK when the communication of safety became a key factor for the beef supply chain. The role of clear rules, inspections and trustworthy inspectors are key

Table 1. Requirements for effective communication of quality signals to customers.

Level of Supply Chain	Requirements
Farm	Clearly written standards, covering all necessary areas in sufficient detail, developed by experts in the industry.
Transport	Transparent inspection protocols with inspections undertaken by independent, competent inspectors.
Abattoir	Strictly upheld rules in case of failure to attain standards
Processing Plant	Assessment of scheme standards and inspectors by an independent, competent third party.
Distribution/Retail	Where necessary, integration of the supply chain to allow full traceability.

Source: Northen, 2000.

points for the assurance scheme. QAS have since become a widespread strategy to differentiate product quality and safety.

QAS also have different aims related to the agent who controls it (Sterns *et al.*, 2001). They proposed three broad categories:

1. Company specific quality assurance scheme – defined and managed at the individual firm level and audited by retailer's quality control staff. Retailers' own brand is a good example of this kind of QAS.
2. Industry quality assurance scheme – these are audited by third party to increase transparency and trust. This industry level can be managed either by the retail, or by the food industry (an example is EUREPGAP).
3. Generic international quality assurance scheme – defined and managed by independent standards organization such as ISO.

Reardon and Farina (2002) described implementation of a QAS for one of Nestlé's coconut suppliers in Brazil. Nestlé obliged this company to be part of its QAS, a combination of ISO (International Standards Organisation), HACCP (Hazard Analysis and Critical Control Points) and GMP (Good Manufacturing Practices already defined). The incentives for compliance were the large-scale operation, the creation of a reputation, and cost reduction from efficiency increase. Fulfilment was a critical to the maintenance of the company as a Nestlé's supplier.

QAS seems an important form of supply chain co-ordination, though it can have different aims, different co-coordinators and different benefits to producers.

2.2 Strategic alliances in supply chains

There are several forms of assuring that a process is accomplished: to develop the activities internally, to acquire companies that have abilities and specialised resources, to use the spot market mechanism and, finally, to develop strategic alliances with other companies. Depending on the resources, capabilities, capital and on products and services available, companies can make their choices. The objective, in the case of developing strategic alliances, is to obtain the advantages of the internal development of the activities without arching with their costs, guaranteeing the necessary conditions that are not found a lot of times in the open market, spot trading transactions.

Strategic alliances are a strategy of explicit co-operation among companies (Hitt *et al.*, 2001) that have interest in to complement their resources or to increase their use in new markets. Strategic alliances allow the companies to overcome entry barriers in markets where it would not be possible to enter in an isolated way.

Gulati (1998) defined strategic alliances as voluntary arrangements among companies involving change, sharing and conjoint development of products, technologies or services. They can happen for several reasons and objectives, to have a variety in the form and to happen in both the horizontal and vertical levels. The type that is of particular interest in this review is the complementary strategic alliance (Hitt et al, 2001), particularly of the vertical type formed by companies that aim to gather their resources and capabilities in different levels of the value chain levels, creating more value to the end customer.

As discussed by Doz and Hamel (2000), alliances can have at least three different objectives: (1) co-option, (2) co-specialization, and (3) learning and internalization. In the co-option, the companies try to establish alliances that are able to increase their competitive capacity, neutralizing possible competitors. Co-specialization refers to the creation of value based on the sharing of resources, positions, capabilities and knowledge for the companies. This kind of alliance gathers partners whose capabilities are complementary for the utilization of a specific opportunity. The third objective of an alliance is the learning and internalization of knowledge. This can be obtained by the conjoint work between the partners, either by the appropriation of existent knowledge, or by the development of a new knowledge.

According to Yoshino (1996), a strategic alliance is defined when it possesses simultaneously three necessary characteristics, as described below:

1. Two or more companies (partners) co-operate to accomplish a group of combined goals remain independent after the establishment of the alliance;
2. Partners share the benefits of the alliance and control the performance of the specified tasks - maybe the most distinctive feature of the alliances and the one that makes it difficult to manage;
3. Partners contribute continually in one or more crucial strategic areas; for instance, technology, and products, among others.

Evaluation of the viability of strategic alliances should contemplate several aspects (Doz and Hamel, 2000; Simchi-Levi *et al.*, 2003): one of the main objectives of the alliances and the fundamental element on which depends their success is the value aggregation to the product. This is possible by several ways, such as quality improvements, reduction in the supply chain time to market, and improvements in the distribution and supply of products and raw materials. In agreement with Stank and Crum (1999), the supply chain management, co-ordinated by companies that are part of alliances, has been important to promote the improvement of service quality provided to the consumers (Ziggers and Trienekens, 1999).

Specifically in the case of food products, where there is a tendency towards differentiation processes, the partnerships within the supply chains have been making possible the value aggregation of raw materials by primary producers (Hobbs and Young, 2000). In this way, in spite of existing problems, the opportunities of closer relationships among the agents of the agri-food chains are growing. That is a consequence of the consumers' growing demand for products more and more refined and adapted to their needs (Jongen and Meulenberg, 1998; Ziggers and Trienekens, 1999).

According to Fearne and Dedman (1999), as far as the fresh produce meat sectors are concerned, there are five benefits which might generally be expected to result from a partnership arrangement:

1. Improved market access.
2. Improved communications.
3. Higher profit margins.
4. Greater discipline.
5. Higher barriers to entry.

Other potential benefits of those alliances, in agreement with a study accomplished by IEL, CNA and SEBRAE (2000) are, for the rural producer 1) increase in the profitability; 2) learning and use of more modern handling techniques, that can provide productivity gains in the medium and long term; 3) sales are guaranteed to the abattoirs that value specific characteristics of the animals (such as steers); 4) product differentiation in the long term. In turn, the abattoir will have benefits from 1) guarantee of inputs regularity; 2) guaranteed sales, 3) differentiation from the competitors. For the distribution segment, the potential benefits refer to 1) guarantee of inputs regularity, 2) to offer an assured quality product to the final consumer.

Another perceived advantage of alliances is the market access improvement, based on the rationalization of the advertising costs, partnerships with companies of the distribution channel and complimentarily, to assist the customers' needs.

In agreement with Oliver (1990), the development of partnerships among organizations has been motivated by several improvements obtained at the internal operations level. Among them, there are reduction of the unitary cost, reduction of the losses and processing, resulting in a significant assets increase. Other advantages in the operation throughout the supply chain refer to the reduction in the transport costs, handling, packing and information processing that also results in a larger operational efficiency.

The improvement of the technological capacity is other advantage found in the alliances, when abilities are shared among the partners (Hitt *et al.*, 2001). Also technological changes can be facilitated when one of the partners dominates a certain technology (Ziggers and Trienekens, 1999). In a general way, the easiness of adoption of new patterns is as facilitated as much closer is the relationship among the links. Within stable relationships, it is possible to promote long-term investments and larger warranties of return.

The profitability increase is one of the primordial objectives of the companies, and it can be reached with a closer relationship among the companies of an alliance (Cooper and Elram, 1993). The increase in profit can be obtained by volume improvements, sharing of the assets, stock reduction and larger stability at the production and sales level (Noordewier *et al.* 1990), among other factors. In this sense, the alliances make possible increments in the organizational abilities by the sharing of management knowledge among the companies. As well, there are also profitability increments, by the reduction of fixed costs and sharing of risks (Ziggers and Trienekens, 1999).

The similarity of strategic objectives is an important factor contributing for the success of the alliance (Cooper and Elram, 1993) as well as the balance between contributions and benefits. These elements should be considered fair for the partners (Ziggers and Trienekens, 1999; Simchi-Levi *et al.*, 2003), in other words, it is important that both parts can obtain benefits, considered relevant for their businesses, in a way that they have a realistic expectation of the potential outcome of the partnership (Lambert *et al.*, 1996).

Another success factor that has an increasing importance in strategic alliances analysis is trust that can be an efficient co-ordination mechanism of economic transactions. If there is trust, the management of the alliance aims to maximize the returns for all of the partners (Simchi-Levi *et al.*, 2003). When trust does not exist, the partners have to allocate more resources and efforts to monitor the activities within the chain (Hitt *et al.*, 2001). In this sense, trust is an important asset, because it is difficult to specify all the operational/relational aspects in a contract (Simon, 1961; Williamson, 1985).

For Degand *et al.* (1998), trust was the main co-ordination mechanism found in the UK fresh fruits supply chain. When there are high market uncertainties, it is difficult to establish the contractual terms that are going to be used, such as the amounts and the prices. In that way, trust among the agents, developed by long-term relationships, replaces the contracts as a supply chain form of governance (Zanquetto Filho *et al.*, 2003).

The main risks associated with inappropriate conditions within the alliances are the establishment of inadequate contracts and difficulties in the sharing of complementary resources (Hitt *et al.*, 2001). If these conditions exist, the management of the partnership should guarantee the establishment of detailed contracts and its monitoring, as well as to promote the development of the relationship based on trust.

Notwithstanding, the main obstacles to the agri-food chain organization refer to the agents' opportunism and the lack of trust among them (Bankuti and Machado Filho, 1999; Pigatto *et al.*, 1999; Bliska *et al.*, 1998). In this case, the contractual break (Williamson, 1985) is frequent and, consequently, interruptions in the operation flow can happen (Zylberstajn, 2000).

Many vertical partnerships are based on tax exemption or reduction, but these elements have not been enough to sustain the investments accomplished in the chain, given the fact that the final price of the differentiated product is quite similar to the common meat (Pigatto *et al.*, 1999; Bankuti and Machado Filho, 1999). The absence of global strategic planning by the chain participants is also one of the main causes of the failure of several alliances in Brazil (Pigatto *et al.*, 1999; Perosa, 1999).

3. Method

The present study is characterised as exploratory, and its objective is to *"offer a picture that can make the deduction process easy of pertinent subjects in the investigation of a phenomenon"* (Tripodi *et al.*, 1981). This kind of study has a descriptive character and phenomenon exploitation, in this case, two vertical partnerships in beef supply chain.

Regarding this exploratory factor, researchers can formulate concepts and hypotheses to be deepened in further studies (Tripodi *et al.*, 1981; Mattar, 1993). At the same time, this research is an in-depth study that aims to analyse several aspects involved in the research subject, which is the strategic alliance concept itself.

The method is a case study, an adequate procedure for the exploratory character of the research (Tripodi *et al.*, 1981). The studied cases are established alliances among companies of the beef supply chain, set by producers (individual or coordinated by a producers association), abattoirs and retailers. One case study was held in Brazil, and the other one, in the UK.

The Brazilian case involves the Brazilian Angus Association, Mercosul Ltd. abattoir and Zaffari Supermarket chain, joined for the production of Certified Brazilian Angus Beef. The UK initiative is composed by an Organic Aberdeen Angus producer (Fordel Angus, Scotland), Dovecote Park abattoir and the Waitrose Supermarket chain.

The purpose of these vertical partnerships is the establishment of beef quality production, involving specific characteristics such as age, breed and carcass fat cover, as well as the quality process accomplished by all of the involved agents. However, there are no direct comparison of the two different alliances which also brings the issue of the different corporate cultures, structures and practices. The cases are presented and analysed individually, and further, implications based on their results are discussed.

One objectives of this research was to identify the producers' perception on the advantages and disadvantages of the alliance. The categories of analysis were established by the literature review. In this sense, producers in Brazil and in the UK were questioned on benefits noticed for their business and for the other partners, as well as on what they considered disadvantages or risks of the partnership. The possible advantages and disadvantages asked are summarised in Table 2.

The categories of analysis considered above were used as a guide for data collection that was accomplished through semi-structured personal interviews. The interviewees were questioned on advantages and disadvantages of the alliances, for their business and for the other partners. In the beginning, the respondents could answer the questions in a broad way and in case the specific analysis elements were not mentioned, the researchers asked them about. The reasons of alliance failures (based on cases that did not have continuity in the beef supply chain) were also questioned to the interviewees, as well as which would be are success factors for the establishment of strategic alliances. Two in-depth interviews were carried out in Brazil in June 2005 and one in the UK in July 2005. The interviews took from 50 minutes to one hour and 30 minutes. After the typing of the interviews, the researchers made the content analysis, trying to identify and to characterise the agents' perception on the vertical partnerships advantages and disadvantages for their business and for each one of the partners.

4. Results and discussion

The results are presented by first characterizing the alliance followed by the perceived advantages and disadvantages for the producers, and finally the causes of success in these alliances.

4.1 Case study 1: Certified Brazilian Angus Beef program

Context

In the mid 90`s in Brazil, there was a flourishing of strategic alliances in the beef sector, motivated by the new globalization scenery. Benchmarking the success and reputation of the Angus beef quality programs worldwide (USA and Argentina, for example), the Certified Brazilian Angus Beef Program was launched in 2001, in Rio Grande do Sul, a Southern Brazilian state. The

Table 2. Possible vertical partnership advantages and disadvantages questioned to beef producers.

Avantages	disadvantages
Sharing of resources, capabilities and competences	Allocation of specific resources
Value aggregation to the product	Investments needed
Improvement of the market access	Reduction of the flexibility
Knowledge of the consumer's demand	Need of exclusiveness
Strengthen of chain operations	Strictness of the organizational structures
Management and control of chain activities	Reduction of management control
Technological improvements	Difficulties in bargaining
Development of organizational capabilities	
Increase in profitability	

Brazilian Angus Association is located in this state, since the temperate climate offers the best conditions for breeding pure breed Angus.

In the initial stage, the Association hired one technician and identified the potential of Angus beef production from a dedicated plant (Mercosul Ltd. Abattoir). The identification was only based on the number of Angus (and crossbreed Angus) cattle that fitted the determined standard, established in conjoint by the Association and the abattoir, and based on market demands (age, weight and carcass fat thickness). When the number of Angus monthly identified became consistent, both Angus Association and the board of directors of the abattoir decided to officially launch the Certified Brazilian Angus Beef Program. They aimed to attract a group of producers, with the same productive goals. The beef produced under this scheme was destined for both the domestic and export markets, allowing the value aggregation necessary to maintain the viability of the program.

Alliance characteristics

The main objective of the alliance was the possibility to add value to the product, based on brand development and consequent differentiation from commodity beef. In that sense, at the beginning of 2003, the program launch was an attempt to create a program based on the production, processing and commercialization of quality beef, in a way to produce a standardised, high quality final product, breed based. Angus cattle are known for their ability to produce high quality, marbled beef if processed to a high standard.

It was established that the association would be responsible for the origin certification, accompanying the whole production process to the final product. In order to obtain conformity of the product, the companies opted for the vertical partnership alliance, valuing the closest relationship and benefiting the animals classified according to the established program standard.

The characteristics that differentiate the product of the alliance are: the animal's standards (age, breed, fat thickness), the certificate of origin assured by the Brazilian Angus Association, and the supply of relevant information to the consumer (nutritional information, validity date, Federal sanitary inspection), besides the brand quality assurance.

Regarding the relationship among the agents, the abattoir paid premium prices to Angus producers for carcass quality and also a bond scheme that foresees financial benefits (*Fidelity Bonus*) to the producers that can plan the slaughtering schedule every trimester. These strategic managerial actions, that differentiate this company from its regional (and even national) competitors, aimed to benefit the most efficient producers and to reduce the raw material supply uncertainties to the industry.

The alliance attracted the local retailers' attention, and six months after the beginning of the certification process, a retail chain showed interest to join the partnership. The subsequent period involved contract negotiations, such as commercialization conditions, brand definition and market positioning, as well as the establishment of product (cuts) standards. In February 2004, the product arrived at the supermarket shelves. In an innovative way in the beef market in Brazil, the new brand development resulted in the first supermarket own brand followed (and certified) by a breed name: *Angus Zaffari Beef*. All the visual identification, name, logotype and communication material were developed in a conjoint effort of all the partners. Nevertheless, it is important to state that although the abattoir was allowed to process and commercialise the beef and the supermarket was allowed to use the "Angus" name, *Angus* is a registered trade mark, legally owned by the Brazilian Angus Association in Brazil.

Up to date, the alliance is considered a successful initiative in the Brazilian market, based on the continuity, regular supply and increasing growth in volume. Since 2003, more than 1,000 tons of Certified Angus Beef was produced. In 2005, 80,337 carcasses were classified, a 43% growth rate compared to 2004. The program started with the inscription of 8 producers, and until December 2005, 250 producers were committed.

Advantages and disadvantages

The Angus Director interviewed declared that one of the main benefits perceived by the producer that participates in the alliance is the premium price paid for the abattoir, around 10% more than other companies. "*The beef price has been the lowest in many years, which makes negotiations tougher, but the Angus producer, committed to the program, has the best deal one can get*". The industry partner is the market leader and traditionally produced high quality beef for domestic and export markets. In that way, it could pay better prices to the producers, since it also sold beef for a more expensive price to its buyers.

Other great advantage, according to the interviewee, was the uncertainty reduction, because the producer has a steady buyer to its animals. There was a previous negotiation and it was not necessary to "*go looking after a buyer for each lot of animals to be sold*". Consequently: "*... the market access is one of the main advantages, because we are working with the best industry and the best retailer. The retailer has a long-term tradition in selling quality products, including beef, what makes it beneficial to associate our brand to theirs. Producers can see their production efforts and results in the shelves. The abattoir is also an industry recognized at the domestic and export beef markets. They have strict audits from the European Union authorities, and the quality standards are quite high*".

In agreement with the interviewee, it was necessary to realise that the customer of the association, or of any producer, is the final consumer, not the abattoir or the retailer. "*What the producer does at the farm level, will have an impact at the consumer*". So, it becomes necessary that the final consumer can approve the product and recognize the association certification as a quality brand. "*For that, to work over brand attachment is fundamental, even if initially we don't win as much as we were able or willing to, but it is important in this stage to consolidate the association brand in the consumer's head. Since our partners have a quality tradition, this helps a lot. That is, even if the brand that the consumers see is not only the association one, it is there, and the consumer is going to learn on how to identify it*".

From the producer's point-of-view, the information transfer could be higher, since the abattoir could make use of the producers' network. They could get an increase in participation throughout word-of-mouth, for example, and consequently, increase production.

For the producers, the alliance facilitated the access to market information allowing them to obtain demand knowledge and understanding of the way the chain operates. The operation knowledge of the other agents of the chain can improve the group efficiency in a significant way. In the interviewee's words: "*We know that for the abattoir, for instance, to get a larger profitability, obtained by costs optimization, it is necessary to have a full load of the truck, in other words, 25 animals. If the producer is organized for that, there is a reduction of the operational costs for the industry and it can revert as an improvement in the remuneration received by the producer. But we only get this information if we start a closer relationship within the chain"*. However, in order to allow this higher commitment, it was necessary that a real predisposition of the agents existed, and that the advantages of such relationship could embody all the participants.

The interviewee also considered that necessary changes happened when the alliance was established, because producers needed to adapt their production system to produce what the market wanted. "*To give you an example...for the export market, the carcass weight must be at least 220 kg, which is a larger carcass than we are used to produce for the domestic market. You spend more to produce it, but you also get more money for it. Therefore, this information indicates that that's the product that gives to the producer and to the industry the biggest profit. This product specification modification has imposed changes in the productive system of some producers"*.

Considering the advantages for the other alliance participants, the perception of the producers was that the largest global advantage was the assured quality of the product, via the certifying process. For the retail and the abattoir this advantage had direct consequences over their operations, since they share the quality control and product responsibilities with other agents.

The knowledge of the demand was definitely an important advantage, but it had also created the need of technological and process improvements. In this regard, based on external market demands, the program is oriented towards a third party accreditation, and the AUSMEAT™ scheme is probably the one to be followed in the near future. EurepGap® and ISO 22000 are also being considered.

In regard to the disadvantages, the interviewees considered that the main disadvantage is the fact of being "in the hand" of a single company, because of the established exclusiveness contract among the parts. The negative consequences involve the reduction of the bargain power in relation to the price negotiation and reduction in the flexibility because of the exclusiveness.

The allocation of specific resources and inputs was not considered a disadvantage, because most of the producers already produced the standard animal type for the programme. In that way, this can even be considered an advantage since this kind of product has a guaranteed market.

Other disadvantages found in the literature, such as the strictness of the organizational structures and reductions of management control were not at all important, according to the interviewee point-of-view. He actually believed that the establishment of organizational structures (governance) helped in the alliance development, and that there was an increase in management control of the supply chain occasioned by the alliance. These are all positive trends that contributed to the success of the partnership.

4.2 Case study 2: Fordel Angus Organic Beef

Context

According to Fearne (1988), an important feature of the UK beef industry in the 90´s has been the emergence of partnerships between producers, abattoirs and supermarkets. In recent years these partnerships have extended further up the supply chain to include breeders and feed suppliers. These partnerships have been difficult to establish and slow to develop – they currently account for approximately one fifth of beef production – but the momentum behind them is gathering and there is a growing acceptance that partnership arrangements are the only sustainable form of trading relationship in the long term.

Alliance characteristics

In the UK, commercialization of niche market beef has been a reality for a long time. Devocote Park is supplying organic beef to the Waitrose for the last 10-15 years, according to the interviewee. Fordel Angus farm joined the alliance motivated by higher profit margins. Since 1985 they were producing organic cereals, and in 1990 experimented with organic beef. The organic lamb certification started in 1999, and beef started in 2000. Since that time, they have been supplying Devocote Park and Waitrose.

All Waitrose beef was reared specially for them, on selected British farms and many farmers have a long-standing relationship with Waitrose. Beef supply comes from animals born, reared and processed in the UK. Waitrose Organic beef is a high standard product in the UK market (supermarket stores in England and Wales, although beef supply comes from all the UK members). It was produced from a variety of continental and native breeds on farms that comply with the UK Register of Organic Food Standards (UKROFS) regulations. Animal welfare was a high priority and the cattle are fed a diet approved by the UKROFS. Organic beef was hung on the bone and matured in vacuum bags to develop flavour and tenderness. Waitrose sourced from farmers who complied with the standards set by Assured British Meat (ABM). This means regular checks were carried out to ensure high standards were maintained. By law, all cattle had a passport which contained information about their breeding and all farm movements.

Waitrose complied with the beef labelling scheme run by the Department of Environment, Food and Rural Affairs (DEFRA). This meant that all their beef could be traced back to the group of farms where the cattle were reared. The cattle exclusively produced for Waitrose were only transported by haulers approved by ABM, which set high standards to minimise stress to the animals during transit. Once the cattle arrived at the processing plant, they were inspected and then rested in an area with bedding and water supply. A meat hygiene inspector supervised throughout.

Advantages and disadvantages

Regarding the advantages, according to the interviewee, the share of competences, capabilities and resources with the other agents was not happening the way it should be. Also there was little knowledge about the consumer's demand. Such important matters should be discussed in regular basis within the partners. He proposed to the abattoir's manager to increase the attention to these topics, since he was aware of the consumer's power in the chain.

A better market access and reduction of uncertainty were very important advantages, in his point-of-view, since he works based on fixed price grid and cattle purchase, what "*reduced dramatically the uncertainty*", according to the producer. He also stated: "*It is very important in the production of organic produce to ensure that it is bought at organic prices. I worked hard to get into this alliance because I could not see any other professional fixed return that was greater*".

The strength of the supply chain operations (like reduction of production costs, increase in productivity of resources) were also a perceived advantage, since he became able to plan the resources and production more efficiently. However, he considered that he could have further co-operation through the chain to assist in further improvements and cost reductions.

He believed that better management and control of supply chain activities (more efficient proceedings in the farm, organizational structures that improve productivity) act the same way as described above, *i.e.*, cattle can be managed more efficiently, but supply chain communication and product development can be improved further. In his view improvement of the technological capacity (new processes implemented in the farm) has not happened yet. There was a discussion with Dovecote in regard to establishing a genetic improvement programme (tenderness). The allocation of resources, the reduction in the flexibility, time constrain because of the audits/specific production proceedings and need of more labour because of production specificities were not considered disadvantages by the producer.

The interviewee considered the cost of the investment (production specifications, information system, and management control) not significant, although the higher cost of the inputs for organic production may create a problem here because of the new EU rule changes for organic farmers. Bargain impossibilities or difficulties were not considered a problem to Fordel Angus, since they are specialist breeders and receive the top valorization offered by the retailer.

Interestingly, the need of exclusiveness was considered an advantage by the UK producer, while Brazilian producers considered it a disadvantage. This is probably due to the high differences in price received by the British producers, *i.e.*, the price received by beef niche market producers in the UK was substantially higher than the one received by producers in Brazil. This certainly stimulates more the producers in the UK to join the alliances (as stated by the producer) than in Brazil.

4.3 Success factors in beef chains

There is a general agree that the success factors in the vertical alliances must involve commitment, trust and the payment of premium prices to producers, according to both producers in the UK and Brazil. The lack of the mentioned conditions could also explain the failure in some previous initiatives.

For the interviewee in the UK, successful alliances demand the producer to believe, to be more optimistic, and to be more participative. It is also necessary that some commercialization warranties might exist, to avoid the distrust. *"I do believe in the initiatives that are linked to cattle breeds, but everyone should be committed. We can adjust the cattle, the diets and the management to suit, if we know that the buyer is going to pay us a profitable price for our work, but also if the buyer is making a profit themselves. If someone in the chain is not making money from the activity then the chain will have a weak link, and a change of management at that organization could require greater profitability and consequently break the chain."*

Improving communication skills and the knowledge about consumers is also an important issue from the producer's opinion. As he stated: "*…however, although the Dovecote Park supply chain is good in that they have built long term trust and dedicated suppliers, they do not integrate their suppliers in to the customer/consumer information flow*".

In that way, we can summarize the factors considered fundamental for the success of the alliances: (1) commitment, (2) constant supply and (3) product standardization. Communication is also an outstanding key factor for the success of the alliances. From the farm to fork, it becomes necessary that everybody can recognise and value the alliance.

5. Managerial implications and conclusions

Regarding the two studied alliances, it was possible to characterise them as complementary strategic alliances (Hitt et al, 2001), established among partners of different vertical links of the beef chain. The alliances were formed on voluntary basis as contractual arrangements between beef cattle producers (organized or not on producer's associations), abattoirs and a retail companies. The main objective focused on the supply of products and complementary services among the partners (Gulati, 1998) for the production of a final product (beef) with increased added value (Hitt et al, 2001). Therefore, they are characterised as alliances whose main objective is the co-specialization (Doz and Hamel, 2000).

Producers perceived very few disadvantages of being part of an alliance. In one hand, specific allocation of resources, the cost of the investments (production specifications, information system, management control), the reduction in the flexibility, time constrains and more labour need because of the audits/specific production proceedings were not significant disadvantages in their view.

These two cases are used here as illustrative and they can not be directly compared due to the differences on the culture, regulation and consumer's preference on the countries where the studies were carried out. However, they bring out important implications for the study of vertical alliances and quality assurance scheme. These are discussed below.

In the case of the Certified Angus Beef Program the governance mechanism was based mainly on trust (Degand *et al.*, 1998; Zanquetto Filho *et al.*, 2003), what could signify emerging risks of failure in the future. Nevertheless, although the lack of trust was one of the pointed causes for the discontinuity of other established alliances in the past, they believed that their alliance have enough strength to survive in the forthcoming years. They also believed that their objectives, product consistency and market acceptance would exempt the alliance from a more rigid governance structure.

It was important to notice, from the obtained results, that the interviewees perceived more advantages than disadvantages in the alliances. This finding helped to explain why, in the literature, there were so few references on the disadvantages. Since this question was an important research question for us, we can consider that it is easier for the participants to see the positive outcomes instead of the negative ones. This may be caused by inexperience or lack of consistent and official results from other established alliances. In this case, we suggest that further studies could go deeper in the understanding of failure motives of beef alliances.

While there are some differences on the institutional environment of these two countries and this may affect the success of the alliance, Angus producers in both countries agree that the main advantage of the alliances was to add value to beef, allowing the producers to receive premium prices for delivered quality at the industry. This value aggregation happened basically through the creation of specific alliance brand. For achieving this brand development level, several operations happened in each one of the productive chain links, characterizing the productive complementarities. In this sense, producers were committed to supply to the abattoir an animal with specific quality characteristics. The abattoir must accomplished the slaughter operation, the de-boning and processing according to established standards, and the retail must distribute the product to the final consumer in a differentiated way from the other products presented at their stores.

Specifically to the Angus producers in Brazil, the main advantage was the market access of a product with their own brand. The access form to the market by itself is a strong differential in the alliance, because in the traditional form of operation of the beef chain, most of the products don't have any identification, they are just commodities. The access to the international market was also a reason of expectation to the association, considering that the industrial partner had access to the external market.

The fact that the product had the brand and certification from the association was a warranty of quality recognition for the product in the future. In this case, the quality assurance provided by the alliance would be independent from the continuity of the current initiative, if the Association could continue to be the responsible entity for the certification process. The fact that the brand was also associated to the retailer (Zaffari in Brazil and Waitrose in the UK) was considered an advantage in both cases, because these retailers were associated to quality by the consumers.

Considering the obtained results it was possible to conclude that producers in Brazil and in the UK are optimistic in relation to the initiative, believing that it can bring increasing profitability for their business. It remains unknown however, if the benefits will happen in a way to satisfy the participants' expectations and that the earnings are fair for all the partners.

The fact that the Angus initiative in Brazil was recent in time (3 years, compared to the American Angus Beef Program, that has 27 years, for example), limits a broader analysis in relation to the noticed disadvantages. A better analysis would be worthy in future stages of the alliance development. It was concerning, however, the apparent lack of risks and disadvantages analysis, mainly in relation to the partners. This can mean a superior expectation than that which can be really reached. This exploratory study also suggests that these two alliances are on different stages of development, showing that as the alliance evolves, there are new gains and better reward to all members. While vertical alliances and quality assurance are relatively new and led by the private sector in Brazil, in the UK there is a strict monitoring of beef quality by the public sector and third-party certifying bodies. The British consumer is more aware of this differentiated product and, consequently, proneness to pay a reward price than the Brazilian consumer.

Vertical alliances refer to the mechanism of chain co-ordination while quality assurance scheme is the transmission of the strictness of the co-ordination to a label or a brand. The quality assurance label or brand was the value perceived by the end consumer.

In relation to the theoretical advances related to new advantages or disadvantages, we found a better evaluation, from the interviewees, of the advantages for the producers who join an alliance. In this sense, although the producers consider themselves the link that has smaller disadvantages, they foresee a positive future for alliances, both in Brazil and in the UK. The main advantage, in our view, was subliminal to the producers (either Brazilian or from the UK), but very clear to the researchers. Producers joining a strategic alliance in beef agri-chain are improving their market access throughout the establishment of quality assurance schemes and rigorous quality standards. This process improvement at farm level can increase their competitiveness and provide the aimed increase in profitability.

We expect to develop, in the future, further analysis of agri-food alliances, in beef specifically, seeking to enlarge the research base and to reduce the limitations of these two case research studies. It also seems necessary to develop a framework to consider the impact of local regulation, consumer habits and culture to the formation and development of the vertical alliance. Nevertheless, we consider the results quite illustrative of the beef chain alliances reality in general.

References

Bankuti, F.I. and C.P. Machado Filho, 1999. Novas alianças no sistema agroindustrial da carne bovina no Brasil. In: I Workshop de Gestão de Sistemas Agroalimentares, PENSA/FEA/USP, Ribeirão Preto.

Bliska, F.M.M., J.L. Parré and J.J.M Guilhoto, 1998. O programa de produção de carne qualificada de bovídeos no Estado de São Paulo sob o enfoque da Teoria dos Jogos. Informações Econômicas, 28, 2, Feb. São Paulo.

Cooper, M. and L. Elram, 1993. Characteristics of supply chain management and the implication for purchasing and logistics strategy. The Int. J. of Logistic Management 4, 412-434.

Degand, J., Pirotte, N. and B. Frahan, 1998. Supply strategies in the European off-season fruit sub-sector. In: International Conference of Chain Management in Agrobusiness and the Food Industry, III. Proceedings... Management Studies Group, Wageningen Pers, Wageningen.

Doz, Y. and G. Hammel, 2000. A vantagem das alianças: a arte de criar valor através das parcerias. Qualitymark, Rio de Janeiro.

Fearne, A., 1998. The evolution of partnerships in the meat supply chain: insights from the British beef industry. Int. J. of Supply Chain Management 3, 214-231.

Fearne, A. and S. Dedman, 1999. Supply chain partnerships for private label products: insights from the UK. In: Conference of the Food Distribution Research Society, Wye College, University of London.

Gulati, R., 1998. Alliances and networks. Strategic Management Journal 19, 293-317.

Hitt, M., D. Ireland and R. Hoskisson, 2001. Strategic management: competitiveness and globalization. South-Western Publishing, London:

Hobbs, J.E. and L.M. Young, 2000. Closer vertical co-ordination in agri-food supply chains: a conceptual framework and some preliminary evidence. Supply Chain Management 5, 131-141.

IEL, CNA and SEBRAE, 2000. Estudo sobre a eficiência econômica e competitividade da cadeia agroindustrial da pecuária de corte no Brasil. Brasília.

Jogen, W.M.F. and M.T.G. Meulemberg, 1998. Innovation of food production systems: product quality and consumer acceptance. Wageningen Pers, Wageningen.

Lambert, D.M.; M.A. Emmelhainz and J.T Gardner, 1996. Developing and implementing supply chain partnerships. The Int. J. of Logistics Management 7,1-17.

Loader, R.J. and J.E. Hobbs, 1999. Strategic responses to food safety legislation. Food Policy 24, 685-706.

Mattar, F., 1993. Pesquisa de Marketing. 1, 2. Atlas: São Paulo.

Noordewier, T.G., G. John and J.R. Nevin, 1990. Performance outcomes of purchasing arrangements in industrial buyer-vendor relationships. Journal of Marketing 54, 80-93.

Northern, J.R., 2000. Farm assurance schemes in the United Kingdom livestock sector: their use as quality signals. Unpublished PhD Thesis, The University of Reading, Department of Agricultural and Food Economics.

Oliver, C., 1990. Determinants of inter-organisational relationships: integration and future directions, Academy of Management Journal 15, 241-265.

Perosa, J.M., 1999. Papel da coordenação em alianças de mercado: análise de experiência no SAG carne bovina. In: II Workshop de Gestão de Sistemas Agroalimentares, PENSA/FEA/USP, Ribeirão Preto.

Pigatto, G., A.L. da Silva and A.M de Souza Filho, 1999. Alianças mercadológicas: a busca da coordenação na cadeia de gado de corte brasileira. In: II Workshop de Gestão de Sistemas Agroalimentares, PENSA/FEA/USP, Ribeirão Preto.

Reardon, T. and E.M.M.Q. Farina, 2001. The rise of private food quality and safety standards: illustrations from Brazil. Int. Food and Agrib. Management Review 4, 413-421.

Rocha, J.C.M., M.F. Neves and R.B. Lobo, 2001. Experiências com alianças verticais na coordenação da cadeia produtiva da carne bovina no Brasil. In: International Conference on Agri-Food Chain/Networks Economics and Management, 3, Ribeirão Preto, SP. Proceedings.... PENSA: Ribeirão Preto, 1-13.

Simchi-Levi, D., P. Kaminsky and E. Simchi-Levi, 2003. Cadeia de suprimentos: projeto e gestão. Bookman, Porto Alegre.

Simon, H., 1961. Comportamento administrativo. USAID, Rio de Janeiro.

Spriggs, J., J. Hobbs and A. Fearne, 2000. Beef producer attitudes to coordination and quality assurance in Canada and the UK. Int. Food and Agrib. Management Review 3, 95-109.

Stank, T. and M. Crum, 1999. Benefits of interfirm co-ordination in food industry supply chain. Journal of Business Logistics 20, 21-41.

Sterns, P.A., 2001.Quality and quality assurance in the fresh produce sector: a case study of European retailers. In: AAEA Annual Meeting, Chicago.

Tripodi, T., P. Felin and H. Meyer, 1975 Análise da pesquisa social. Petrópolis, RJ.

Williamson, O., 1985. The economic institutions of capitalism. The Free Press, New York.

Yoshino, M., H. Rangan and U. Srinivasa, 1996. Alianças estratégicas. 1, 3-29, Makron Books, São Paulo.

Zanqueto Filho, H., Pizzolato and A. Fearne, 2003. Os benefícios das parcerias e suas correlações com os recursos investidos: uma pesquisa amostral na indústria de frutas e vegetais do Reino Unido. In: Encontro Anual da Associação Nacional dos Programas de Pós-Graduação em Administração (ENANPAD), 28, Atibaia, SP. Proceedings...Atibaia: ANPAD, 40-41.

Ziggers, G.W and J. Trienekens, 1999. Quality assurance in food and agribusiness supply chains: developing successful partnerships. Int. J. of Prod. Economics 60-61, 271-279.

Zylberstajn, D., 2000. Conceitos gerais, evolução e apresentação do sistema agroindustrial. In: D. Zylbersztajn and M. Neves (eds), Economia e Gestão dos Negócios Agroalimentares, Pensa, Pioneira, São Paulo.

Vertical coordination, quality management and firm performance of the pork processing industry in China

J. Han, J.H. Trienekens, T. Tan, S.W.F. Omta and K. Wang

Abstract

With rapid economic development in China, food safety and quality is increasingly becoming an issue of concern to its consumers. This poses challenges to the organization of the pork supply chain. Pork processing companies are endeavoring to establish closer cooperative links with its suppliers and downstream customers in order to be competitive in the market. This study focuses on the exploration of the links between vertical coordination, quality management and firm performance within the framework of transaction cost analysis and quality management. Data collected from 140 pork processors in China will be used to study the causal relationships between these variables with a focus on the upper stream of the pork supply chain. Analysis leads to a number of provisional conclusions regarding the development of closer supply chain relationships between the suppliers and processors in the pork supply chain in China.

Keywords: vertical coordination, quality, performance, pork processing industry, China

1. Introduction

The organization of supply chains in agriculture is changing, requiring more customer orientation. Impersonal spot markets have evolved towards closer vertical co-ordination between the stages in the chain (Boger, 2001). As pork sector is concerned, the vertical coordination systems between livestock producers and processors have dramatically shifted toward long-term contract coordination or vertical integration (Lawrence and Hayenga, 2002). Among a number of driving forces for this shift, level of risk faced by agricultural producers, stringent quality requirements for processing, changes in technology have been the main ones (*e.g.* Fearne, 1999; Hobbs and Young, 2001). It has been indicated in their studies that consumer demand for food quality and diversity has been a pivotal factor in increasing vertical coordination in the food industry.

Many companies have invested heavily in quality management to gain competitiveness in the market place and retain consumers. Will quality investment contribute to firm performance? Empirical studies showed a mixed performance implications accrued from QM practices. Therefore one of the motivations of this research is to investigate the relationship between quality management and firm performance. We would like to find out whether quality improvement contributes to firm performance so that pork processors are willing to invest in quality improvements.

A variety of theoretical approaches contribute to the understanding of vertical coordination (Hobbs, 1996, 2001), for example, transaction cost economics (TCE), resource based view and agency theory. Although all are important, TCE attaches special emphasis to the asset specificity involved in trading the good or service in question. The basic question to be analyzed in TCE is what kind of governance structure provides an efficient framework for transactions with varying levels of asset specificity (Boger, 2001). Although TCE has been criticized for its difficulty in measuring, a lot of empirical studies have been conducted since the early 1980s to investigate either the influence of transaction costs themselves or the attributes of the transaction on organizational efficiency and found out the causal link between the degree of vertical coordination and the transaction cost attributes (*i.e.* Frank and Henderson, 1992; Fahlbeck, 1996). To our best knowledge, little empirical studies have been done to explore the relationship between the various transaction cost attributes and vertical coordination in the pork supply chain and the relationship between vertical coordination and firm performance of the pork processing firms in China. This is the other motivation for this research.

The article begins with a description of the industry profile. This serves as the background for this research. To understand the mechanism of the selection of vertical coordination modes, the coordination in the pork supply chain will also be described. As hog production and slaughtering and processing are the two most important stages in pork supply chain management, attentions on the coordination between the suppliers and processors will be focused on in the research. Then theoretical framework will be developed by addressing transaction cost economics as well as quality management and hypotheses will be drawn. Afterwards, the research method for the empirical study on Chinese pork processors will be described and analysis will be made on the basis of a survey conducted in two provinces in East China. Finally it is concluded with some implications for the development of more effective and efficient pork supply chain in China.

2. Research background

Fundamental changes have taken place in the Chinese pork industry since the government removed state procurement quotas and price control in 1985. Since 1990, China has become the largest pork production and consumption country in the world. The total output of pork production reached 51.19 million tons in 2003. Slaughtered heads reached 592 million in the year (National Bureau of Statistics, 2004). The pork consumption in China accounted for 50.8% of the total pork consumption in the world, with an average per capita pork consumption of 33 kg. Although pork consumption has the tendency of decreasing in the last 5 years, it is still the most popular meat for Chinese people. The outbreak of bird flu and swine streptococcus in Sichuan province, south west of China, made pork consumption rebounded in 2005. Despite of the development of processing sector, only 15% of the pork is processed into packaged cuts and chilled meat, with the rest sold as fresh and frozen meat in the retail market. Out of that 15%, less than 27% is cooked meat, including traditional ham, cured meat, Western flavoured ham, sausage and bacon, *etc.* (Pan and Kinsey, 2002).

After China's accession into the WTO, more pork imports can be expected. The entrance of more foreign companies and joint ventures are already foreseen in China. Competition will depend on meeting the quality and safety standards of both domestic and international consumers. In recent several years, leading meat-processing companies like Shineway and Yurun Co. Ltd. have invested heavily to develop cold chain to provide the consumers with brand products. These companies establish closer vertical coordination with their retailers through franchise and long-term contracts (Zhou and Dai, 2005). However, loose coordination still prevails in the current pork supply chain in China. And quality problems are still of great concern to Chinese consumers. The reasons are mainly due to:

(1) Large number of small household hog production: although commercial operations and specialized households are growing, they provide only about 20% hogs all over China. The remaining 80% of the hogs are provided by backyard farms (Pan and Kinsey, 2002). Lower quality and sanitation standards from these backyard productions prevent pork industry from gaining competitive advantage and entering into the world market. And it also induces high transaction

cost. Slaughtering plants or processors have to depend on the traders to collect hogs from the small farm households into certain lot size. The lack of monitoring on the small producers also produces safety problems. It is believed by many people that the chemical residues tend to be high in China.

(2) Short-term spot market transaction between buyers and producers: Previous empirical study has indicated that traditional spot market transaction was still the most popular market channel that farmers used in selling their hogs (Zhou and Dai, 2005). This brought uncertainty in quantity, quality and price to the processors on one hand, difficulty in making decisions on production and marketing to the producers on the other hand.

(3) Large numbers of illegal slaughterhouses: One of the main sources of safety and quality problem also originates from the large numbers of illegal slaughterhouses. There are altogether 3,728 meat-processing firms in China. Among them, only 200 are recognized as large and medium-size (Jiang *et al.*, 2005). To regulate pork slaughtering, Chinese government issued Regulation on Hog Designated Slaughtering on Jan. 1, 1998. Although the total capacity of designated plants is less than half of the total slaughtered hogs per year, overall utilization of the capacity can be as low as 15%. This is because commercial firms can't compete with the low costs of individual butchers (Pan and Kinsey, 2002). The existence of these unregistered individual butchers was one of the sources for quality problems. It was widely reported by the media that the occurrence of swine streptococcus in Sichuan province in April 2005 was from one of the butchers in the area.

With the development of supermarket chains and the advent of higher incomes in China, some modern pork companies have developing more close relationships with their upstream suppliers and downstream customers. They have invested to have their own pig production farms and retail stores. And contracts are more often applied in doing business. The close coordination of activities along the supply chain from breeding and genetics, to production, slaughter, processing, further processing and exporting was recognized as the success factor to the Danish pork industry (Hobbs, 2001). How does close vertical coordination contribute to the firm performance? Does the investment in quality management contribute to the firm performance of pork industry? To our best knowledge, no empirical study has been done to examine the critical factors of quality management and their casual linkages to the performance of pork processing industry in China. This article will try to answer these questions through an empirical survey in East China.

3. Theoretical framework and hypothesis

3.1 Vertical coordination and transaction cost

One of the most popular cited definitions of vertical coordination was provided by Mighell and Jones (1963): "*...includes all the ways of harmonizing the vertical stages of production and marketing. The market-price system, vertical integration, contracting and cooperation single or in combination are some of the alternative means of coordination*". Within this succinct definition is the notion that vertical coordination encompasses a continuum of possibilities, from open market spot transactions at the one end, through to full vertical integration at the other end and including strategic alliances, joint ventures, contracting, *etc.* (Hobbs and Young, 2001). Transaction cost economics (TCE) helps us to understand many of the recent changes in vertical coordination in the agro-food sector. While there have been many discussions and elaborations to the theory, we focus only on one of the important propositions made by Williamson (1975, 1985, 1991a, b) regarding the governance of transactions and the relationship between the governance structure and the two principal attributes of transactions, asset specificity and uncertainty. In TCE, the transaction is the unit of analysis. It focuses on *"transactions and the costs that attend completing transactions by one institutional mode rather than another"* (Williamson, 1975). The theory's central claim is that transactions will be handled in such a way as to minimize the costs involved in carrying them out (David and Han, 2004). Williamson (1991a , b) identifies three alternative forms of transaction governance: market, hybrid and hierarchy. TCE maintains that there are 'rational economic reasons' for choosing the means of governing transactions (Williamson, 1985). It is what Williamson (1991b) called the 'discriminating alignment hypothesis,' which holds that transactions, which differ in their attributes, are aligned with governance structures – *i.e.* market, hybrid, or hierarchy – in a discriminating (*i.e.* transaction-cost-economizing) way. Therefore the governance mode that minimizes transaction costs is the preferred option. In the Williamson version of TCE, asset specificity is the most important variable influencing transaction costs among the three principal attributes of transactions. The other two are uncertainty and frequency. As asset specificity increases, greater coordinated adaptation is warranted in order to coordinate productive resources effectively (Williamson, 1991a). The second important dimension of transactions is uncertainty. The effect of uncertainty on the choice of governance form, however, is conditional. When asset specificity is low, market governance should be preferred whatever the degree of uncertainty, since continuity matters little and new transaction arrangements can easily be arranged by both parties if necessary (Williamson, 1985). According to the theory, transaction costs are presumed to increase with an increase in asset specificity. When aspects of the transaction are uncertain, spot markets may result in higher information and monitoring costs, consequently closer forms of vertical coordination such as long-term contracts, strategic alliances or full integration are predicted, As asset specificity increases, however, we move to more formal vertical coordination alternatives.

Governance structure is not equivalent to vertical coordination. Coordination is just one of the functions of a governance structure. Safeguarding the investments in the transaction is actually the most important function of a governance structure (Williamson, 1991a, b). But from the above description on vertical coordination and transaction cost economics, we can find that vertical coordination and governance structure are similar in the way that they both encompass a continuum of possibilities, with spot market transaction at one end and vertical integration (hierarchy) at the other end. Between these two polar extremes, there are various hybrid forms of transaction, *i.e.* strategic alliances, joint ventures, and contracting. The control degree between the supplier and the buyer is from weak to strong in the continuum. Under this context, the two terms 'vertical coordination' and 'governance structure' that we use in this research are referring to three transaction modes, namely spot market, contracting and vertical integration.

The fundamental assumption underlying the TCE framework states that the alignment of governance mode does affect governance performance and serves to reduce transaction costs. Empirical work based on transaction cost analysis has produced modest evidence of the governance efficacy of hybrid governance (vertical coordination) in inter-firm relationships (Buvik, 2002). This study will explore the relationship between the vertical coordination and the firm performance of the pork supply chain.

3.2 Transaction cost economics and quality management

TCE offers one perspective on the relationship between market organization and product quality. One class of transaction costs are measurement, or information, costs (Hobbs, 1996). These include costs of searching for information about buyers or sellers in the market, inspecting goods prior purchase, and assigning a price. Markets may be organized to reduce measuring costs that are associated with assuring a closer correspondence between product value and price, or actions and rewards (Williamson, 1985). When hog quality attributes are difficult to measure the producer may engage in opportunistic behaviour to exploit private information by failing to perform as agreed, such as shirking or cutting corners on quality, also referred to as moral hazard. This is expected to lead to contracts with added security features to mitigate the hazard (Martinez and Zering, 2004). Contract advantages over spot market related to hog quality and consistency are reflected in two packer survey in the USA in 1996 and 2001 respectively (Martinez and Zering, 2004).

To cope with the competitive environment, many companies have applied quality management systems. The objective of quality management efforts should be focused on achieving customer satisfaction. The importance of quality and its associated benefits such as improvements in customer satisfaction and in the bottom-line have been well acknowledged (*e.g.* Hendricks and Singal, 1997). Madu *et al.* (1995) studied QM practices in Taiwan's manufacturing firms. They found a significant causal relationship between the quality dimensions (*i.e.* customer satisfaction, employment satisfaction, and employee service quality) and organizational performances. Mohrman *et al.* (1995) found that 83% of the surveyed companies had a "positive or very positive" experience with QM, and 79% planned to "increase or greatly increase" their QM initiatives in the next 3 years. Kuei and Madu (2001) note that the focus of quality-based paradigm has shifted from the traditional company-centred setting to complete supply chain system. A number of articles offer insights on the critical success factors in a traditional quality or a broader supply chain context. Saraph *et al.* (1989) reported that eight critical factors could be used for QM assessment, namely the role of the quality department, training, product/service design, supplier quality management, process management, quality data and reporting, and employee relations. However, a survey by A.T. Kearney of 100 British firms and a survey of the executives in US manufacturing and service firms revealed that only 20-30% believed TQM made them more competitive (Economist, 1992; Mathews and Katel, 1992). Dooyoung *et al.* (1998) also reported estimates of QM failure rates as high as 60-67%. These mixed findings put forward the necessity to study the QM practices-performance link in Chinese pork processing industry. Based on research done by Saraph *et al.* (1989) and Kaynak (2003), the following quality management practices will be used in this research: management leadership, supplier quality management, product/service design and process management.

3.3 Firm performance

In the literature of performance much attention has been devoted to three main streams: financial, organizational and strategic. Organizational theory offers three approaches to measuring organizational effectiveness or performance (Murphy *et al.*, 1996), namely goal-based approach, system approach and multiple constituency approach. After comparing different measures of performance, they suggested that multiple dimensions of performance should be considered where possible, including both financial and non-financial measures. Accounting-based indicators, with efficiency, sales growth rate and profitability (*e.g.* return on sales or on investments) are the indicators most commonly used (Murphy *et al.*, 1996). In addition, it is important to examine operational (non-financial) performance measures, such as product quality, customer satisfaction, new product introductions and market shares. The set of non-financial measures involves qualitative as well as quantitative indicators (Claro, 2004). By taking these into consideration, our research use both financial and non-financial indicators to measure performance. As it is difficult to get the exact market share and profitability, we use qualitative measures. The indicators we use to measure the performance of the pork processing firms are: growth rate, market share, profitability and productivity.

3.4 Conceptual model and hypothesis

Based on the above literature review, the following conceptual model (Figure 1) can be drawn. The model has five constructs: asset specificity, uncertainty, vertical coordination, quality management practices and firm performance. Firm performance will be measured by tangible and intangible variables. They are used as latent dependent (endogenous) variables. The other four variables will be used as latent-independent (exogenous) variables. For operationalization, each construct contains a set of indicators (See Appendix).

From this conceptual model, we can develop the related hypotheses in the next part. Although all three dimensions of transaction cost are expected to affect the choice of governance mode, asset specificity is the most important. Numerous empirical studies have investigated the effect of asset specificity on governance, and the hypothesis that asset specificity leads to vertical integration is largely supported (*e.g.* Mahoney, 1992; Rindfleisch and Heide, 1997). However, empirical studies have not been able to find a positive relationship between transaction frequency and vertical integration (*e.g.* Rindfleisch and Heide, 1997), and the effect of uncertainty on governance is mixed. We would like to let the empirical results speak for themselves regarding the foundations of the theory (David and Han, 2004). Therefore it is hypothesized:

H_1: There is a positive relationship between the degree of asset specificity and the degree of vertical coordination in the pork producer-processor transaction.

H_2: There is a positive relationship between the degree of uncertainty and the degree of vertical coordination in the pork producer-processor transaction.

With regard to the relationship between vertical coordination modes and pork firms' performance, empirical work has confirmed that investments in relation-specific assets are often correlated with superior performance (Parkhe, 1993; Dyer, 1996). One fundamental assumption underlying the transaction cost analysis perspective is that the alignment of governance forms does reduce transaction costs (Williamson, 1985). Therefore, it is hypothesized:

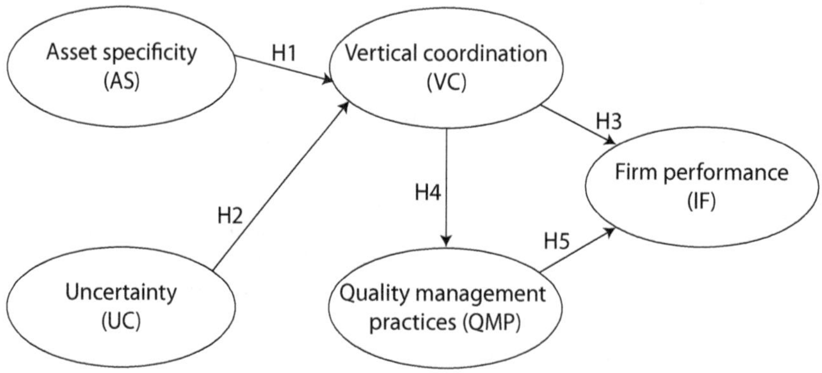

Figure 1. Conceptual model.

H_3: There is a positive relationship between the degree of vertical coordination and the performance of the pork processing firms.

The empirical study of Zhou and Dai (2005) showed contract was not commonly used in the current pork supply chain management. However large meat processing firms tended to use contracts more often to guarantee the quality of meat products and develop more high quality brand products to the market. In the U.S. pork industry, hogs sold through marketing contracts account for approximately 69% of total hog sales, compared with 11% in 1993. This suggests possible contract advantages related to changes in quality emphasis (Martinez and Zering, 2004). Therefore, it is hypothesized:

H_4: There is a positive relationship between the degree of vertical coordination and the quality management practices.

Kuei *et al.* (2002) reported that the organizational performance of enterprises could be differentiated by the following quality-related variables: processing management, supplier selection, and training on statistical methods. Tracey and Vonderembse (1998) also confirm that improved supplier performances on delivery time, shipping damage and inbound component quality have positive impacts on manufacturing performances. Their results indicate that the levels of QM practices and supplier involvement do relate to the organizational performance. However, as previously stated, paradoxes are shown on the relations among quality management practices, firm performance and customer satisfaction.

H_5: There is a positive relationship between quality management practices and the performance of pork processing firms.

4. Empirical assessment

4.1 Population and sample

As China is very large, we concentrated on two provinces in East China to make the survey applicable. They are Shandong province and Jiangsu province. Visits were paid to the government agencies with relevance to pork production, supply and distribution in order to get the name lists of pork slaughtering plants and processors. The effort in trying to get the list from the Provincial Administration of Industry and Commerce was proved to be futile as the files were not updated. In this situation, the list of the members joining into the meat association in the two provinces is used. It was learned from the Jiangsu Meat Association that the members were usually of large and medium size. Special attention was also given to the smaller ones in the actual survey due to the fact that majority of the processors are quite small in China.

4.2 Measures

To test the five hypotheses without the potential confounding effects of variation in the variables, we use two multi-item scales of measurements. One is 7-likert scale measuring system. The other is 5 point measuring system with specific questions that are easy to cause variation. For example, our mid-level managers inspect the work floor on a regular basis to check all operational processes. Different companies can interpret the regular check differently. Therefore we developed five items for the respondents to select: every day, twice one week, once every week, one half a month and once/more than one month. By doing so, we can reduce the variation to minimum.

The measurement items in this study are, to a great extent, developed from previous theories and empirical studies for the high degree of validity and reliability. For example, the quality management practices are mainly derived from Saraph *et al.* (1989), the firm performance mainly from Claro (2004). A survey questionnaire was designed in English with measurement items developed. It was translated into Chinese. To validate the equivalence of the translation, one person translated the questionnaire into Chinese and another back-translated into English. The original and back-translated versions were compared for conceptual equivalence and refined where necessary (Lai *et al.*, 2005). The questionnaire was sent to 10 academics from agricultural economics, meat sciences and animal husbandry at Nanjing Agricultural University for pre-test. Visits to 8 pork processing companies of different sizes in Jiangsu province were paid and panel discussions were made to the managers of these companies for the operation of the questionnaire items. Further refinement of the survey questionnaire was conducted on the basis of all the feedback from the academics and professionals. The revised questionnaires were then sent to 10 local pork designated slaughtering plants in Nanjing. Telephone calls were made by Nanjing Administration of Pork Designated Slaughtering Plants to remind them of doing the questionnaire. After about one month, all the 5 questionnaires returned and the item and scale reliability of the questionnaire was high in this pilot test. The finalized measurement items are provided in Appendix A.

4.3 Data collection

During pre-test period, it was very difficult to get the questionnaires done by the meat processing companies. The companies are still not used to do the mail questionnaire yet in current situation. After several trials, it was deemed impossible to send the questionnaires by mail to the pork processing firms, as the response rate would be too low. Experts from Nanjing Agricultural University also hold the same opinions. Therefore survey was carried out by students majoring in marketing, management and animal sciences during the winter and summer vacations in 2005. As China is a big country, only Jiangsu and Shandong provinces were selected to do the survey to make the study applicable. Four training sessions with each lasting for two hours were organized for the students. They were divided into small groups to improve the effectiveness of the training. A guideline on how to do the survey was distributed to the students. After explaining the research background and all the questions, students were asked to work in pairs to practice the roles as respondent and the interviewer. Two rounds of telephone contacts were made when the students were doing the survey, monitoring the process and answering questions. The survey lasted for one month for each time and the two surveys produced 140 valid questionnaires after checking the result. The 8 useless questionnaires were not completed by the processors.

4.4 Results

Structural equation modelling, a multivariate statistical technique, is one of the most applied and consolidated means of testing relations and causality in the field of buyer-supplier relationships (Malhorta *et al.*, 1999). The advantage of SEM over standard regression analysis (*i.e.* OLS) is its explicit consideration of the measurement error in the indicators and simultaneously estimation of a system of structural equations. Moreover, SEM is a powerful method for testing causal models, because it enables the simultaneous evaluation of the individual paths constituting the model, total effects and the complete model's goodness of fit (Hair *et al.*, 1998). However due to the potential problem brought by the moderate sample size, we directly regard each set of measurement indicators as latent variables in the analysis, that is to say the five constructs will be analyzed as component items.

Prior to evaluating the structural equation model, the validity of the measurement models was tested (Byrne, 1998). The scales in exploratory factor analysis were evaluated and refined by a confirmatory factor analysis before testing the full latent variable model (Gerbing and Anderson, 1988). After estimating the measurement model for all factors without constraining the covariance matrix of the factors, the SEM for the factors together with the measurement models was estimated. It was done by examining the standard errors, t-value and a number of goodness-of-fit statistics (Kaynak, 2003). AMOS 4.01 version was used to conduct structural equation modelling analysis. The Maximum likelihood was used to evaluate the parameters. Due to the absence of 5% data in the questionnaire, the Max likelihood is also used to adjust the missing data in model estimation. Therefore the goodness-of-fit index (GFI) and the adjusted GFI (AGFI) are unavailable in the result of model estimation. Table 1 shows the result of the confirmatory factor analysis, including factor loading and path coefficiency. The evaluation of the goodness-of-fit has several indicators. The common ones are CMIN/df, probability value, Normal Fit Index (NFI), Comparative Fit Index (CFI), Root Mean Square Error of Approximation (RMSEA) and Tucker-Lewis Index (TLI indicator).

A clear understanding of these fit indices will be useful for the explanation of the result of analysis. We follow Bagozzi and Yi (1988) on statistics of fit (except GFI), namely CFI, NFI, TLI and the RMSEA. Table 1 shows the result of the confirmatory factor analysis.

Table 1. Result of the confirmatory factor analysis.

Goodness-of-fit statistics	Indicator	Value	Result
Discrepancy	94.184		
P	0.034	<0.05	Significant
Discrepancy / df	1.327	<2	Significant
GFI	-	>0.8~0.9	
AGFI	-	>0.8~0.9	
TLI	0.993	>0.9	Significant
CFI	0.995	>0.9	Significant
NFI	0.981	>0.9	Significant
RMSEA	0.048	<0.05	Significant
Unstandardised estimation	Estimate	S.E	C.R. (t-value)
GS ← AS	0.54	0.856	1.631*
GS ← UC	0.94	1.248	1.843*
QM ← GS	0.305	0.052	5.868**
FP ← QM	0.446	0.145	3.065**
FP ← GS	0.037	0.556	0.578

** P<0.01; * P<0.05

These statistics suggested that the data fit the measurement model reasonably well (df=1.327; TLI=0.993; CFI=0.995; NFI=0.981; RMSEA=0.048). It indicated that all of the measurement items in the theoretical model also significantly loaded on the constructs on which they were loading. Table 2 shows the result of path verification and explains the proposed hypotheses:

From the table, we can see that positive relationships have been found in all the hypotheses except the relation between the degree of vertical coordination and firm performance of the Chinese pork processing firms. That is to say only H_3 is rejected.

5. Implications and discussion for pork supply chain management

This study represents one of the first empirical studies with regard to upstream pork supply chain management in China. It was aimed at investigating the relationships among transaction cost attributes, vertical coordination, quality management and firm performance of the pork supply chain. Based on literature review, a hypothesized structural model was developed and tested. The research result generates some significant theoretical and managerial implications. It is discussed in the next part.

The empirical assessments contribute to the transaction cost economics and quality management. Positive relationship has been found between the two transaction cost attributes and the degree of vertical coordination (H_1 and H_2). The two attributes under investigation were asset specificity and uncertainty. And the relationships between the degree of vertical coordination and quality management practices (H_4) and between quality management practices and firm performance (H_5) were analyzed. These findings supported the results of previous research done by Frank and Henderson (1992) and Boger (2001) with regard to the relations between transaction cost attributes and vertical coordination. As for the relation between quality management and firm performance, the research findings were consistent with many previous studies, *i.e.* Douglas and Judge (2001) and Kaynak (2003).

The confirmed four hypotheses in this research have significant implications on the processing firms in China in selecting appropriate modes of vertical coordination and implementing quality management practices. To improve quality of the product and reduce the uncertainty in hog supply, the companies should develop close relationships with their suppliers. In the survey, we

Table 2. Result of the proposed hypotheses.

Description	Path coefficiency	Result
H_1: the degree of asset specificity and the degree of vertical coordination	0.54	Accepted
H_2: the degree of uncertainty and the degree of vertical coordination	0.94	Accepted
H_3: the degree of vertical coordination and the performance of the pork processing firms	0.037	Rejected
H_4: the degree of vertical coordination and the quality management practices	0.305	Accepted
H_5: quality management practices and the performance of pork processing firms	0.446	Accepted

found more contracts were used by the processors with their suppliers. By doing so, they could reduce the costs of searching for adequate suppliers offering the most favourable prices and the availability of hogs with specific attributes such as specific breeds. Brand is one of the six kinds of asset specificity described by Williamson (1985). To safeguard brand, companies should use contracting and vertical integration as preferable governance modes than spot market. As mentioned earlier in this paper, majority of the hogs are produced by small household farmers in China. It is simply difficult to sign contracts with a large number of small farmers. Spot market is still the most common way of transaction used by many processing firms. It was found in the survey that hogs supplied through vertical integration only accounted for small part of the slaughtering capacity. One-third was regarded as heavily integrated in the supply management. This can explain the rejection of the hypotheses with regard to the relation between the degree of vertical coordination and firm performance.

The confirmed positive effects quality management practices on firm performance in this research is very encouraging for the practitioners. Firms that wish to improve their performance should invest in quality management. And quality management practices have been elaborated in this study, namely, management leadership, supplier quality management, product/service design and process management. Clear information has been provided to those companies who are endeavouring to implement total quality management. The research done by Kaynak (2003) indicated management leadership is directly related to training, employee relations, supplier quality management, and product design, and indirectly related to quality data and reporting, and process management. The various quality management practices were computed as component items in our study, further research is needed as what practices are directly related with management leadership and what indirectly related with management leadership.

6. Conclusions and further research

This study contributes to the literature of TCE and quality management by validating the relationships among the two transaction cost attributes (asset specificity and uncertainty), vertical coordination, quality management and firm performance for the pork processing industry in China. It is deemed valuable as not much empirical study has been done in the sector so far. The research result generates several important implications to the various stakeholders in the pork supply chain. Firms should forge strategic partnerships and develop closer coordination relationship with their suppliers. Investment in quality management is critical to improve custom satisfaction and bring benefit to the company.

This study focuses on the relationship between the upstream parts of the pork supply chain. Further empirical research should be conducted to gain more insight into the relationship between processors and retailers. Processing firms' strategic decision has a great impact on their performance. Although the causal relation has been found between quality management practices of the companies and their performance, further research is needed to find out what practices have direct relations with the firm performance and what have indirect relations. The result will be useful for the decision-making. Furthermore, we need to investigate the relationships among vertical coordination, quality management and firm performance in the downstream parts of the pork supply chain to give a clear picture of the pork supply chain management.

References

Bagozzi, R.P. and Y. Yi, 1988. On the evaluation of structural equation models. Journal of the Academy of Marketing Science 16, 74-94.

Boger, S., 2001. Agricultural Markets in Transition, an empirical study on contracts and transaction costs in the Polish hog sector. PhD thesis, Shaker Verlag, Aachen.

Buvik, A., 2002. Hybrid governance and governance performance in industrial purchasing relationships. Scand. J. Mgmt. 18, 567-587.

Byrne, B.M., 1998. Structural equation modeling with LISREL, PRELIS, and SIMPLIS: Basic concepts, applications, and programming. Lawrence Erlbaum, Mahwah, NJ.

Claro, D.P., 2004. Managing business networks and buyer-supplier relationships. PhD thesis, Wageningen University and Research Center, the Netherlands.

David, R.T. and S.K. Han, 2004. A systematic assessment of the empirical support for transaction cost economics. Strat. Mgmt. J. 25, 39-58.

Dooyoung, S., J.G. Kalinowski, and G. El-Enein, 1998. Critical implementation issues in total quality management. SAM Advanced Management Journal 63, 10-14.

Douglas, T.J. and W.Q. Judge, 2001. Total quality management implementation and competitive advantage: the role of structural control and exploration. Academy of Management Journal 44, 158-169.

Dyer, J.H., 1996. Specialized supplier networks as a source of competitive advantage: evidence from the auto industry. Strat. Mgmt. J. 17, 271-292.

Economist, 1992. April 18, pp. 67.

Fearne, A., 1999. Building partnerships in the meat supply chain, the case of the UK beef industry, Agricultural Economics Research Institute (LEI), the Hague, the Netherlands.

Fahlbeck, E., 1996. Transaction cost determinants in co-operatively organized slaughter business in Sweden. In: E. Fahlbeck (ed.). Essays in transaction cost economics. Uppsala: Swedish University of Agricultural Sciences.

Frank, S.D. and D.R. Henderson, 1992. Transaction costs as determinants of vertical coordination in the U.S. food industries. American J. of Agr. Eco. 74, 941-950.

Gerbing, D.W. and J.C. Anderson, 1988. An updated paradigm for scale development incorporating unidimensionality and its assessment. J. of Marketing Research 25, 186-192.

Hair, J.F., R. Anderson, R.L. Tatham and W.C. Black, 1998. Multivariate data analysis. Prentice Hall, Upper-Saddle River.

Hendricks, K.B. and V.R. Singhal, 1997. Does implementing an effective TQM program actually improve operating performance? Empirical evidence from firms that have won quality awards. Management Science 43, 1258-1274.

Hobbs, J.E., 1996. A transaction cost approach to supply chain management. Supply Chain Management 1, 15-27.

Hobbs, J.E., 2001. Against all odds, explaining the exporting success of the Danish pork cooperatives. Center for the Study of Cooperatives, University of Saskatchewan, Canada.

Hobbs, J.E. and L.M. Young, 2001. Vertical linkages in agri-food supply chains in Canada and the United States. Research and Analysis Directorate Strategic Policy Branch, Agriculture and Agri-Food Canada.

Jiang, Y.Y., Z.Q. Liu and S.W. Wang, 2005. The status quo of Chinese meat production and processing and development trends. Meat Research, 38-40.

Kaynak, H., 2003. The relationship between total quality management practices and their effects in firm performance. Journal of Operations Management 21, 405-435.

Kuei, C., C.N. Madu, C. Lin and W.S. Chow, 2002. Developing supply chain strategies based on the survey of supply chain quality and technology management. The International Journal of Quality & Reliability Management 19, 889-901.

Lai, K.H., T.C.E. Cheng and A.C.L. Yeung, 2005. Relationship stability and supplier commitment to quality. Int. J. Production Economics 96, 397-410.

Lawrence, J.D. and M.L. Hayenga, 2002. The U.S. pork and beef sectors: divergent organizational patterns, paradoxes and conflicts. In: Paradoxes in Food Chains and Networks. Proceedings of the 5th International Conference on Chain and Network Management in Agribusiness and the Food Industry, Noordwijk, 6-8 June 2002, J.H. Trienekens, S.W.F. Omta (eds.), Wageningen Academic Publishers, the Netherlands, pp. 512-521.
National Bureau of Statistics, 2004. China statistical yearbook 2004. China Statistics Press.
Madu, C.N., C. Kuei and C. Lin, 1995. A comparative analysis of quality practice in manufacturing firms in the U. S. and Taiwan. Decision Sciences 26, 621-635.
Mahoney, J.T., 1992. The choice of organizational form: vertical ownership versus other methods of vertical integration. Strategic Management Journal 13, 559-584.
Malhotra, N.K., M. Peterson and S.B. Kleiser, 1999. Marketing research: a state-of-the-art review and directions for the twenty-first century. Journal of the Academy of Marketing Science 27, 160-183.
Martinez, S.W. and K. Zering, 2004. Pork quality and the role of market organization, Agricultural Economics Report number 835, www.ers.usda.gov.
Mathews, J. and P. Katel, 1992. The cost of quality, Newsweek 120, 48-49.
Mighell, R. and L. Jones, 1963. Vertical coordination in agriculture. USDA, ERS, Agricultural Economics Report No. 19, February.
Mohrman, S.A., R.V. Tenkasi, E.E. Lawler III and G.G. Ledford Jr., 1995. Total quality management: practice and outcomes in the largest US firms. Employee Relations 17, 26-41.
Murphy, G.B., J.W. Trailer and R.C. Hill, 1996. Measuring performance in entrepreneurship research. Journal of Business Research 36, 15-23.
Pan, C. and J. Kinsey, 2002. The supply chain of pork: U.S. and China. Working paper 02-01, the Food Industry Center, University of Minnesota, USA.
Parkhe, A., 1993. Strategic alliance structuring: a game theoretic and transaction cost examination of interfirm cooperation. Academy of Management Journal 36, 794-829.
Rindfleisch, A. and J.B. Heide, 1997. Transaction cost analysis: past, present and future applications, Journal of Marketing 61, 30-54.
Saraph, J.V., P.G. Benson and R.G. Schroeder, 1989. An instrument for measuring the critical factors of quality management. Decision Sciences 20, 810-829.
Tracey, M.A., and M.A. Vonderembse, 1998. Building supply chains: A key to enhanced manufacturing performance. Proceedings of the Decision Science Institute, 1184-1186.
Williamson, O.E., 1975. Markets and hierarchies: analysis and antitrust implications, New York: Free Press.
Williamson, O. E., 1985. The Economic Institutions of Capitalism, Firms, Markets, Relational Contracting. The Free Press, London
Williamson, O.E., 1991a. Strategizing, economizing, and economic organization. Strategic Management Journal, Winter Special Issue 12, 75-94.
Williamson, O.E., 1991b. Comparative economic organization: the analysis of discrete structural alternatives. Administrative Science Quarterly 36, 269-296.
Zhou S.D. and Y.C. Dai, 2005. Selection of vertical coordination forms by the hog producers under the supply chain management context. China Rural Economy No. 6, 30-36.

Appendix: Summary of the measurement items

AS Asset Specificity; 1=strongly disagree to 7=strongly agree
1 If we switch to other products, we would lose a lot of investment in facilities and human resources
2 If we switch to new suppliers, we would lose a lot of investments in time and effort in establishing relationship with our former key suppliers.

UC Uncertainty; 1=strongly disagree to 7=strongly agree
1 The amount of hogs supplied by our most important suppliers is always stable.
2 The quality of hogs supplied by our most important suppliers is always stable.

VI Vertical coordination; 1=strongly disagree to 7=strongly agree
1 We trade with our most important suppliers on the basis of spot market.
2 We trade with our most important suppliers on the basis of contracts.
3 We invest to establish our own farms in order to ensure the quality of hog supply.

QMP Quality management practices; 1=strongly disagree to 7=strongly agree
1 The quality goals of our company are based on 5 years projecting.
2 Our mid-level management team is well trained in quality management practices.
3 Our employees are well trained in the implementation of quality management practices.
4 We pay our most important suppliers a premium for good quality pigs.
5 We provide our most important suppliers with feed and technology in order to get good quality hogs.
6 We emphasize quality of new product/service in relations to cost in new product development
7 We have clear product/service specifications and procedures for employees
8 Our mid-level managers inspect the work floor on a regular basis to check all operational processes.
9 We inspect the quality of hogs purchased from our most important suppliers on a regular basis.

FP Firm performance; 1=strongly disagree to 7=strongly agree
1 Over the last three years, the total sales volume of our company is growing faster than that of our most important competitors.
2 Over the last three years, the market share of our company is increasing faster than that of our most important competitors.
3 Compared with our most important competitors, we achieved higher profitability on pork production during the past three years.
4 Due to the shortage of hog supply, the production capacity of our firm can't be met.

Improving quality and ecological sustainability of natural resources in international supply chains: The role of market-based incentives

E. Kambewa, P. Ingenbleek, A. van Tilburg and I. van der Lans

Abstract

Demands for food quality and safety as well as concerns over the degradation of natural resources are increasing. On one hand, international organizations and agro-business chains promote the use of practices and procedures such as the HACCP (Hazard analysis and critical control points) to ensure food safety. On the other hand, both public and private policies try to collaborate to protect and conserve natural resources. They try to promote sustainable practices such as those outlined in the FAO Code of Conduct for Responsible Fisheries. There is little understanding, however, on how actors from the downstream part of supply chains can collaborate with those in the upstream, particularly in developing economies who face technological and socio-economic constraints to implement the required practices and procedures. This study explores the use of marketing approach – use of incentives for primary producers to implement practices that may improve quality and protect natural resources. To achieve this practices of the fresh Nile perch chain from Lake Victoria are examined. The incentives include access to special investments, price information, selection of transacting partner, and enforcement mechanisms. Data were collected through a conjoint design from 275 fishers who rated their preferences to contract to implement better practices. Results show that access to specific resources was most important followed by price information. Select a trading partner and enforcement were the third and last in importance respectively. These results imply that to promote quality and sustainability in the upstream, downstream actors and corporate businesses should give appropriate incentives.

Keywords: ecological sustainability, quality, market-based incentives, small-scale producers

1. Introduction: problem statement and goal

This study addresses the willingness of primary producers to implement practices that may improve quality and sustainability if given particular incentives. As such the study contributes to literature on quality management and protecting natural resources (*e.g.* Adede, 1979; Alam *et al.*, 2005; Taylor and Kane, 2005) that face implementation challenges among primary producers especially, in developing economies.

The rapidly changing consumer lifestyles, increased demands for food safety and quality (Leeflang and van Raaij, 1995; Angulo *et al.*, 2005; Buckley *et al.*, 2005) continue to drive major structural changes in the global marketing arena. While changing consumer demands affect a cross section of supply chains, those that depend on ecological products such as fisheries face additional challenges of declining production. On one hand, the agro-food sector is closely coordinating the food supply chains by implementing various codes of conduct for quality and safety (*e.g.* the HACCP, Eurrep-GAP) to meet safety and quality demands. On the other hand, international organizations develop other codes of conduct (*e.g.* FAO Code of Conduct for Responsible Fisheries; Convention of Biological Diversity) to conserve and protect natural resources (NR).

The implementation of these standards for sustainability of NR and product quality faces challenges in the upstream part of the supply chain. For example, the implementation of the codes of conduct for quality and safety (*e.g.* HACCP) require relatively huge financial and organizational resources beyond what most developing economies can afford (Henson *et al.*, 2000; Henson and Loader, 2001). The codes of conduct for conserving NR face enforcement and non-compliance problems. Some studies, for example in fisheries resources (*e.g.* Nielsen, 2003; Nielsen and Mathiesen, 2003) attribute non-compliance to a number of reasons such as 1) failure to ensure fair distribution of socio-economic benefits; 2) incompatibility with the socio-economic situation of the implementers; 3) failure to institute justice when non-compliance is detected; and 4) failure to involve implementers in the formulation process. As public concerns over the degradation of NR continue to grow, some corporate businesses try to address these issues in their corporate social responsibility programs (Maignan *et al.*, 2002). For example, protection of fish supply through promotion of responsible, environmentally appropriate, socially beneficial and economically viable fishing practices, was an important motivation for Unilever and the World Wildlife Fund to invest in the Marine Stewardship Council in 1997 (www.msc.org). At the moment these codes of conduct are however not widely implemented among primary producers who face institutional, technological and socio-economic constraints especially, in developing economies.

The goal of the present study is therefore to test, from a marketing perspective, the use of incentives for primary producers to implement practices that promote sustainability and quality. Using fisheries resources as a case study, better practices for sustainability refer to use of recommended fishing gears. Practices to improve quality refer to the use of quality management tools and practices such as ice to keep fish fresh, and good handling of fish to enhance texture. The paper is organised as follows. Immediately, we introduce the case study channel *i.e.* the fresh Lake Victoria Nile perch channel. Next, we outline our research framework followed by the research methodology. Results and discussions conclude the paper.

2. The Lake Victoria fresh Nile perch channel

Lake Victoria, the largest fresh water body in Africa is bordered by Kenya (6%), Uganda (45%) and Tanzania (49%) (LVFO, 1999). Nile perch (*Lates niloticus*), the case study fish, was introduced into the Lake in mid 1950s to improve the fishery's productivity and commercial value (Geheb and Crean, eds. 2000). The Nile perch production boomed about two decades later and triggered unprecedented socio-economic benefits through employment, food security and foreign exchange earning through export markets (see Abila and Jansen, 1997; Jansen, 1997; Geheb, 2002). About 80% of the fishers earned their primary income from fishing and the income was fairly and evenly distributed (Abila, 2000; Henson and Mitullah, 2003). In mid 1990s fish catch and species diversity rapidly declined. The decline was attributed among other factors to over-fishing, habitat degradation due to siltation, encroaching aquatic weeds, pollution and predation by Nile perch (LVFO, 1999). Following increased competition between export and domestic markets, domestic fish consumption dropped since mid 1990s. Subsequently, lakeshore areas, where most fishing and fish processing take place, are among the areas that registered highest levels of food insecurity, malnutrition and absolute poverty in Kenya (Geheb, 2002).

The European Union remains the main international outlet taking up about 80% of Nile perch export volume (see FAO Globefish, 2005, www.globefish.org). Previously, between 1997 and 1999 the Lake Victoria fresh and frozen fish products had been banned into EU market due to alleged contamination by salmonella, cholera outbreak and poisoning (Mwaura, 1998; Omondi, 2000).

In response, the channel restructured to implement the HACCP procedures to meet the EU food safety requirements. The primary sector, dominated by small-scale fishers and middlemen, has however, continued to face technological constraints. According to Henson and Mitullah, (2003) out of the 300 landing sites in Kenya only 30% have weighing shades, 9% have electricity, 5% have a fish store, 0.3% have a cold room, 20% have access to all-weather roads.

In summary, Lake Victoria Nile perch channel faces threats to ecological sustainability and increasing challenges to meet quality demands in international markets. Whereas regional co-operation through the Lake Victoria Fisheries Organization (see www.inweh.unu.edu/lvfo) admit that environmental pressure and over-fishing must be minimised, existing governance mechanisms often based on spot market transactions motivate actors to maximize their short term profitability and overexploit the fishery. Whereas the responsibility to protect the fishery remains in the hands of public institutions, cooperation with private sector is critical. For example, at primary stages actors are not only custodians of the fishery but also have private interests to fulfil. Furthermore, as the fish catch continues to decline, desperation for survival among the poor to meet livelihood needs increases (Odada *et al.*, 2004). For the little fish that they catch, lack of quality management tools means that that they lose fish to spoilage and hence limit the monetary income they obtain from fish sales. Hence the channel presents a suitable case to follow on behalf of other NR facing similar challenges.

3. Theoretical framework and approach

This study is founded on a number of theoretical concepts due to its nature and field –common pool resources. Transaction cost economics and agency theory (Williamson, 1985) enable us to understand the choice of use of contracts under high risks and uncertainties. Given the fisheries resources are unique in their own respect, we use property rights theory (Foss and Foss, 2005) and resources based theory (Hunt and Morgan, 1995) to understand the value that could be realised from protecting common pool resources in view of the risks associated with common resources. According to transaction costs economics and agency theory, contracts are incentives deployed to minimise opportunism (Williamson, 1985). In a principal –agent relationships deployment of specific investments creates hold up problems for the principal and conducive conditions for opportunism for the agents. In this case contracts are used to safeguard the investment (Williamson, 1985).

The resource based and property rights theories exemplify the importance of development, acquisition and deployment of unique resources from the value creation perspective (Srivastava, *et al.*, 2001; Frels *et al.*, 2003). Some studies (*e.g.* Ghosh and John, 1999; Rokkan, *et al.*, 2003) show that deployment of specific resources may not always be associated with opportunism. Instead, the value arising from the deployment of resources may benefit the channel. Applying these perspectives to common pool resources raises some fundamental questions. For example, would resource deployment in the upstream really enhance food safety and quality, and sustainability? In view of declining production would further resource deployment be warranted? And in view of the resources being common pool, would the producers be willing to implement the codes of conduct in view of the risks of opportunism from other resource users? This study establishes the incentives that may influence producers' willingness to improve sustainability and quality.

This brings us to our conceptual model explaining the producers' preference to contract to improve sustainability and quality. We offer them incentives in return. These incentives constitute the terms of contracts or contract attributes. These attributes are access to specific resources, market price information, selection of contract partner and sanctions for non-compliance (Figure 1).

3.1 Research approach

This study is undertaken under real life situation among fishers in Kenya. The research model was operationalised through a conjoint design (Hair *et al.*, 1998). The conjoint design enables us to examine how fishers may develop preferences for the incentives in return for sustainability and quality. Like previous studies (*e.g.* Wathne *et al.*, 2001; Van der Haar *et al.*, 2001), focus group discussions (Stewart and Shamdasani, 1990) involving 51 fishers were used to explore and solicit key decision factors to motivate fishers. Sustainability was defined from the perspective of using recommended fishing methods outlined in the legal framework (Kenya Government, 1991). Fish quality was defined from the use of good post harvest handling practices such as avoiding beating and throwing of fish, timely landing and icing of fish. In the design sustainability and quality were fixed because actors could not be asked to choose between sustainability or lack of it and between giving good or bad quality. Four decision factors with two level each were selected based on how they were favoured or divided participants during group discussions; their criticality for the intended contribution of the study, their actionability and communicability (Hair *et al.*, 1998) (see Table 1).

Access to technologies varied between acquiring recommended fishing gears through credit and quality management tools (*ice*). This was based on reports that the use of destructive fishing gears was increasing due to the fact that recommended fishing gears were more expensive than the bad gears. Furthermore, fishers who could not afford to buy gears (good or bad) either rent or obtain fishing gears through informal loans from buyers. Such loans are then tied to fish supply. Drawing from the theory of interlinked factor markets (Bardhan, 1980; Platteau and Abraham, 1987), fishers could be exploited. For quality fishers lose fish to spoilage due to lack of ice.

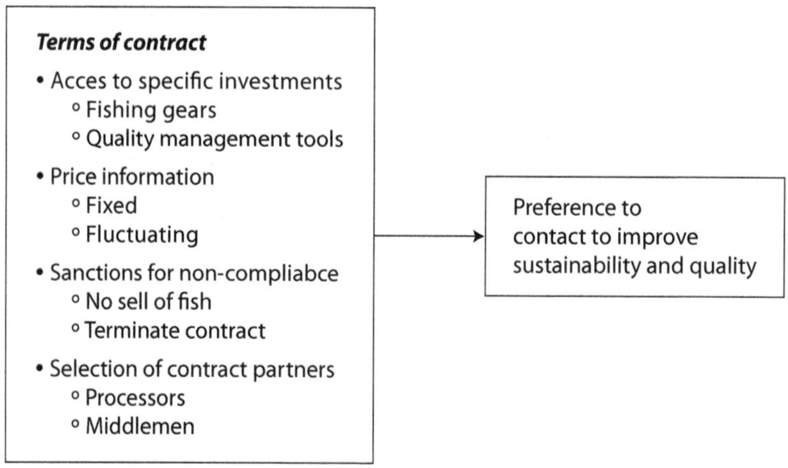

Figure 1. Research framework for producers' preference to contract to improve sustainability and quality.

Table1. Conjoint attributes and attribute levels.

Attribute	Attribute levels
Sustainability	Use of recommended fishing methods
Quality	Supply of fresh and non bruised fish
Access to technologies	1. Improved access to recommended fishing gears on credit 2. Improved access to quality management tools on credit
Price information	1. Fluctuating price (no information) 2. Fixed price (advance price information)
Choice of partners	1. Middlemen (small-scale buyers) 2. Processing factories (industrial buyers)
Sanctions for non-compliance	1. Withdrawing fishing gears if fish supplied with less than 1.5 kg 2. No payment for fish if supplied with fish less than 1.5 kg

Price information varied between fluctuating and stable price. Stable price implies that fishers get price information in advance and fluctuating price means no price information which is leading to unannounced and unpredictable price changes.

Contract partner varied between having contracts with small-scale middlemen and large scale processors. Fishers were not trading directly with processors at the time of the study. This was alleged to contribute to information asymmetries and price disparities to the benefit of middlemen. The only feasible option for fishers was thus to maintain status quo *i.e.* contract middlemen or contract processors.

Enforcement and sanctions mechanisms to enforce compliance with contractual obligations are often part of the terms of contract. Sanctions varied between withdrawing fishing gears and not paying for fish when under-size fish (juveniles less than 1.5 kg) are detected. Fish size followed the minimum sizes from the Fisheries Act (Kenya Government, 1991). We considered withdrawing gears a severe sanction since fishers would lose their fishing gears. Withholding payment was considered lenient as the fisher would only lose income for a particular consignment.

4. Sampling, data collection and validation

Data was collected from fishers that at least land Nile perch. Fishers were selected from 18 beaches. Beaches were selected from a list of landing sites obtained from the Kenya Marine and Fisheries Research Institute. Beaches were selected based on four criteria namely, 1) landing of Nile perch; 2) ease of access; 3) number of fishers landing Nile perch at the beach (at least 10); and 4) directly linked to middlemen who were buying and selling Nile perch to processing factories. While at the beach, there was no sampling frame. As a result we sampled fishers on the basis of opportunity or convenience (Schuman and Kalton, 1985; Poate and Daplyn, 1993) to get them, *i.e.* we positioned ourselves at the selected beaches waiting for and interviewing fishers as they landed their catch. A total of 275 fishers rated twelve conjoint tasks, four of which were holdout tasks (Hair *et al.*, 1998). The conjoint tasks were administered through personal interviews (Yin, 1984). A full-profile presentation method using a fractional-factorial design (Hair *et al.*, 1998) was used to estimate main effects only. Respondents were asked to consider the degradation of fisheries resources and Nile perch in particular, and quality losses they incur. Then they were being offered incentives to stop further degradation of fisheries resources and to improve the quality of their fish. Respondents rated contracts on a Likert scale from 1 to 5, where 1 was the least preferred and 5 the most preferred.

Conjoint scores were validated by calculating Kendall's tau (Hair *et al.*, 1998) for the holdout tasks. One questionnaire was dropped due to missing data. The remaining questionnaires had average Kendall's tau of 70.37% which shows high correlation between actual and predicted scores.

5. Results and discussion

Although we do not offer apriori hypotheses about relative importance of one factor over the other, our conjoint research design enables us to undertake such comparisons *ex post*. The aggregate factor importance (Figure 2) shows that access to technologies was the most important factor in influencing preference to contract followed by price information. The high importance attached to access to technologies over price information can be explained by their importance in the fishers' livelihoods. Although the nominal monetary income is dependent on the daily price, fishers' livelihood (welfare) is dependent on fishing with or without selling fish. It is not surprising therefore, that access to technologies was more important for contracting than price information. Whether or not a fisher has price information, it may not guarantee both food and money. Of course having price information is essential for fishers to plan their businesses and survival strategies.

In that respect price information only becomes an important factor after one had produced. However, given that price information is equally decisive in preference to contract means that nominal price may not always be the most important factor for sustainable economic development. Although fishers may not influence the price, having information is equally good enough to change their behaviour. The choice of contract partner was slightly more important than enforcement. The low importance of selecting contract partner can be explained by the fact

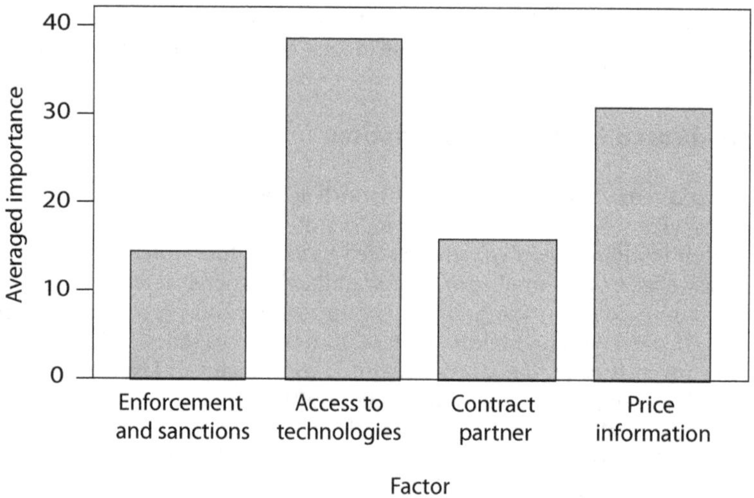

Figure 2. Conjoint factors importance.

it may not necessarily be a matter of who the partner is but rather what he offers like price or other relational factors that would matter most in deciding who to transact with.

The fact that enforcement was the least in importance means that given appropriate incentives fishers may not need any enforcement on their part because they are much more willing to improve sustainability of the fisheries and quality of their product. This is sensible in that fishers will gain from both fronts *i.e.* from long term sustainability in the interest of both welfare and economic gains and in the short run generate direct economic gains from say, improved quality and price information.

Table 2 shows partworth utilities that show how factor levels contributed to preference to contract. The table shows that fishers preferred having improved access to fishing gears compared to access to ice. Fishers also preferred having price information *i.e.* fish prices that are periodically negotiated and fixed over fluctuating prices. Although enforcement and sanctions were the least in importance fishers preferred to have no payment for all fish supplied with juveniles compared to termination of withdrawing fishing gears. On the choice of contract partners, fishers preferred to have direct transaction with industrial processors unlike the middlemen.

6. General discussion and conclusion

This study brings to light both conceptually and practically the incentives that may motivate primary producers to implement activities that promote sustainability and quality. While sustainability is not a substitute for quality, the impact of catch decline and lack of quality management tools on fishers' livelihoods is reflected in their differential importance on preference to contract. Like fishers argued, (during focus group discussions) acquiring ice to keep fish fresh without the ability to fish would be meaningless. But having fishing gears without ice is still productive; after all not all fish gets spoilt. Moreover, the fish that is rejected from export channel is still consumed in the domestic market. These findings underline the realities for improving quality to meet consumer demands and practical choices that small-scale producers may make. The latter should not be undermined. So quality management in fresh fish cannot start from the "boat" if production technologies are not secured first and foremost. Production technologies and quality management tools should be considered together but certainly not as a choice between them.

Table 2. Partworth utilities.

Factor	Average Importance	Factor levels	Utility
Access to technologies	38.67	1. Improved access to recommended fishing gears	0.5098
		2. Improved access to quality management tools	-0.5098
Price	30.91	1. Fluctuating price	-0.4179
		2. Stable price information	0.4179
Choice of partners	15.95	1. Agents	-0.0666
		2. Processing factories	0.0666
Enforcement and sanctions	14.47	1. Withdrawing fishing gears if fish less than 1.5 kg supplied	-0.0542
		2. No payment for all fish supplied with fish less than 1.5 kg	0.0542

Fishers' preference for lenient punishment might be seen as quite obvious. But considering that willingness to protect NR, especially those under common property rights, is more voluntary than mandatory, the importance of lenient measures may not be as obvious. Evidently arresting fishers and burning their gears to induce compliance has not always been effective in combating use of illegal gears in Lake Victoria fishery. If anything, it only creates more resentment among implementers, animosity between law enforcers and the implementers, and retaliatory behaviour not beneficial for the sustainability of NR. Moreover, as Nielsen (2003) notes, compliance with regulations is determined by the economic gains from breaking the law compared to the risks of being detected and punished. Importantly, actors weigh the costs and benefits of contributing to - and conforming with - the existing governance mechanisms (Nielsen, 2003; Nielsen and Mathiesen, 2003) to decide their next course of action. But as Grewal and Dharwadkar (2002) argues the more attractiveness the incentives are, the higher the likelihood that the actors would conform to the requirements of the institution offering the incentives. So better incentives for the responsible behaviour could, in our view, be more rewarding and self enforcing, than threats of severe sanctions which are anyhow not effectively monitored and enforced.

In conclusion, this study raises the possibility of using marketing-based incentives to promote sustainability of NR and product quality. We are of course mindful of the possibility that the consultative process used in the identification of the incentives (contracts attributes) and tested in this study might have contributed to the positive results. In fact it is for the same reason that we are optimistic that market-based incentives may not only be acceptable but their benefits could outweigh the often overemphasized transaction costs associated with contracts. The only critical issue though is identifying and giving the appropriate incentives. We therefore recommend that actor in downstream supply chains and corporate businesses should identify and give appropriate incentives to upstream partners in order for them to implement codes of conduct for ecological sustainability and product quality.

However the use of contracts (incentives) depends on the context in which they are applied (Cannon *et al.*, 2000). While this study has not addressed this aspect, we recognise its need as a subject of further research. Furthermore, as transactions are always bilateral, we also recognise the need for determining how and whether buyers at different stages of the channel may support the implementation of recommended practices at primary level. This is another topic for further research.

References

Abila, R.O., 2000. The development of the Lake Victoria fishery. A boon or bane for food security? Socio-economics of the Lake Victoria Fisheries. IUCN, Report No. 8.

Abila, R.O. and E.G. Jansen, 1997. From local to global markets. The fish exporting and fishmeal industries of Lake Victoria -Structure, strategies and socio-economic impact in Kenya. Socio-economics of the Lake Victoria fisheries. IUCN, Report No. 2.

Angulo, A.M., J.M. Gil and L. Tamburo, 2005. Food safety and consumers' willingness to pay for labelled beef in Spain. J. Food Products Marketing 11, 89-105.

Adede, A.O., 1979. Utilization of shared natural resources: towards a code of conduct. Environmental Policy and Law 5, 66-76.

Alam, S.M.N., C.K. Lin, A. Yakupitiyage, H. Demaine and M.J. Phillip, 2005. Compliance of Bangladesh shrimp culture with FAO code of conduct for responsible fisheries: A development challenge. Ocean & Coastal Management 48, 177–188.

Bardhan, P.K., 1980. Interlocking Factor Markets and Agrarian: A Review of issues. Oxford Economics Papers, New Series 32, 82-98.

Buckley, M., C. Cowan, M. McCarthy and C. O'Sullivan, 2005. The convenience consumer and food-related lifestyles in Great Britain. J. Food Products Marketing 11, 3-25.

Cannon, J.P., R.S. Achrol and G.T. Gundlach, 2000. Contracts, Norms, and Plural Form Governance. J. Academy of Marketing Science 28, 180-194.

Foss, K. and N.J. Foss, 2005. Resources and transaction costs: How property rights economics furthers the resource-based view. Strat Mgmt J. 26, 541–553.

Frels, J.K, T. Shervan and R.J. Srivastava, 2003. The integrated networks model: Explaining resource allocation in network markets. J. Marketing 67, 29–45.

Geheb, K. (ed.), 2002. Report of the LVFRP nutrition survey. LVFRP Technical Document No. 18. LVFRP/TECH/00/18.

Geheb, K.and K. Crean (eds.), 2000. The Co-management survey: Co-managerial perspectives for Lake Victoria's fisheries. LVFRP Technical document No.11. LVFRP/TECK/00/11.

Ghosh, M. and G. John, 1999. Governance value analysis and marketing strategy. J. Marketing 63, 131–45.

Grewal, R. and R. Dharwadkar, 2002. The role of the institutional environment in marketing channels. Journal of Marketing 66, 82-97.

Hair, J.F., R.E. Anderson, R.L. Tatham and W.C. Black, 1998. Multivariate data analysis. Fifth edition, Prentice Hall International Inc.

Henson, S. and R. Loader, 2001. Barriers to agricultural exports from developing countries: The role of sanitary and phytosanitary requirements. World Development 29, 85-102.

Henson, S. and W. Mitullah, 2003. Kenyan exports of Nile perch: Impact of food safety standards on an export-oriented supply chain. University of Guelph and University of Nairobi.

Henson, S., A.M. Brouder and W. Mitullah, 2000. Food safety requirements and food exports from developing countries: The case of fish exports from Kenya to the European Union. American Journal of Agriculture Economics 82, 1159–1169.

Hunt, S.D. and R.M. Morgan, 1995. The comparative advantage theory of competition: Dynamics, path dependencies, and evolutionary dimensions. J. Marketing 59, 1–15.

Jansen, E.G., 1997. Rich fisheries – poor fishfolk. Some preliminary observations about the effects of trade and aid in the Lake Victoria fisheries. Socio-economics of the Lake Victoria fisheries. IUCN, Report No. 1 September 1997.

Kenya Government, 1991, The laws of Kenya: The fisheries act. CAP 378 (Revised 1991), Government Printer, Nairobi, 90 pp.

Lake Victoria Fisheries Organization (LVFO), 1999. Strategic vision for Lake Victoria (1999-2015), Jinja, Uganda.

Leeflang, P.S.H. and W.F. van Raaij, 1995. The changing consumer in the European Union: A "meta-analysis". Intern. J. of Research in Marketing 12, 373-387.

Maignan, I., B. Hillebrand and D. McAlister, 2002.Managing socially-responsible buying: How to integrate non-economic criteria into the purchasing process. Europ. Management J. 20, 641–648.

Mwaura, J., 1998. EU bans Kenya Fish. Electronic Mail and Guardian, Johannesburg, South Africa.

Nielsen R.J., 2003. An analytical framework for studying: compliance and legitimacy in fisheries management. Marine Policy 27, 425-432.

Nielsen R.J. and C. Mathiesen, 2003. Important factors influencing rule compliance in fisheries lessons from Denmark. Marine Policy 27, 409-416.

Odada, E.O., D.O. Olago, K. Kulindwa, M. Ntiba and S. Wandiga, 2004. Mitigation of environmental problems in Lake Victoria, East Africa: Causal chain and policy options analyses. Ambio 33, 13-23.

Omondi, V., 2000. Will Kenya let go EU fish market again? The East African, The Nation Group.

Platteau, J.-P. and A. Abraham, 1987. An inguiry into quasi-credit contracts: The role of reciprocal credit and interlinked dear in small-scale fishing communities. J. of Dev. Sudies 23, 461-490.

Poate, C.D. and P.F. Daplyn, 1993. Data for agrarian development. Cambridge University. Cambridge University Press.

Rokkan, A.I., J.B. Heide and K.H. Wathne, 2003. Specific investments in marketing relationships: Expropriation and bonding effects. J. Marketing Research 40, 210–224.

Schuman, H. and G. Kalton, 1985. Survey methods. In: G. Lindzey and E. Aronson (eds.), Handbook of social psychology, Volume 1New York: Random House, pp. 635-697.

Srivastava, R.K., L. Fahey and H.K. Christensen, 2001. The resource-based view and marketing: The role of market-based assets in gaining competitive advantage. J. Management 27, 777–802.

Stewart, D.W. and P.N. Shamdasani, 1990. Focus group: Theory and practice. Applied social research methods series, Vol. 20, 153 pp.

Taylor, E. and K. Kane, 2004. Reducing the burden of HACCP on SMEs. Food Control 16, 833–839.

Van der Haar, J.W., R.G.M. Kemp and O.S.W.F. Omta, 2001. Creating value that cannot be copied. Industrial Marketing Management 30, 627-636.

Wathne, K.H., H. Biong and J.B. Heide, 2001. Choice of supplier in embedded markets: relationship and marketing program effects. J. Marketing 65, 54–66.

Williamson, O.E., 1985. The economic institutions of capitalism. The Free Press, MacMillian, Inc., New York.

Yin, R.K., 1994. Case study research design and methods. Second edition: Applied Social Research Methods Series, Vol. 5, Sage Publications.

Assessing and stimulating environmental pro-activeness in food supply chains

D.J. Haverkamp, H.J. Bremmers and S.W.F. Omta

Abstract

This paper investigates the factors that stimulate environmental pro-activeness of food companies. The goals are: (1) to assess the relative importance of factors that stimulate this and (2) to give advice on governmental policy change, to stimulate companies going further than mere compliance with public prescriptions. A distinction is made between two types of pro-activeness: *over-compliance* and *beyond-compliance*. The first is linked with the adoption of a process-oriented environmental management system (EMS) using end-of-pipe environmental technologies, whereas the latter is linked to a chain-oriented EMS which assesses *supply-chain and network relationships*. Data were gathered using survey questionnaires in 2002 and 2005 including, Dutch agri-food companies with at least 50 employees. Public stakeholders (*i.e.* government, environmental action groups, and local inhabitants) appeared to provoke over-compliance (P<0.01). The impact of both public and private stakeholders (buyers, suppliers, and competitors) is affecting beyond-compliance behavior (P<0.01). Furthermore the 2002 and 2005 data show that pro-activeness is influenced by *public-private correspondence* of environmental policies. In order to further stimulate environmental pro-activeness, we recommend the government to grant environmental permits for products instead of individual business locations. A second recommendation is to anticipate the limited individual resources of especially smaller companies, environmental information exchange (knowledge and experiences) between chain and network parties should be encouraged.

Keywords: environmental management system (EMS), environmental pro-activeness, environmental policy

1. Introduction and goal

The attention paid to safeguarding the natural environmental (called "the environment" hereafter) has been growing rapidly over the past few years. Because a considerable part of the environmental pollution is caused by industry, companies have experienced growing pressure from society and government to take care of their environmental impact. Ideally, companies would act pro-actively with respect to their environmental management, because a pro-active company *initiates* changes in its environmental policy on a voluntary basis rather than *reacts* to events, *e.g.* new legislative demands (Aragón-Correa, 1998: 557). In practice, however, some companies develop beyond legislative requirements, whereas others lag behind. How can this be explained? An answer to this question is important for both public policy makers and the companies themselves. Pro-active behavior is beneficial for the government, because it stresses self-responsibility and it reduces opportunism and agency costs. Three reasons can be mentioned why companies would act pro-actively including: customer preferences, anticipating/influencing (future) environmental legislation and because 'pollution prevention pays' (Reinhardt, 1999). In addition, a company might act pro-actively to meet environmental demands from especially the local society in order to maintain a social "license-to-produce" (Kagan and Gunningham *et al.*, 2003). The goal of this paper is to get insight in the determinants of environmental pro-activeness using data from the Dutch agri-food industry originating from 2002 and 2005.

So, environmental pro-activeness is expected to be influenced by different stakeholders. A distinction is made between private and public stakeholder groups. Private stakeholders such as buyers could, for instance, impose environmental requirements on a company (Holt, 2004), or consumers could prefer "green" product characteristics (Prakash, 2002). Examples of public stakeholders with interest in the environmental business practice are already mentioned, *e.g.* local inhabitants living in the neighborhood of the company. Also government is an important public stakeholder which imposes environmental standards on companies. Of special interest in this paper is the role of *correspondence* of public and private environmental policies. We define it as the extent to which the governmental environmental policy is perceived as effective and efficient. If correspondence is lacking, a pro-active investment in for instance an environmentally friendly production technology might turn out to be very costly, because it has to be adapted to new and unexpected (future) legislative standards (Walley and Whitehead, 1994; Rugman and Verbeke, 1998). Therefore, the degree of correspondence is expected to be an important determinant of environmental pro-activeness. The results of in-depth interviews with environmental coordinators indicated a lack of perceived correspondence for many companies in the Dutch agri-food industry (Haverkamp, 2005).

In the Netherlands, companies are confronted with a governmental environmental policy that is focused on the implementation of a *process-oriented* environmental management system (EMS)(Wätzold *et al.*, 2001). An EMS provides managerial support to companies to implement environmental measures in their organization (Netherwood, 2004). Process-oriented measures aim at the reduction of environmental pollution by using end-of-pipe technologies that are integrated in the production process at the level of the single firm or business unit. So, the Dutch government focuses on the optimalization of the environmental performance of isolated business units. A *chain-oriented* EMS aims at the reduction of environmental impacts beyond this level focusing at the supply chain and/or network perspective (De Bakker *et al.*, 2002). Because environmental impacts surpass the boundaries of the individual firm, a chain-oriented EMS is believed to be crucial to enhance a sustainable and environmental sound future production (Canning and Hanmer-Lloyd, 2001; Holt, 2004; Handfield *et al.*, 2005). Unfortunately, environmental measures including the promotion of the necessary supply chain and network policy and procedures are not stimulated by the government. We expect nevertheless that such a policy would be synergetic with existing business policy and practice in food supply chains in particular for reasons of food quality, hygiene and risk reduction. Moreover, the agri-food industry is well known for innovations that surpass the individual business unit in terms of transportation logistics and recycling of packaging material (Batterink *et al.*, 2005).

In the next section, the research model and hypotheses are addressed (Section 2). Then, the material and methods are presented (Section 3) as well as the results (Section 4). Finally, conclusions are drawn (Section 5) and public and private policy recommendations are sketched (Section 6).

2. Research model and hypotheses

The following clusters of variables are expected to influence the extent to which companies act pro-actively:

- Impact of public stakeholders.
- Impact of private stakeholders.
- Correspondence of governmental and company's environmental policy.

These elements are included in the research model (see Figure 1).

Environmental pro-activeness

The level of pro-activeness can be measured using the extent to which companies have implemented an EMS (Christmann, 2000; González-Benito and González-Benito, 2005). A distinction can be made between two categories of environmental pro-activeness: *over-compliance* and *beyond-compliance* (Prakash, 2001). The first implicates that a company does more than legally required using process-oriented environmental measures at the level of the individual business unit. Therefore, the degree of over-compliance shows a linkage with the adoption of a process-oriented EMS. The second (beyond-compliance) focuses at chain-oriented measures and policies assessing supply-chain and network relationships. It aims at cooperation with buyers and/or suppliers in order to reduce the environmental impact related to, for instance, the production of the raw materials and transportation activities between companies. It is therefore linked with the implementation of a chain-oriented EMS.

A process-oriented EMS consists of the stages in the Deming plan-do-check-act cycle, see for instance the ISO 14,001 guidelines (Netherwood, 2004). These stages can be referred to using the following 4 C's: commitment, compliance, control, and communication (Bremmers, 1995). Commitment is the inclusion in the organization of an environmental strategy and action program. Compliance covers the setting of environmental standards in concordance with preferably external norms. Control refers to the measurement of environmental impacts, registration of environmental data and (external) auditing of the environmental performance. Communication contains the feedback of the environmental performance to both internal and preferably also external stakeholders.

Chain-oriented environmental measures can be introduced in concordance with the stages in the Deming cycle (Rocha and Brezet, 1999; Hervani and Helms, 2005). Chain measures aim at reducing environmental impacts that are occurring outside the company. They include for instance collaboration with buyers and/or suppliers on environmental issues such as transportation logistics, product (re)design, and recycling of packaging material.

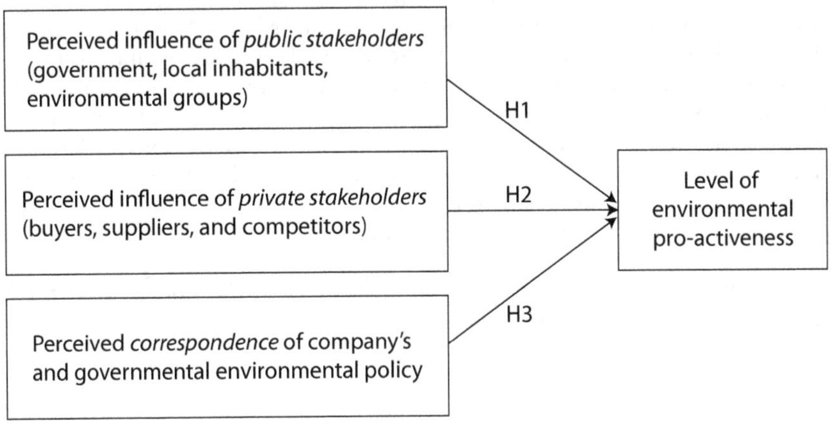

Figure 1. Research model and hypotheses.

Public stakeholders

The present paper focuses at the following three public environmental stakeholders: government, local inhabitants, and environmental action groups. We expect that they are important to enhance over-compliance, or the implementation of a process-oriented EMS. The importance of government is already mentioned in the introduction: they stimulate the implementation of process environmental measures in particular (Wätzold *et al.*, 2001). Local inhabitants often have interests in the environmental practice of companies because of smell and noise problems which are typical environmental issues to deal with in a process EMS (Bremmers *et al.*, 2006). Environmental action groups can have a severe negative impact on the company by bringing for instance negative environmental messages about the company's environmental practice in public (Van Tulder and Van der Zwart, 2003).

H1: Public stakeholders are important to enhance over-compliance behavior.

Private stakeholders

The following three private stakeholders are distinguished: buyers, suppliers, and competitors. We expect that they are important to stimulate beyond-compliance, or the adoption of a chain-oriented EMS. In the buyer-supplier relationship, companies can impose environmental claims on each other and/or opt for environmental collaboration. Companies can for instance apply environmental selection criteria on their suppliers (Holt, 2004). Companies might also choose for voluntary adaptation of their environmental strategies and activities in a cooperative way (Canning and Hanmer-Lloyd, 2001). This might for instance include working together on product (re)design (Handfield *et al.*, 2005). Pressure from competitors to pay attention to environmental management might include the introduction of environmental standards on an industry-level (Porter and Van der Linde, 1995).

H2: Private stakeholders are important to enhance beyond-compliance behavior.

Correspondence of public-private environmental policies

Environmental goals of government versus companies can be either conflicting or corresponding of a kind (Rugman and Verbeke, 1998; Clemens and Douglas, 2005). This correspondence includes the extent to which the governmental environmental standards are perceived as effective and efficient. Tension can for instance arise due to imposed environmental measurements that are costly, whereas companies have only a limited financial budget (Del Brío and Junquera, 2003). Adaptation to new and/or unexpected governmental standards could easily be perceived as costly only (Walley and Whitehead, 1994) and therefore hamper environmental pro-activeness (Rugman and Verbeke, 1998). As the Dutch government policy is biased towards process-oriented environmental measures, we expect that it is perceived as critical to companies that aim at the adoption of a chain-oriented EMS (*i.e.* beyond-compliance behavior).

H3: Correspondence of public-private environmental policies is more critical for beyond-compliance than over-compliance behavior.

3. Material and methods

3.1 Study sample

The data are obtained from the Dutch agri-food industry in 2002 and 2005 by means of a questionnaire survey. In order to increase the content validity, the questionnaires of both years were pre-tested with experts from business practice. If necessary, the questions were changed in order to increase a uniform interpretation of the used concepts. Only companies with more than 50 employees were included in this study. Their addresses and the number of employees were obtained from the Dutch Chambers of Commerce. In 2002, a questionnaire was sent to 355 companies of which 106 were returned. This equals a response rate of 29.9%. In 2005, 418 questionnaires were sent of which 100 questionnaires were returned, equaling a response rate of 23.9%. All the returned questionnaires could be used for analysis. In total, 23 companies were included in both years. In order to enable independent comparison between the sample of 2002 and 2005, they are included in the sample of 2002 (N=106) only and excluded from the sample of 2005 (N=77).

3.2 Constructs and reliability

Influence of public and private stakeholders

The influence of the different stakeholders on the environmental management of companies is measured using 5-point Likert scales (Friedman and Amoo, 1999) with answer categories ranging from "No impact at all" to "Very high impact". We have included the following private stakeholders: suppliers, buyers, and competitors. The reliability of this construct is satisfactory with a Cronbach alpha coefficient of 0.67 in 2002 and 0.60 in 2005. The following stakeholders are included as public stakeholders: government, environmental action groups/organizations, and local inhabitants. For 2005 only, the influence of the government on the corporate environmental policy was measured at the EU, national, and local level. Again, the reliability of the construct was satisfactory although relatively low in 2002: the Cronbach alpha coefficient equaled 0.56 in 2002 and 0.77 in 2005.

Correspondence

Correspondence is measured using 5-point Likert scales with answer categories ranging from "I completely agree" to "I completely disagree". The following question statements are included:

- "the environmental policy of the national government is perceived as effective";
- "the environmental policy of the lower governmental authorities is perceived as effective";
- "the environmental legislation contributes to the realization of our corporate environmental goals";
- "the lower governmental authorities are suitable for collaboration".

The reliability of this construct is satisfactory although relatively low in 2002: the Cronbach alpha coefficient was 0.56 in 2002 and 0.80 in 2005.

Environmental performance measurement

The measurement of process-oriented EMS-performance is based on the plan-do-check-act stages in the Deming-cycle. A number of different question statements (yes/no) were selected per stage to measure the environmental management performance at both levels. The statements referred to the availability of an environmental strategy and action plan, the implementation of various environmental measures, the collection of environmental information, and regular internal environmental reporting, see Table 1. So, for each company a process-oriented score is calculated for both 2002 and 2005. Each stage in the Deming-cycle is considered equally important. Also within each stage, the included questions are rated as equally important.

The chain-oriented performance is measured based on the implementation (yes/no) of environmental information collection:

1. on product (re)design;
2. for information exchange with buyers and suppliers;
3. for chain-oriented environmental care.

And the implementation of the following elements which were asked on a 5-point Likert scale (not at all – very much):

4. active collaboration with suppliers on environmental issues;
5. active collaboration with buyers on environmental issues.

Each element of the chain-oriented EMS is rated equally important. After rescaling, both the process- and chain-oriented EMS-performance are expressed on a continuous scale with five being the maximum score.

3.3 Data analyses

Significant differences between 2002 and 2005 are tested using the two independent samples (Anova t-test) as well as the longitudinal sample (Paired t-test). Spearman's rank correlation coefficients for non-parametric data are included in the correlation matrices. The data of the two

Table 1. Included question statements (yes/no) in the process-oriented EMS.

	Implementation of:
Plan	Formal environmental strategy
	Environmental action program
Do	Environmental audit
	Environmental education of and information for employees
	Environmental information system/database
Check	Information collection for internal environmental care
	Environmental information collection on emissions
	Regular check of environmental impact
Act	Information collection for internal environmental evaluation
	Regular internal environmental reporting

independent samples of 2002 and 2005 are combined in the regression analysis. In concordance with the research model (see Figure 1), the following regression model is applied.

$$EMS = \beta_0 * Intercept + \beta_1 * Log(Size) + \beta_2 * PublicStake + \beta_3 * PrivateStake + \beta_4 * Correspondence + \beta_5 * Year$$

Where:

EMS	= Environmental management performance at process- and chain-level, respectively
Intercept	= Intercept of the model
Log(Size)	= Log value of the number of employees
PublicStake	= Perceived influence of the public stakeholders
PrivateStake	= Perceived influence of the private stakeholders
Correspondence	= Perceived correspondence of environmental policy of government and the company
Year	= Dummy including the year of measurement (2002 and 2005)

The dummy parameter is used to measure possible influences of the year of measurement. Linear regression analysis (*Forced entry mode*) is performed using the statistical package of SPSS version 12.0.1.

4. Results

4.1 Size and sub-sectors

Table 2 includes the distribution of the samples over the different size-categories. From the table it can be noticed that both samples contain relatively many companies with 50-150 employees. They are even stronger represented in the longitudinal sample. In time, the average company size of the longitudinal sample has grown: fewer companies are included in the middle-size category in 2002 than in 2005.

Considering the sample distribution over the different branches, most of the companies belonged to the following sub-sectors: meat (2002: 23.6%, 2005: 18.2%), bakery (2002: 16.0%, 27.3%), and vegetables and fruit (2002: 10.4%, 2005: 13.0%). In the longitudinal sample, the following sub-sectors were well represented: meat (26.1%), bakery (21.7%), and beverage (13.0%).

Table 2. Distribution of the sample populations over the different size-categories (%).

Number of employees	Independent samples		Longitudinal sample	
	2002 (N=106)	2005 (N=77)	2002 (N=23)	2005 (N=23)
50-150	54.7	53.2	91.3	73.9
151-250	20.8	20.8	8.7	21.7
>250	24.5	26.0	-	4.3

4.2 Baseline statistics

The mean score on the process-oriented EMS has increased significantly from 2.90 ± 1.20 (± standard deviation) in 2002 to 3.26 ± 1.23 in 2005 (P<0.05), see Table 3. Over the same period, no significant differences are found in the chain-oriented EMS-performance. Its mean score is relatively low with 1.39 ± 1.01 in 2002 and 1.37 ± 1.00 in 2005.

From Table 3 it can be seen that the perceived influence of both public and private stakeholders has significantly increased from 2002 to 2005 (P<0.01). No significant differences are found in the perceived level of correspondence comparing 2002 to 2005.

Table 4 presents the mean scores for the longitudinal sample. For 2005, we compared the mean scores of the companies included in the longitudinal (see Table 4) with the independent sample (see Table 3). A significant difference was found for company size only, but of minor importance (P<0.10).

Within the longitudinal sample, an increase in the process-oriented EMS-performance is measured from 2002 to 2005 (P<0.10), see Table 4. In accordance with a previous notice (see Table 1), the average size of the companies in the longitudinal sample has grown significantly (P<0.01). Remarkably, no significant differences are found in the perceived influence of public and private stakeholders between 2002 and 2005. Also the perceived level of correspondence does not show a significant difference over this period.

Table 3. Mean scores given for the independent samples (Anova t-test).

	2002 (N=106)	2005 (N=77)
ProcessEMS	2.90 ± 1.20	3.26 ± 1.23*
ChainEMS	1.39 ± 1.01	1.37 ± 1.00
Log(Size)	2.18 ± 0.31	2.20 ± 0.35
PublicStake	3.08 ± 0.73	3.37 ± 0.67**
PrivateSta	2.21 ± 0.70	2.67 ± 0.77**
Correspondence	3.03 ± 0.52	3.08 ± 0.71

Significant differences compared to 2002 are indicated with: ** P<0.01, * P<0.05

Table 4. Mean scores given for the longitudinal sample (N=23; Paired t-test).

	2002	2005
ProcessEMS	2.81 ± 1.20	3.25 ± 1.36#
ChainEMS	1.29 ± 1.02	1.30 ± 1.08
Log(Size)	1.98 ± 0.15	2.07 ± 0.22**
PublicStake	3.12 ± 0.62	3.24 ± 0.69
PrivateStake	2.15 ± 0.72	2.39 ± 0.72
Correspondence	2.93 ± 0.56	2.85 ± 0.72

Significant differences compared to 2002 are indicated with: ** P<0.01, * P<0.05, # P<0.10

4.3 Correlation matrices

Table 5a and 5b represent the correlation matrix including the sample of 2002 and 2005, respectively. A relatively high correlation is found between a process- and product-oriented EMS in both years (P<0.01).

As can be seen in both tables, company size is positively correlated with a process EMS (P<0.01). Public stakeholders show a positive relationship with a process and product EMS in both years. In contrast to 2002, private stakeholders do not have a significant correlation with a process EMS in 2005.

Considering the perceived level of correspondence, significant results are found in 2005 only: it is significantly correlated with both a process (P<0.01) and product-oriented (P<0.05) EMS as well as with the impact of public (P<0.01) and private (P<0.10) stakeholders.

4.4 Regression results

The results of the regression analyses are included in Table 6. The explanatory power of the process (Adj. R^2=25.0%) and chain (Adj. R^2=24.2%) EMS are both relatively high. Table 6 shows that the process EMS-performance is significantly determined by company size (P<0.01) and the perceived influence of public stakeholders (P<0.01). Furthermore, correspondence appears to play a role, but of minor importance (P=0.10).

Table 5a. Spearman rank correlations in 2002 (N=106).

		1	2	3	4	5
1	ProcessEMS					
2	ProductEMS	0.44**				
3	Log(Size)	0.33**	0.06			
4	PublicStake	0.35**	0.32**	0.01		
5	PrivateStake	0.21*	0.49**	0.01	0.38**	
6	Correspondence	0.14	0.16	0.13	0.05	0.03

Significant correlations are indicated with: ** P<0.01, * P<0.05

Table 5b. Spearman rank correlations in 2005 (N=77).

		1	2	3	4	5
1	ProcessEMS					
2	ProductEMS	0.43**				
3	Log(Size)	0.46**	0.20			
4	PublicStake	0.27*	0.33**	0.17		
5	PrivateStake	0.10	0.31**	0.12	0.25*	
6	Correspondence	0.33**	0.28*	0.01	0.32**	0.21#

Significant correlations are indicated with: ** P<0.01, * P<0.05, # P<0.10

The implementation of a chain EMS is significantly effected by the impact of both public and private stakeholders (P<0.01). Also the year of measurement has a significant influence on the chain-oriented EMS-performance (P<0.05). This is probably related to the fact that some correlations changed in significance from 2002 to 2005, see Table 5a and 5b (*e.g.* between public-private correspondence and chain EMS). Company size and the perceived level of correspondence play a role, but of minor importance (P=0.10).

5. Conclusions and discussion

In this sub-section we refer to the hypotheses (see Section 2). A distinction was made between two types of pro-activeness: over-compliance and beyond-compliance. In concordance with the first hypothesis (H1), the influence of public stakeholders groups appeared to be important to enhance over-compliance behavior (*i.e.* the adoption of a process-oriented EMS). This implicates that companies that perceive higher levels of impact of these groups (*i.e.* government, environmental action groups, and local inhabitants) are indeed more sensitive to adopt a process EMS (Henriques and Sadorsky, 1999). Furthermore, company size plays a significant role in this (P<0.01). Many other studies show also that larger compared to smaller companies have a better EMS-performance (Aragón-Correa, 1998; Halkos and Evangelinos, 2002). In comparison with the smaller companies, they have more resources (money and knowledge) available to invest in environmental measures (Del Brío and Junquera, 2003). Moreover, they frequently attract more attention from external stakeholders such as environmental action groups (Henriques and Sadorsky, 1999).

In concordance with our second hypothesis (H2), private stakeholders did show to enhance beyond-compliance environmental behavior (*i.e.* the implementation of a chain-oriented EMS). However, the influence of public stakeholders was significant too (P<0.01), as well as the perceived level of correspondence of public and private policies (P<0.10). This implicates that in order to stimulate a chain-oriented EMS, adaptation of environmental policies of both public and private stakeholders is important. Cooperative "green" initiatives of supply-chain and network actors could for instance be stimulated by government, rather than focusing on individual companies only (Madsen and Ulhøi, 2001; Buysse and Verbeke, 2003).

Table 6. Regression analysis of the combined sample of 2002 and 2005 (N=177).

	Process-oriented EMS R^2=27.2%; Adj. R^2=25.0%; P<0.01 Standardized coefficients			Chain-oriented EMS R^2=26.4%; Adj. R^2=24.2%; P<0.01 Standardized coefficients		
	Beta	Std. Error	Sig.	Beta	Std. Error	Sig.
Intercept		0.703	0.00		0.585	0.00
Log(Size)	0.355	0.244	0.00	0.124	0.203	0.06
PublicStake	0.291	0.123	0.00	0.242	0.102	0.00
PrivateStake	0.032	0.116	0.66	0.326	0.096	0.00
Correspondence	0.109	0.132	0.10	0.121	0.110	0.07
Year	0.047	0.169	0.50	-0.171	0.141	0.02

Our third hypothesis (H3) is not supported by the results. It proposes that public-private correspondence is more critical for beyond-compliance than for over-compliance. But, such a difference was not found in the analysis. Interestingly, the correlation matrices did indicate that the importance of correspondence has increased from 2002 to 2005 (see Table 5a and 5b). In 2005 only, it showed a positive and significant correlation with both a process-oriented ($P<0.01$) and product ($P<0.05$) EMS. How can this be explained? Our answer on this question is speculative. During interviews with different environmental coordinators in the Dutch agri-food industry, many informants indicated that the government is not sufficiently anticipating (future) EU-legislation (Haverkamp, 2005). Over the past few years, the integration of EU-standards in national environmental policies has increased considerably (Knill and Lenschow, 2005). We observed that some large companies in particular are more confident about the eventual effects of these processes on their business practice and how to anticipate them (compare: Buysse and Verbeke, 2003). Therefore, probably a big group of companies will remain that is relatively uncertain about (future) environmental legislation and thus likely to perceive relatively low levels of public-private correspondence.

6. Public and private policy recommendations

This section focuses on how to enhance environmental pro-activeness through stimulating the implementation of a process and a product-oriented EMS. Two propositions can be made:

1. Enhancing chain and network relationships in environmental management, via granting *environmental permits for products* rather than individual business units, and
2. Anticipating the effect of company size via stimulating *environmental cooperatives* in chains and networks.

Ad 1.

As indicated in the introduction, the Dutch government focuses strongly on individual companies. In line with this focus, environmental permits are granted to individual business units. This policy strategy obviously ignores the potential contribution of chain and network relationships of companies in greening their business. One solution to encourage environmental pro-active behavior would be that government links environmental permits to products rather than single business units (Haverkamp *et al.*, 2005). This will contribute to especially chain-oriented environmental targets of beyond-compliance companies (*e.g.* Sharma, 2001). In addition, it will stimulate individual companies to impose environmental requirements on other companies upwards in the supply-chain. This is called the "green multiplier effect", meaning that companies forward environmental requirements that are imposed on them to their suppliers (Holt, 2004: 79). It will enhance the influence of private stakeholders on company's environmental management, which will consequently contribute to the self-responsibility of the agri-food industry to develop into a more sustainable and environmentally friendly direction.

Ad 2.

Our results shows, that size is an important determinant for both over- and beyond-compliance. In general, smaller companies have limited resources (knowledge and money) and therefore they try to limit their environmental management to compliance with legislation only (Del Brío and Junquera, 2003; Hillary, 2004). To overcome limitations of company size, we recommend companies to search for environmental information exchange with supply chain and network partners to share knowledge and experiences on environmental management. This will enrich the available environmental knowledge in the organization and enhance efficient and effective adoption of an EMS and stimulate environmental pro-activeness (Cramer, 1998; Verheul, 1999). In order to stimulate the dissemination of environmental information in the agri-food industry, branch-organization can play an important role. They are advised to take responsibility in for instance organizing discussion platforms on environmental management issues on a regular basis and to assist companies with questions on their environmental management directly, but also by bringing them into contact with relevant other parties (governmental bodies, colleague companies, consultancies, research institutes, *etc.*). We belief that this kind of (active) support is important to provide companies the necessary insight in alternatives to act beyond mere compliance with environmental legislation.

References

Aragón-Correa, J.A., 1998. Strategic proactivity and firm approach to the natural environment. Academy of Management Journal 41, 556-567.

Batterink, M.H., S.W.F. Omta and E.F.M. Wubben, 2005. Managing postharvest innovation: a review of developments and trends. Stewart Postharvest Review 1(2).

Bremmers, H.J., 1995. Milieuschade en financieel verslag. Deventer, Kluwer.

Bremmers, H.J., S.W.F. Omta, R. Kemp and D.J. Haverkamp, 2006. Do stakeholder groups influence environmental management system development in the Dutch agri-food sector? Business Strategy and the Environment (in press).

Buysse, K. and A. Verbeke, 2003. Proactive environmental strategies: a stakeholder management perspective. Strategic Management Journal 24, 453-470.

Canning, L. and S. Hanmer-Lloyd, 2001. Managing the Environmental Adaptation Process in Supplier-Customer Relationships. Business Strategy and the Environment 10, 225-237.

Christmann, P., 2000. Effects of "best-practice" of environmental management on cost advantage: the role of complementary assets. Academy of Management Journal 43, 663-680.

Clemens, B.W. and T.J. Douglas, 2005. Understanding strategic responses to institutional pressures. Journal of Business Research 58, 1205-1213.

Cramer, J., 1998. Environmental management: from 'fit' to 'stretch'. Business Strategy and the Environment 7, 162-172.

De Bakker, F.G.A., O.A.M. Fisscher and A.J.P. Brack, 2002. Organizing product-oriented environmental management from a firm's perspective. Journal of Cleaner Production 10, 455-464.

Del Brío, J.A. and B. Junquera, 2003. A Review of the Literature on Environmental Innovation Management in SMEs: implications for public policies. Technovation 23, 939-948.

Friedman, H.H. and T. Amoo, 1999. Rating the rating scales. Journal of Marketing Management 9, 114-123.

González-Benito, J. and O. González-Benito, 2005. Environmental proactivity and business performance: an empirical analysis. Omega 33, 1-15.

Halkos, G.E. and K.I. Evangelinos, 2002. Determinants of environmental management systems standards implementation: evidence from Greek industry. Business Strategy and the Environment 11, 360-375.

Handfield, R., R. Sroufe and S. Walton, 2005. Integrating environmental management and supply chain strategies. Business Strategy and the Environment 14, 1-19.

Haverkamp, D.J., 2005. Environmental care in the Dutch agri-food industry (in Dutch). Wageningen, Wageningen University.

Haverkamp, D.J., H.J. Bremmers and M. De Rooij, 2005. Milieuzorg geketend of milieuzorg in ketens? (in Dutch). Bestuurskunde 14, 29-38.

Henriques, I. and P. Sadorsky, 1999. The relationship between environmental commitment and managerial perceptions of stakeholder importance. Academy of Management Journal 42, 87-99.

Hervani, A.A. and M.M. Helms, 2005. Performance measurement for green supply chain management. Benchmarking: an international journal 12, 330-353.

Hillary, R., 2004. Environmental management systems and the smaller enterprise. Journal of Cleaner Production 12, 561-569.

Holt, D., 2004. Managing the interface between suppliers and organizations for environmental responsibility - an exploration of current practices in the UK. Business Strategy and the Environment 11, 71-84.

Kagan, R.A., N. Gunningham and D. Thornton, 2003. Explaining corporate environmental performance: how does regulation matter? Law & Society Review 37, 51-89.

Knill, C. and A. Lenschow, 2005. Compliance, communications and competition: patterns of EU environmental policy making and their impact on policy convergence. European Environment 15, 114-128.

Madsen, H. and J.P. Ulhøi, 2001. Integrating Environmental and Stakeholder Management. Business Strategy and the Environment 10, 77-88.

Netherwood, A., 2004. Environmental management systems. Corporate environmental management: systems and strategies. R. Welford. London, EarthScan.

Porter, M.E. and C. Van der Linde, 1995. Green and competitive. Harvard Business Review 73, 120-134.

Prakash, A., 2001. Why do firms adopt "beyond-compliance" environmental policies? Business Strategy and the Environment 10, 286-299.

Prakash, A., 2002. Green marketing, public policy and managerial strategies. Business Strategy and the Environment 11, 285-297.

Reinhardt, F., 1999. Market failure and the environmental policies of firms: economic rationales for "beyond compliance" behavior. Journal of Industrial Ecology 3, 9-21.

Rocha, C. and H. Brezet (1999). Product-oriented environmental management systems: a case study. The Journal of Sustainable Product Design(July): 30-42.

Rugman, A.M. and A. Verbeke, 1998. Corporate strategies and environmental regulations: an organizing framework. Strategic Management Journal 19, 363-375.

Sharma, S., 2001. Different strokes: regulatory styles and environmental strategy in the North-American oil and gas industry. Business Strategy and the Environment 10, 344-364.

Van Tulder, R. and A. Van der Zwart 2003. Reputaties op het spel; maatschappelijk verantwoord ondernemen in een onderhandelingssamenleving (in Dutch). Utrecht, Uitgeverij Het Spectrum BV.

Verheul, H., 1999. How social networks influence the dissemination of cleaner technologies to SMEs. Journal of Cleaner Production 7. 213-219.

Walley, N. and B. Whitehead, 1994. It's not easy being green. Harvard Business Review May-June 1994, 46-52.

Wätzold, F., A. Bültmann, M. Eames, K. Lulofs and S. Schucht, 2001. EMAS and regulatory relief in Europe: lessons from national experience. European Environment 11, 37-48.

Part 8. Case studies

Mapping and quantification of the milk chain in São Paulo State, Brazil

M.F. Neves, M.A. Consoli, F.F. Lopes, E.M. Campos and H.A.B. Paiva

Abstract

The paper presents the results of the "Mapping and Quantification of the Milk Chain in Brazil", a PENSA project, which was financed by the São Paulo Milk Sector Council, to analyze the milk chain, focusing on competitiveness of São Paulo State. The research emphasizes the importance of this chain, as one of the biggest chains in Brazil, in terms of turnover, number of agents, employment, taxes, and social relevance. In 2004 the whole chain had a turnover of USD 22.6 billion, producing 23.5 billion litres of milk, the equivalent of USD 4.2 billion in only milk production. The study analyzes the main sectors within the milk chain and the main destinies of the milk production, such as cheese, fluid milk, powdered milk, yogurts, and farm self-consumption. The study also presents a SWOT analysis of the chain and highlights the fact that São Paulo State is the biggest milk consumer in Brazil, commercializing more than 7.6 billion litres, but producing only 1.7 billion litres. Moreover, despite the importance of the chain, São Paulo has decreased production and increased purchases from other states due to coordination problems, lack of price polices, dispersion of producers, taxes, cooperatives, and other factors. This research was the first step of a bigger study that will provide the basis for the "strategic planning of the milk chain", which will be take place throughout 2006 to propose and implement collective action to improve the competitiveness of the milk chain in São Paulo State, Brazil.

Keywords: agro-industry systems, dairy products, competitiveness, chain coordination

1. Introduction and objectives

The milk sector is an important economic segment in Brazil because it generates employment, wealth and taxes, and it plays a social role by keeping small producers on the farms. Dairy production also fuels various segments of the food industry in the form of fluid milk, cheese, yogurt, chocolate, ice cream, and cookies, among others. Furthermore, dairy farmers, cooperatives and the milk and derivates' processing industry are important consumers of agricultural and industrial inputs, such as veterinary products, genetic material, research, animal feed, breeding, milking equipment, seed, fertilizer, and agrochemical products (from the agricultural input sector). The milk processing companies also buys packaging materials (plastic, paper and glass), aluminium lamina, sugar, industrial equipment, and electric energy (from the industrial inputs sector).

The milk chain is one of the biggest Brazilian chains, with a turnover of about USD 22.63 billion in 2004 and with a milk production of 23.5 billion litres, representing a total of USD 4.25 billion (see Appendix 1). The sector expected to produce about 25 billion litres in 2005 and increase production by 6% a year in the coming years. Brazilian milk production represents about 5% of the world's production, making the country the sixth major producer. Brazil is the second-largest producer in the Americas (just behind the USA), with 16% of the continents' production.

São Paulo State is the biggest milk consumer in Brazil, commercializing more than 7.6 billion litres, but it produces only 1.7 billion litres, thus creating a shortage of products that need to be bought from other states in Brazil. Moreover, São Paulo has decreased milk production and increased transfers from other states in the last decade. These movements and tendencies have made the agents and producers in São Paulo worry about the future and the competitiveness of the state's production.

As the state with the largest economy and population, São Paulo is the biggest consumer of dairy products and milk derivatives, but in the last years the economic geography has changed and the milk production has moved to other states due to production costs, taxes and competition with other agro-chains (mainly sugar cane and citrus, which cover an increasing area of cultivated land in the state).

The results of these changes can be analyzed in terms of São Paulo's share and production compared with the country and other states. The historical data show that from 1995 to 2004 Brazil increased its total production by 42.5%, at the same time that São Paulo decreased its production by 12.1%, thus declining from 12% to 7.4% of the Brazilian total production in this period. Meanwhile, other states increased their production, including Minas Gerais – by 45.5%, Rio Grande do Sul by 45.6%, Paraná by 44%, and Goiás by 81.4%. Together, these states increased their share of total Brazilian production from 57.5% to 61%.

The objective of this paper is to analyze the competitiveness of the milk chain in São Paulo State compared to other producer states in the country and to present the mapping and the quantification of the milk chain in Brazil with some details concerning São Paulo. As secondary objectives, the study research seeks to identify the main sectors in the milk chain, to identify and discuss problems and opportunities for small and big producers in São Paulo State, and the main problems of coordination among the producers, the industry, and the distribution agents present in the system.

2. Literature review

Two different approaches, developed in different places and time, show some similarities in the chain's structures. The first, called the commodity system approach (CSA), was developed by Goldberg (1968) in the USA in a study of, the citrus, wheat, and soy production systems. The CSA approach, beyond analyzing the traditional buyer-seller relationship, analyzes the institutional influences, concluding that the final destiny of agricultural products was not the final consumer but the agricultural industry. This influences the analysis of the sub-system that composes the agro-industrial system.

The theoretical basis of the CSA is derived from the neoclassic theory of production and the Leontief input-output matrix. This approach was the basis for the introduction of inter-sector dependence and also expressed concern with the measurement intensity of inter-sector linking (Zylbersztajn, 2000). The CSA methodology emphasizes the sequence of product transformations in the system.

Goldberg's research had its merit in changing the centre of the analysis from the farm to the system; the agricultural sector was thus no longer dealt with in isolation from the remaining economy. The idea of agribusiness reinforces the existing inter-sector bonds, as it sees agricultural production as part of the commodity system, emphasizing its relations with the world of businesses (Silva, 1991). Goldberg (1968) emphasised the importance of the systemic approach's use to support corporate decisions.

The second approach, proposed by Morvan (1985), considers a chain ("*filière*") as linked operations for the transformation of a good. They are influenced by technology and have complementary interdependencies, according to Batalha (2001). The *filière analysis* applies to the sequence of activities that transform one commodity into a product for final consumption and represents a tool of the French industrial economy's school. Among the authors who used the concept, Louis Malassis emphasized its historical dimension, pointing out the agro-industrial complexity as a characteristic of the capitalist development and industrialization's stage of agricultural production. According to Morvan (1985), the *filière analysis* is an important instrument to describe systems, to define the technology role in the framing of productive systems, to organize integration studies and to analyze industrial polices, firms and association strategies.

Additional theories can also be applied in these analyses (Omta *et al.*, 2001). The *supply chain* is viewed as a system that integrates raw material suppliers, factories, distribution services and consumers. Furthermore, the network concept refers to when organizations are directly involved in different processes that add value in the elaboration of goods and services up to the final consumer.

Lazzarini *et al.* (2001) integrate chain and network concepts in a new concept: the *netchain*. According to these authors, the integration of these approaches allows the consideration of existing organizational interdependencies in a network, as well as the different mechanisms of coordination (managerial plans, processes standardization and adjustments) and sources of value (production and operations optimization, transaction costs reduction, diversity and "co-specialization" of knowledge).

Hardman *et al.* (2002) demonstrated the possibility to increase the competitiveness of the South African apple chain exports through cooperation among producers, packers and exporters. From the ideas of CSA and *filières*, it is possible to develop tools and managerial activities to improve the chains' efficiency. Thus, *Supply Chain Management* (SCM), *network* and *netchain* are the most applied theoretical concepts and empirical notions for the development of agro-industrial systems (Batalha and Silva, 2001).

3. Procedures and operationalization

The first step to characterize and to analyze a system is to define its objectives as well as boundaries and scope, participant subsystems of the system, and its environment (Malhotra, 2001). Batalha (2001) comments that for a chain analysis, the researcher must define some conditions and those are consequences of the objectives to be reached. The most important and difficult definitions are related to the analysis scope and levels that should be detailed. Zylbersztajn (2000) also comments that the definition of the Sag's boundaries depend on the researcher's purpose and the focus on a product.

In this study, the established scope was the milk chain in Brazil, focusing particularly on milk and milk products in São Paulo State. However, in some cases we present data that surpasses this scope, in order to analyze the dependence and importance of some agents and sectors in the system.

This work will present, by means of a descriptive survey, the characterization of the milk chain in Brazil. The methodology used in this study was developed and applied by Neves *et al.* (2001, 2002) in the PENSA study of the Brazilian orange and wheat agro-industry systems. The sequence of this methodology can be summarized in six stages (see Figure 1).

The first step consisted of elaborating a preliminary design of the chain based on theory and the researchers' experience. It was also necessary to identify which segments would be studied, keeping the focus in the central axle of the system, due to the objective of the research. We opted to study milk as a raw material and central object of the system, adopting Goldberg's (1968) commodity system approach (CSA), as well as emphasizing a product as the starting point for the system analysis. The second step was to submit the design to sector specialists and to interview them to obtain their input for possible adjustments. In this way, a realistic picture of the current system could be created. The third step consisted of secondary data research, which, according to Malhotra (2001), is collected for ends that differ from the problem of the research. For this step, we looked for data from sources that have academic and statistical credibility, a good reputation, and integrity with data collected until the year of 2004. Additional efforts were made to not capture unnecessary data, by following Godoy's (1995) suggestions for document choice.

After collecting the available secondary data, we started collecting primary data (fourth step), that is, the research data produced by the researcher for the specific purpose of solving the problem in question (Mattar, 1993; Malhotra, 2001). This included in-depth interviews with representatives of several organizations in the milk sector. In preparing for the interviews, we first identified which data was not found through the secondary research, and selected these topics to focus on in the interviews with agents in the chain. Agents were selected based on a number of characteristics; *i.e.*, access to the information and data related to the sector being studied, knowledge and experience about the system, willingness to collaborate with the research and maintain a communication channel for futures contact, and contacts with other agents who might be able to contribute unavailable data.

The quantification (fifth step) determined the turnover of each sector in the chain, through the companies' revenues and estimates, given in Brazilian Real currency. The values were converted to US dollars, using the 2004 average exchange rate of R$ 2.93 per USD. Because currency values fluctuate it is important to identify the period of the research's evaluation and the currency used.

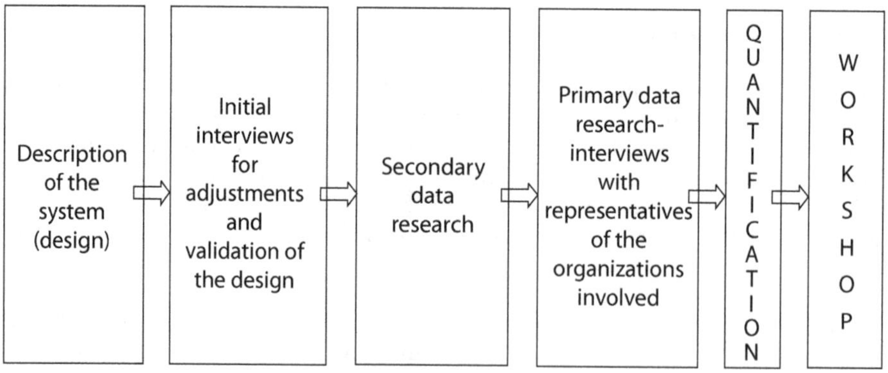

Figure 1. Sequence of steps to mapping and quantification of agro-industrial systems. Source: Neves et al. *(2004).*

To guarantee the data's reliability some secondary and primary data were compared, in an attempt to find incongruous information. In this process we tried to use at least two different data sources to check the results, and conducted additional interviews with similar agents when needed.

In the sixth step a workshop was held. There was great concern that all links in the chain be represented to make ensure a good discussion and valid data. It is important to note that some agents who participated in the workshop have little involvement or experience in the milk chain given their specific business activities more on the periphery of the milk chain. They were nevertheless also invited to develop agents' networks and to discuss future collective actions. After the researcher's presentation, the participants were invited to meet in randomly chosen groups of 8 to 12 people, to compose a discussion table and data validation using the focus groups technique. Each group discussed and elaborated a proposed list of actions that were presented together with the data in a plenary session for all the participants. With the results, a "milk chain agenda" was compiled to be developed and implemented.

4. Results: description and analysis of the milk chain

In the next sections we will present the main information derived from the milk chain quantification. The final report on the whole research project comprises more than 600 pages, including an executive summary, PowerPoint presentations and spreadsheets.

4.1 Agricultural inputs

Within the cattle milk chain in Brazil, the veterinary products sector had a turnover of USD 82.9 million in 2004, the equivalent of 11.8% of the cattle market (it represents 55.6% of the veterinary market). The main segment in this market is products related to the *Aphthovírus* (known as foot-and-mouth disease), which represent about 25% of this market. It is a concentrate sector in Brazil, where CR4 was 36% in 2004. Since 1997 this sector has increased sales by 25% and it has a projected growth of about 11% until 2008. Other important agricultural inputs are supplied by genetics companies, which increased their turnover by 30% in the last five years and have great growth potential in the coming years. Sales in 2004 reached USD 15 million, of which USD 1 million for São Paulo State. According to specialists, the sector will increase sales by 5% each year in the coming years, due to increased profits from the use of artificial insemination.

About 342,000 ha in Brazil are used for the production of crops (mainly corn) that are to be converted into ensilage. In São Paulo 21,800 ha are used for the production of 327,500 tons of corn for ensilage, which requires USD 1.12 million in corn seeds. The animal feed sector sold almost USD 731 million to the milk chain, of which São Paulo is responsible for USD 58 million, almost 8% of the country's consumption. Furthermore, the cattle for milk production consume 73% of the Brazilian cattle feed. In a global context, Brazil is the third largest animal feed producer, just behind the USA and China.

The milk sector also had a turnover of USD 1.4 billion in animals sold, with USD 116 million in São Paulo. Other inputs, like equipment for milking and refrigeration, amounted to USD 1.46 billion, reaching about 10% of farmers and milk producers in Brazil with 13,750 units of cooling equipment sold in 2004.

4.2 Industry inputs

In our chain analysis, we also analyzed some inputs used in the processing industry of milk and milk products. Some data were not completely available and were thus calculated using industry indicators and consumption ratios. The most important industry input, packaging materials, had a larger turnover in 2004 than in the previous three years, with USD 9.75 billion. The industry worked at 85.6% of its capacity to achieve this level (Abre, 2005). About USD 49 million of this total amount represented packages sold to the milk chain companies.

4.3 Milk production

The value of Brazilian milk production was USD 4.25 billion in 2004 (23.5 billion litres). The main producers are concentrated in the south-eastern states (SP, MG, RJ, ES), whose production was evaluated at USD 1.7 billion, followed by the southern region (USD 1.1 billion). In the Southeast, Minas Gerais is the largest producer (USD 1.26 billion), followed by São Paulo (USD 300 million). With respect to international trade, the milk chain is expected to have a surplus trade balance of USD 2.73 million in 2005 (in 2004, the balance was also positive with USD 3.92 million).

In the last 10 years, Brazil's milk production grew faster than its population (production increased by 48% from 1994 up to 2004). Per capita consumption reached 135 litres/inhabitant, but this was still only about half of the UN's recommended consumption. However, it is still possible to promote and improve both consumption and production. For example, at 45% above the national average, São Paulo State's per capita consumption is very close to the recommended one.

The current level of national production and its recent growth have been accomplished in large part by informal producers (about 24% of the national production) without government inspections. To reduce the informality and improve the quality and quantity of milk production, important steps are needed such as professionalization of management (costs, health, traceability and quality), adequacy of the load tax, increase of investments, and improvement in government regulations.

These factors have affected the competitiveness of São Paulo State over other regions. Milk production has increased (in the last four years) mainly in the states of Paraná, Rio Grande do Sul, Goiás and Minas Gerais. Moreover, São Paulo has decreased its production, due to the expansion of sugar cane cultivation, effectively pushing milk producers to areas were productivity is lower (São Paulo is currently in 9th position in productivity). Due to this loss of production (12% in the last five years) and increasing consumption (32.4%), São Paulo needs to "import" about 5.87 billion litres per year. The states of Minas Gerais and Goiás are the main generators of surplus for to São Paulo State (Pensa, 2005). Figure 2 illustrates this scenario and Table 1 shows detailed internal transferences in Brazil.

The milk producers in Brazil (and the chain as a whole) have received little attention from the responsible agencies for the national statistics. The last national rural census dates back to 1995. According to IBGE (Brazilian Institute of Geography and Statistics), the last census showed that Brazil had about 1 million of establishments involved in cattle and milk production. In São Paulo there were about 78,000 establishments.

Currently, the 16 largest milk receivers buy from 117,494 milk producers, with a daily average of 205 litres/producer. The total number of producers (formal and informal) is bigger than this.

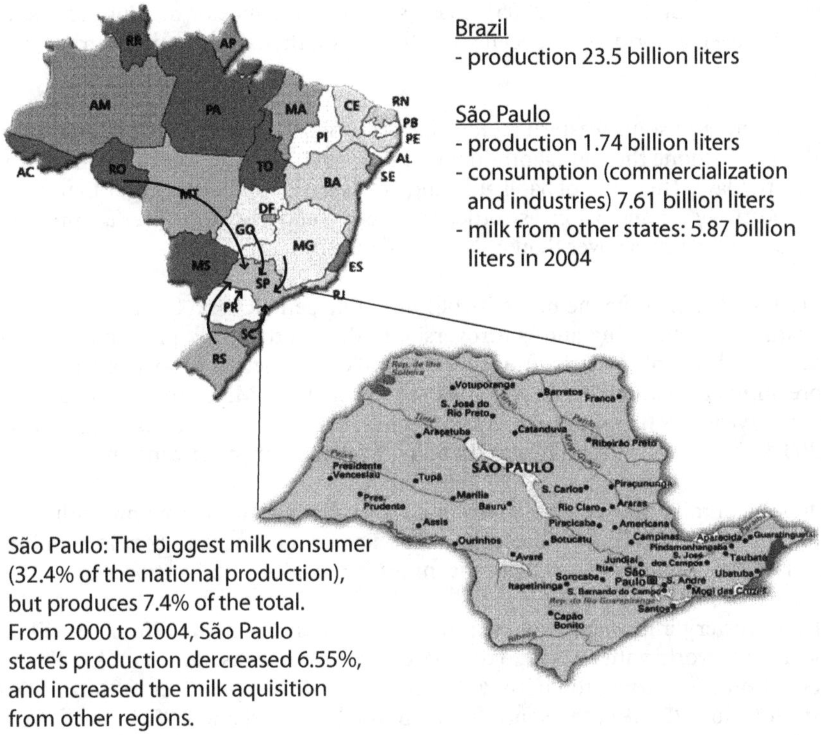

Figure 2. Brazilian production and internal transfer to São Paulo. Source: Pensa (2005).

Table 1. Main producer and consumer states – internal transferences.

Balance between Production and Consumption (internal transfers) [1]			
State	Production* (l)	Estimated Consumption (l)	Balance (l)
São Paulo*	1,739,397,000	7,615,689,393	-5,876,292,393
Minas Gerais	6,928,184,894	2,195,099,093	4,733,085,801
Goiás	2,634,168,173	718,223,806	1,915,944,367
Rio Grande do Sul	2,488,144,614	1,699,251,468	788,893,146
Paraná	2,260,777,038	1,575,452,868	685,324,170
Santa Catarina	1,471,292,187	1,022,880,334	448,411,853
Bahia	829,569,917	997,107,088	-167,537,171
Pará	610,812,612	416,577,736	194,234,876
Rondônia	582,969,141	125,233,691	457,735,450
Mato Grosso	512,409,211	368,664,193	143,745,019

[1]PENSA Estimation; *IBGE/SCOT Consultoria/PENSA
Source: IBGE, Pesquisa Trimestral do Leite (2005) and Pensa (2005).

What's more, according to IBGE (2005) there is a predominance of medium and small-scale producers. However, according to specialists in the sector, there is a trend toward concentration of production.

Only the southern and south-eastern regions have a production above the national average (5.6 litres/cattle/day). Among the biggest producing states, Santa Catarina has the highest production (8.6 litres/cattle/day). The states of São Paulo and Goiás have production lower than the national average (respectively 5.2 and 5.4 litres/cattle/day). As a comparison, the average production per cow in Brazil is three times lower than in New Zealand.

These indicators demonstrate the necessity of improving genetics and feeding conditions of the Brazilian cattle. An interesting and controversial finding is that milk production has a higher profitability than many other agricultural activities, like production of soybean, corn and sugar cane or breeding and fattening of cattle for beef production. Milk production provides USD 512.47/hectare/year (3.9 times more than sugar cane). However, it demands a higher investment - USD 7,303.45/hectare (2.5 times bigger than the investment in sugar cane crops).

The profitability of milk production has declined for some producers (mainly with small farms that use lower technology), due to the lack of a policy for minimum milk prices and increased prices of agricultural inputs such as fuels, machines, animal feed, and fertilizers.

Cooperatives are very important players in the agribusiness of Brazilian milk. Brazil is ranked sixth ranked in the world with respect to the share that cooperatives have in milk collection (40% of the production). According to OCB (2002), there are 353 cooperatives in Brazil (47% in the south-eastern region, 27.5% in the South, 15.3% in the Northeast and 9.9% in the Centre-West). Of these, 288 are milk cooperatives, most of which collect more than 55,500 litres/day.

In summary, the factors that negatively affect the competitiveness of São Paulo's milk production industry are efficiency of the pastures, production per cow and logistics.

4.4 Industry

The dairy industry in Brazil (second largest food industry) had a turnover of about USD 4.94 billion (from formal market), which represents 8% of the food industry in Brazil. These numbers show the importance of this industry for the economy. The country has about 1,973 milk collectors. Minas Gerais State represents 34.4% of the establishments, São Paulo 13% and Goiás 10.4%. More than half of these collectors have the capacity to process less than 10,000 litres/day. Only 5.3% of the companies have a capacity to process more than 100,000 litres/day. Among these, 28.8% are located in Minas Gerais and 20.2% in São Paulo. A steady decrease in the total number of producers and the increase in the daily production average indicate that small milk suppliers are being substituted by larger ones.

In terms of international markets, the exported volumes of milk, cream, and powdered milk increased almost 12 times from 2000 to 2004 representing 80% of total exports. Iraq, Algeria and Angola are the main Brazilian milk purchasers and account for 48.5% of the country's exports. Imported volumes of milk and UHT milk declined consistently from 2000 to 2004, thereby falling from the second-highest position to the last in the list of imported dairy products.

Uruguay and, mainly, Argentina are the two largest exporters of dairy products to Brazil representing about 79% of the Brazilian's imports. São Paulo buys 36% of the imported milk and was responsible, in 2004, for almost 100% of the volume of yogurts, curdled, and fermented dairy imports. São Paulo received 44.7% of total Brazilian imports.

The great majority of the milk production is sold in the form of derivatives, and only 6.8% is sold as pure milk in grades A, B or C. Table 2 shows the milk's destiny in Brazil.

4.5 Distribution channels

Distribution agents, mainly retailers, had a turnover of USD 5.29 billion with milk and milk products (the total self-service turnover was USD 33.34 billion), involving about 72,000 stores in the country that get an average gross margin of 17% over the consumer's price. More than half of the milk sold in Brazil (55%) passes through self-service channels (supermarkets), followed by the bakery channel. In the supermarkets, dairy products are responsible for 13% of sales. The bakery channel had a turnover of USD 8.53 billion in 2004, including USD 1.1 billion represented by milk (9%) and other dairy products (5%). In Brazil, 80% of long-life milk, condensed milk and cream are sold in supermarkets. In São Paulo this channel is even more important with 85% of these products' sales. Liquid yogurts are among the categories that showed the greatest growth, as sales increased by more than 19% from 2003 to 2004. Worldwide, this category grew by 4%.

4.6 Workshop discussions and SWOT analysis description

The workshop discussions led to a list of strengths, weaknesses, opportunities and threats to the milk chain, in an attempt to initiate a strategic plan for the chain. The results of the analysis conducted by various groups were merged and are shown in Tables 3 and 4.

Table 2. Destiny of milk production in Brazil – 2004.

Products/ destiny (uses)	Volume	Units	Production Value (USD 1000)	Milk Consumption (1000 l)
Cheeses	808,112	thousand kg	1,892,031	7,919,500
UHT (long life milk)	4,403,000	thousand litres	2,004,642	4,403,000
Powder milk	420,000	thousand kg	1,247,099	4,379,130
Fluid milk A, B, C	1,590,000	thousand litres	488,396	1,590,000
Yoghurts	545,897	thousand kg	869,925	419,904
Butter	75,000	thousand kg	150,512	308,918
Others *				1,138,549
Farms self-consumption	3,341,000	million litres		3,341,000
			6,652,605	23,500,000

* Involves milk production used for products like cream milk, condensed milk, ice-cream, "dulce de leche", exportats and other derivatives.
Source: Pensa (2005).

Table 3. Milk chain strengths and weaknesses discussed during the workshop.

Strengths	Weaknesses
Farm facilities are used in various ways (cattle with other agricultural activities), which amplifies the milk production's profitability.	Milk production dispersed all over the country, which makes better coordination difficult among the agents and increases the dependence of the industry and/or cooperatives.
Low production costs, due to the good climate, land availability, agro-input availability (large production of grain, sugar cane, oranges, tropical foragers, mineral supplements and feed) and genetic adaptation of cattle.	Lack of experience and contact with the international market, which inhibits exporting attempts.
Diversity of efficient production systems (extensive/intensive, and pasture/ration cattle breeding).	Profit losses in the sector distract the attention of private enterprises and the government.
Milk buyers' competition and aptness (cooperatives and industries).	Lack of investments in bovine genetics and disease prevention, shown by the cattle's low productivity and by the incidence of diseases like the Aphthovírus (foot-and-mouth disease).
Strong food's industrial park distributed for all country territory allied to the facility of industrial production and commercialization of food.	Low hygiene control and/or traceability, demonstrated by low quality of milk.
Big internal market and low milk consumption.	Lack of differentiation in milk supply for distinct uses (cheese production, pasteurized milk and shelf stable milk, cakes, biscuits, etc).
Supply of (technical and economic) information provided by presence of research groups in Brazil (EMBRAPA Milk Cattle, ITAL, Esalq, IEA, IAC, CATI, PENSA).	Inadequate labour and rigid environmental laws lead to fraud in production reporting.[?]
	Low professionalization of the production sector and, consequently, low levels of technology on farms, worker capability, and farmer's managerial knowledge.
	Deficient technical veterinary and agronomical assistance and little incentive to develop technically specialized personnel.
	High seasonal variation in prices and price differences among regions in Brazil, which generates instability in the profits of the activities.
	Limited access to credit high interest rates for existing sources of credit.
	Precarious basic infrastructure, very bad road conditions, energy shortage and water scarcity.
	Strong protectionist policies of importing countries – subsidies, duties and sanitary barriers.

Source: Pensa (2005).

Table 4. Milk Chain opportunities and threats discussed during the workshop.

Threats	Opportunities
Lack of bilateral agreements with promising importing countries	Potential of increasing exports and secondary effects on the price balance, especially for condensed milk.
Permanency and continuity of the sanity barriers enforced by the developed countries for the different kinds of cheese.	Possibilities of reaching markets historically protected by subsidies.
Economic instability, mainly in the national interest rate and in value of national currency.	Improvement in the economy , which may translate into national and international investments, profit growth, internal consumption growth, increased imports and stable exchange rate.
National priority on exporting other agricultural products, not milk. That is evident when we note that there are not specific government policies related to the sector.	Opportunities to improve milk quality (genetics and cattle improvement) and to explore niche markets.
Argentinean milk can be imported at a low cost. The future of the Mercosul is still uncertain, especially due to the plea intricacy and the usage of arbitrary establishment's change.	Greater supply of feed due to increased grain production .
Investment possibility, especially of global farmers.	Excellent research centres and increase use of internet provide greater access to technical knowledge.
In general, the Brazilian consumer buying potential is low, which is a reflection of the historical inequality of income distribution.	Development of new milk-based beverages and, in particular, of products in classes C, D, E, to compete with soft drink market.
Lack of public and private investments in the energy sector, contributing to probable new rationing of electric energy and water.	Learn from the developed countries and develop institutional marketing of Brazilian milk supply chain, such as introduction of milk in the social programs and school lunch program .
Restrictive monetary politics with maintenance of high interest rates, which brings high credit costs to the sector.	
Food retail's growth and concentration and, consequently, buying power is concentrated in the hands of large net retailers and retailer's own brands.	
Disrespect of property rights, mainly by governmental carelessness.	
Possible international dispute (litigation) for the registered brand expropriation without royalties' payment.	

Source: Pensa (2005).

The SWOT analysis showing strengths and weaknesses in the chain, allowed us to point to the actual position of the companies, relationships, and contractual features within the chain. The opportunities and threats were discussed in order to gain insight into the milk chain and its agents, analyzing external factors that influence this important chain. Unfortunately, the discussions show that the agents involved perceive more weaknesses and threats than strengths and opportunities in the milk chain.

5. Managerial implications and discussion

Despite the relevance of the milk chain in Brazil and its importance to São Paulo's economy, a lot of challenges remain for the future of the chain, mainly in São Paulo State. Historical developments, trends in other countries and the Brazilian situation have created issues that need to be resolved in order to guarantee the future and development of the milk chain, allowing it to continue to performing the economic and social roles required of it in a big country like Brazil with its thousands of small producers.

Using the quantitative information and the list derived from the SWOT analysis, the workshop participants proposed a list of activities that should be developed by the government and private agents involved in the chain, in order to continue increasing Brazilian milk production and improve São Paulo's competitiveness, in addition to guaranteeing the future of the chain and its agents.

The research helped agents to define action to be taken and priorities to improve the chain's future performance. The main activities discussed involve:

1. Development of institutional marketing activities, like the "milk in the schools" programme and vertical/horizontal associations.
2. Milk quality improvement programs, with implementation of the new Brazilian regulations (Normative Instruction # 51), investments in genetics, cattle wielding technologies, sanitary and traceability systems.
3. A system that sets payment according to quality, providing price differentiations to producers, awarding best practices and establishing annual contracts.
4. Programs to increase capacity of producers, improving managerial skills, cost and control techniques, and technical knowledge.
5. Development of a network of "intelligence centres" to collect, gather, process and share information and benchmark good initiatives from other chains in Brazil and worldwide.
6. Make available more credit lines specific to milk producers, and incentives to invest in technologies for milking, storage, transportation and industrialization.
7. Developments and improvements in the Brazilian cooperative system, improving the management, involvement of producers, and strategic planning, strengthening cooperatives through mergers and acquisitions or cooperative joint ventures. Other strategic movements could involve broadening the cooperatives' boundaries, for example, industrializing and distributing dairy products with under brand names.
8. Review of taxes that affect dairy products, mainly milk, removing tax differences among states and overtaxing.

In summary, the paper presented the mapping and quantifications of the milk chain in Brazil, an initiative of the São Paulo Milk Sector Council and PENSA, as the first step of a continuing program intended to increase the country's and São Paulo's competitiveness through the "strategic planning of the milk chain".

References

Abre, Associação Brasileira de Embalagens, 2005. http: //www.abre.org.br.

Batalha, M.O. (ed.), 2001. Gestão agroindustrial. 2ª ed. São Paulo, Atlas.

Batalha, M.O. and A.L. Silva, 2001. Gerenciamento de sistemas agroindustriais: definições e correntes metodológicas. In: Batalha, M.O. (ed.). Gestão agroindustrial. 2ª ed. São Paulo, Atlas, pp. 23-63.

Godoy, A.S., 1995. Pesquisa qualitativa, tipos fundamentais. Revista Administração de Empresas, São Paulo 35, 20.

Goldberg, R.A., 1968. Agribusiness coordination: a system approach to wheat, soybean and Florida orange economies. Division of Research. Graduate School of Business and Administration. Harvard University, 256 pp.

Hardman, P.A., M.A.G. Darroch and G.F. Ortmann, 2002. Improving cooperation to make the South African fresh apple export value chain more competitive. Journal on Chain and Network Science 2, 61–72.

IBGE, Pesquisa Trimestral do Leite, 2005. http://www.ibge.com.br.

Lazzarini, S.G., F.R. Chaddad and M.L. Cook, 2001. Integrating supply chain and network analyses: The study of net chains. Journal on Chain and Network Science 1, 7–22.

Malhotra, N.K., 2001. Pesquisa de marketing: uma orientação aplicada. 3.ed. São Paulo, Bookman.

Mattar, F.N., 1993. Pesquisa de marketing - edição compacta. Atlas.

Morvan, Y. 1985. Filière de production. In: Fondaments d'economie industrielle, Economica. pp. 199-231.

Neves, M.F., A.M. Val and M.K. Marino, 2001. The orange network in Brazil. Fruit Processing 11, 486-490.

Neves, M.F., D. Zylbersztajn, C.P. Machado Filho and R.T. Bombig, 2002. Collective actions in networks: the case of beef in Brazil. In. J.H. Trienekens and S.W.F. Omta (eds.). Paradoxes in food chains and networks. Wageningen, Wageningen Academic Publishers. pp. 742–750.

Neves, M.F., R.M. Rossi, L.T. Castro, F.F. Lopes and M.K. Marino, 2004. A framework for mapping and quantifying value chains towards collective actions. European Marketing Academy, Murcia. International Food Agribusiness Management Association, Montroeux.

OCB - Organização das Cooperativas Brasileiras, 2002. Anuário do Cooperativismo Brasileiro. Brasilia, OCB.

Omta, S.W.F., J.F. Trienekens and G. Beers, 2001. The knowledge domain of chain and network science. Journal on Chain and Network Science 1, 77–85.

Pensa, 2005. Tomografia do sistema agroindustrial do leite no Brasil. Workshop apresentado na Faculdade de Economia e Administração da USP. Realizado em 26 de outubro de 2005 em Ribeirão Preto.

Silva, J.G., 1991. Complexos agroindustriais e outros complexos. Associação Brasileira de Reforma Agrária 21, 5-34.

Zylbersztajn, D. and M.F. Neves, (eds.), 2000. Economia & Gestão de Negócios Agroalimentares. São Paulo: Pioneira, 2000. 428 pp.

Appendix 1. Quantification of the Brazilian milk chain in million dollars (USD 22.63 billion)

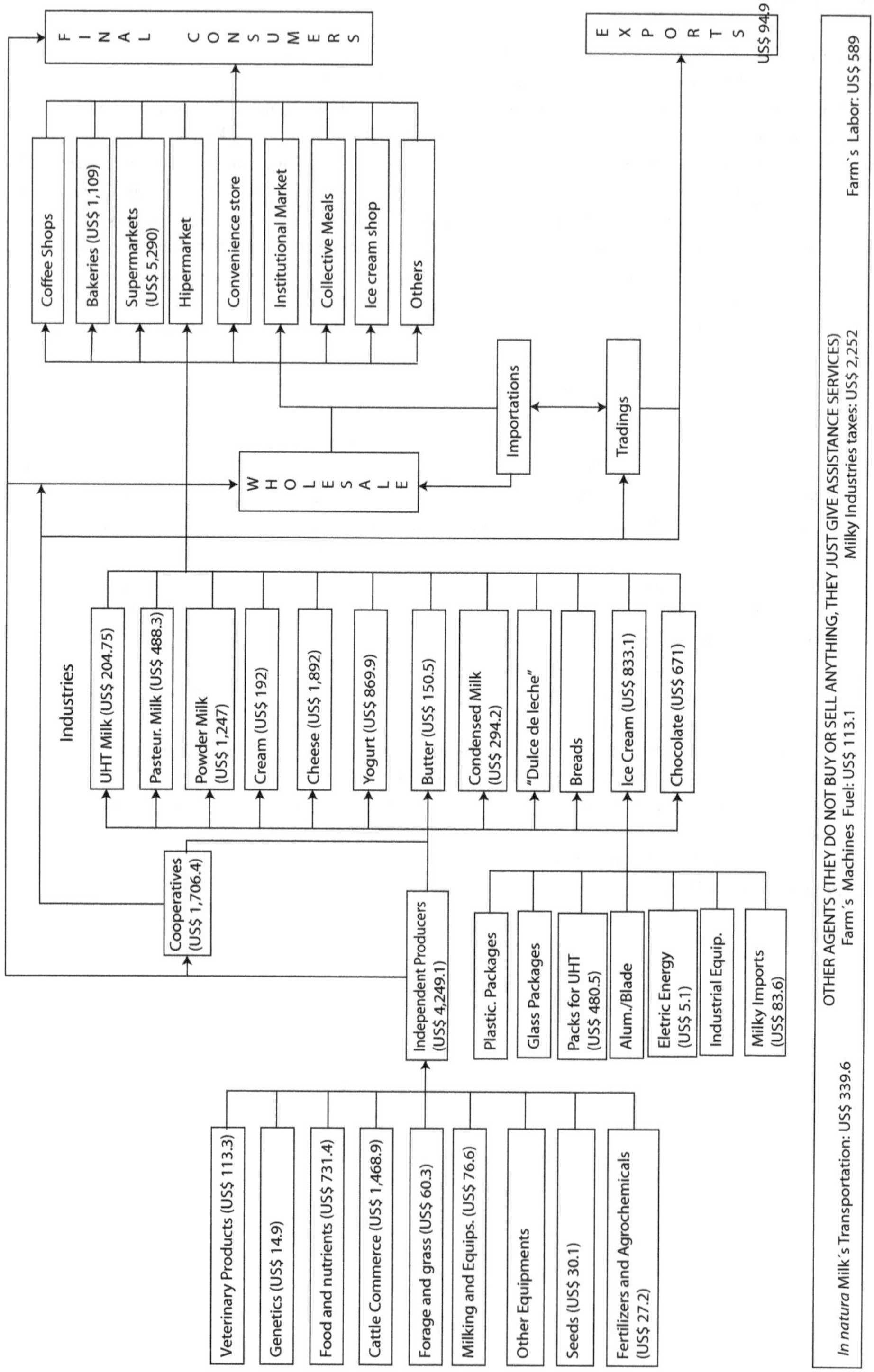

Impacts of the rapid development of supermarkets on China's dairy industry

D. Hu, F. Fuller and T. Reardon

Abstract

The emergence of supermarkets along with rapid development of the economy is quickly changing the traditional dairy industry in China. As supermarkets swiftly replace various traditional dairy product distribution channels, this change is promoting the formation of a national unified dairy product market. The development of supermarkets has created objectively favourable conditions for the movement of milk products from the low-cost production bases in the western pastoral provinces to the major consumption markets in the coastal region, and supermarkets have encouraged the consolidation of dairy firms into large-scale dairy-processing enterprises with centralized management. Along with the improvement of dairy product distribution, supermarkets have created opportunities for consumers in remote townships and rural areas to gain access to a greater variety of dairy products. The gradual increase in the number of rural children consuming dairy products may lead to tremendous growth in dairy product demand by the next generation. China, with a population of 1.3 billion, may one day become the largest consumer of dairy products in the world.

Keywords: dairy processing, dairy products, supply chain.

1. Introduction

Since the beginning of China's economic reforms and the opening of its economy, the dairy industry in China has developed at an unprecedented rate. Milk production has increased from 1.14 million tons in 1980 to 17.4 million tons in 2003, a fifteen-fold increase over 23 years. Milk production growth accelerated in the latter half of the 1990s, and since 2000, milk production has grown in excess of 20% annually. Studies examining the development of dairy production in China cite favourable government policies, foreign investments, growing availability of production inputs, and rising per capita income as the primary sources of growth (Zhou *et al.*, 2002; China Dairy Development Strategic Study Research Group, 1990). While these factors have been crucial to the industry's growth, most authors overlook the important contributions of supply chain development and, in particular, the advent of supermarkets to the rapid expansion of China's dairy supply in the late 1990s. In this paper we seek to address this omission by examining how the growth of supermarket chains in China helped overcome key hindrances to the distribution of dairy products and facilitated a rapid broadening of dairy product consumption.

The next section provides a brief overview of particular challenges that slowed the growth of China's dairy industry in the early reform period. Section 3 quickly recaps the rapid growth of supermarket chains in China, and section 4 connects the two discussions by highlighting the direct and indirect influences supermarkets have on the growth of the China's dairy sector. Based on our observations, we conclude with some thoughts about future developments in the industry and research priorities.

2. Hindrances to the development of China's dairy industry

Even after the start of economic reforms, the expansion of China's dairy industry lagged behind the development of other livestock industries for several years. This changed in the mid-1990s, and both production and consumption began growing at double-digit rates. Three important reasons for the initial sluggish growth of the dairy sector are: (1) prohibitive transportation costs from low-cost production regions to primary consumption regions, (2) monopolization of local retail supply channels for pasteurised milk, and (3) a narrow range of dairy products that appealed to only a limited demographic of consumers.

2.1 Regional distribution of dairy production and consumption in China

One of the greatest challenges facing China's dairy industry in the last two decades has been placing adequate supplies of dairy products on the markets with the greatest demand. The problem arises because the fastest growing markets for dairy products are not located in areas with a comparative advantage in dairy production. From a geographical perspective, dairy product supply and demand in China can be separated into 3 regions: the coastal region, the central agricultural region, and the western pastoral region. The coastal region includes Beijing, Tianjin, Shandong, Jiangsu, Shanghai, Zhejiang, Fujian, Hainan and Guangdong provinces. The key characteristics of the coastal region include the existence of dense population centres, well-developed industries and commerce, high per capita income, and a large demand for dairy products. This region constitutes the main consumption area for dairy products.

The central agricultural region includes Liaoning, Jilin, Hebei, Henan, Anhui, Jiangxi, Yunnan, Guizhou, Sichuan, Chongqing, Shanxi, Shaanxi, Gansu provinces and autonomous regions. More than 60% of China's population resides in these provinces. The economy in the central agricultural region is dominated by crop production, and average per-capita income levels are low compared to the coastal region. Correspondingly, the demand for dairy products is also weaker. The western pastoral region encompasses Heilongjiang, Inner Mongolia, Ningxia, Qinghai, Xinjiang and Tibet provinces and autonomous regions. This region has substantial pasture area with rich feed resources in many locations, and the people in this region have a long history of milk consumption. However, only 7.4% of China's population resides in the pastoral region.

The regional imbalance between supply and demand for milk in China is clearly illustrated in Figure 1. Over 57% of the demand for dairy products comes from the coastal region, but roughly 23% of the production occurs in those provinces. Obviously, milk from the pastoral region has to flow into the other two regions to clear the national market. The evolution of milk production shares shown in Table 1 indicates that achieving this redistribution of milk has not occurred smoothly. Until the recent widespread adoption of ultra high temperature (UHT) pasteurization, lack of adequate cold storage transportation chains prohibited the flow of fluid milk from the pastoral region to the coastal region. Consequently, much of the growth in milk production in the pastoral region in the 1980s was for local consumption or was converted to milk powder. In the early and mid-1990s, some dairy firms began investing in milk production in the agricultural region in provinces adjacent to Shanghai, Beijing, and other large population centres in the coastal region. Thus, some of the growth in milk production in the agricultural region has been channelled into the coastal region since the mid 1990s, accounting for the increase in the central region's share of total milk production from 1990 to 2000. Between 2001 and 2003, milk production in Heilongjiang and Inner Mongolia, respectively, increased by 59% and 90% (a combined total of 3.1 million metric tons (mmt)), making these two provinces the top two milk producing regions. Over

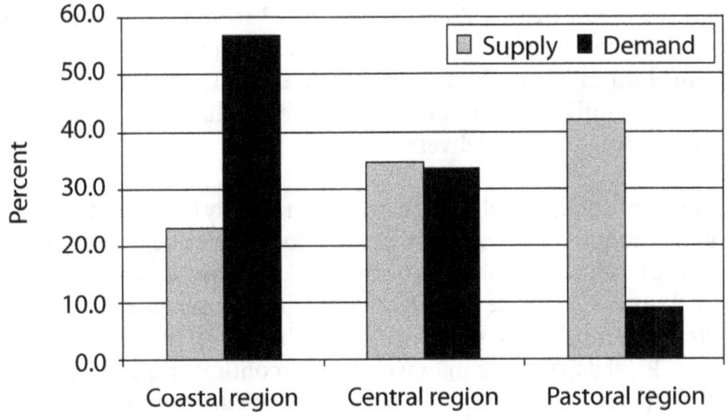

Figure 1. 2002 regional supply and demand shares for dairy products.

the same period, UHT milk production rose from 0.71 mmt to 3.2 mmt. The growing market for UHT milk is promoting greater efficiencies in dairy production by enabling an increasing share of production to shift away from the high-cost coastal and agricultural regions to the low-cost pastoral region (see Table 1). As we discuss below, the rapid rise of supermarkets in China has played an important role in the expansion of UHT milk.

2.2 Monopolization retail supply chains for pasteurised milk

Among all dairy products, fluid milk is consumed in the greatest quantities in China. Since 1949, the retail sector for fluid milk in China has gone through 4 major of phases of development: with

Table 1. Production characteristics of China's dairy regions.

	Coastal Region	Central Agricultural Region	Western Pastoral Region
Share of Dairy Production[1] (%)			
1980	24.5	38.2	37.3
1990	23.7	30.3	46.1
2000	21.7	36.6	41.7
2002	23.3	34.7	42.0
2002 Milk Production Cost per 100 kg[2] (US Dollars)			
Material	18.97	17.40	13.41
Labor	1.09	2.05	1.45
Total	20.06	19.45	14.86
Milk Output per Cow[2] (kg)			
	6541	5210	6265

[1]Source: Dairy Industry Yearbook, 2003.
[2]Source: State Development and Reform Commission

Phase I occurring from 1949 to 1972, Phase II from 1973 to 1985, Phase III from 1986 to 1995, and Phase IV from 1996 to the present. During Phase I, household delivery systems were the primary channels for fluid milk distribution in urban areas. The local milk companies replaced household deliveries with milk distribution stations during Phase II. Consumers picked up milk from these depots according to a set delivery schedule.

During Phase III, the supply of fluid milk in urban cities gradually became more abundant and direct rationing was no longer required. Private vendors and food stores began selling milk, constituting the first challenge to the retail monopoly held by the local state-owned milk companies. At the same time, many milk companies reinstituted household distribution services and maintained deliveries to some milk depots. With most households tied into either direct household or depot distribution systems, the dominant local dairy companies were able to continue restricting competition for fluid milk sales. The lack of interregional competition inhibited the development of a national market for dairy products and slowed the growth of the industry.

However, with the growing presence of multinational dairy firms in the early 1990s and the development of supermarket chains, real competition for milk and dairy product sales appeared. Since 1995, supermarkets and other private retail venues, such as the growing number of convenience stores, have become major forces in the distribution and sale of fluid milk and an expanding palette of dairy products. Most state-owned milk companies have been privatized and have adopted market-oriented business practices. In this current phase, the dairy product retail sector in China allows multiple retail supply chains to coexistence (Liu, 2000), and competition in most urban dairy markets is intense.

2.3 Consumer acceptance of dairy products

Lactose intolerance and the lack of a tradition of dairy product consumption are often cited as primary reasons for the low level of milk consumption in China (Wu, 2003; Garnaut and Ma, 1993). People who suffer from lactose intolerance often have the symptoms of abdominal distension, even abdominal pain and diarrhoea after eating milk products. The disease is caused by the lack of the enzyme lactase, which breaks down the lactose in the digestive process. In addition to hereditary factors, the failure to regularly consume dairy products after the age of 2 can result in the decline of the level of lactase produced by the body. Statistics show that more than 80% of the children (5 years old and above) and adults of the Han people are lactose intolerant, and most of them live in rural areas (Song, 2000). The rural residents rarely drink milk because it is not readily available or they lack the income to purchase it. Thus, many of them display symptoms of lactose intolerance (Liang and Zhou, 1999). As a result, expansion of fluid milk sales in rural areas is very gradual.

It is worth noting that yogurt, ice cream, and cheese contain roughly 50% or less lactose by weight than fluid milk, and people who suffer from lactose intolerance often have less severe reactions to consumption of these products. By focusing on fluid milk production, the domestic dairy industry in China limited the market for dairy products primarily to those in urban areas, where lactose intolerance is less prevalent. The variety and quality of dairy domestic dairy products has increased greatly in the last decade. As products with lower lactose content become more readily available, particularly in rural areas, there is greater potential for expanding the demographic of consumers purchasing dairy products.

3. Development of supermarkets in China

Supermarkets first emerged in China's coastal area, focusing primarily on the large population centres. Once supermarket chains had established a strong presence in the large cities in the coastal region, several large-scale supermarket franchises began expanding into middle and small cities and even into some township cities. Moreover, this expansion into smaller population centres has become increasingly obvious since 2000. Below, we briefly chart the development of China's two largest and oldest supermarket chains to illustrate the rapid growth of this sector.

In 1992, Shanghai Hualian Supermarket Company was established. By 1993, Hualian had opened six stores, and it has been active in developing its chain stores and expanding its franchising network since 1995. As shown in Table 2, there were just 400 supermarket companies in China in 1995, with 6000 stores and a sales volume of 3 billion Yuan. Supermarkets accounted for just 0.18% of the total retailing volume.

Shanghai Lianhua Supermarket Group was created in 1995, and that year it opened 41 stores. In 1996, Lianhua began a joint venture with the French retailer, Carrefour, and formed Shanghai Lianhua Ltd. Co., opening its first sales centre in Shanghai. In 1997, after obtaining investment capital from the Japanese Suzuki company and other foreign investors, Lianhua began expanding into the suburbs of Shanghai and beyond. The company built stores the neighbouring cities of Yangzhou and Nanjing in Jiangsu province, and then in Ningbo and Hangzhou in Zhejiang province. Lianhua had opened 230 stores by the end of 1997, generating 2.4 billion Yuan in total sales. By 1998, the number of supermarket companies in China had reached 1150, totalling 21,000 stores and realizing 10 billion Yuan in sales (3.43% of total retail sales).

Table 2. Development of supermarkets in China (1994-2003)[1].

	Number of supermarket chains		Number of supermarket chain stores		Total sales of supermarket		Rate of total sales of supermarket within total retail sale of consumer goods (%)
Year	Number	Growth rate (%)	Number	Growth rate (%)	Total sales (billion Yuan)	Growth rate (%)	
1994	150	-	2,500	-	0.30	-	0.18
1995	400	166.7	6,000	140.0	0.80	166.7	0.38
1996	700	75.0	10,000	66.7	3.00	275.0	1.21
1997	1000	42.9	15,000	50.0	4.20	40.0	1.54
1998	1150	15.0	21,000	40.0	10.00	138.1	3.43
1999	1800	56.5	26,000	23.8	15.00	50.0	4.82
2000	2100	16.7	32,000	23.1	22.00	46.7	6.50
2001	2250	7.1	40,500	26.6	30.83	40	12.36
2002	2400	6.7	53,100	31.1	45.82	49	17.64
2003	2472	3.1	74,000	39.4	59.57	30	22.48

[1]Source: China Chain Store & Franchise Association, 2003.

By the end of 1999, Hualian Group had opened 500 chain stores in eight cities and provinces including Shanghai, Jiangsu, Zhejiang, Jiangxi, Henan and Shandong. They continued to expand their network, venturing into Shanxi, Inner Mongolia and Beijing. In 2002, Hualian had 1,240 stores, covering 12 provinces and cities and generating 15.06 billion Yuan in sales. In 1999, the number of stores operated by Lianhua reached 606, with a sales volume of 7.3 billion Yuan. By 2002, Lianhua's operations had expanded to 1,921 chain stores, with revenues of 18.3 billion Yuan.

As shown in Table 3, the China's top 10 supermarket franchises operated an average of 570 stores in 2002, 46% more than the previous year. The average sales volume for these 10 franchises was 8.8 billion Yuan, an increase of 39% over 2001. In 2003, the number of supermarket companies was more than double the number in 1998, and the number of stores had more than tripled. Total supermarket sales volume in 2003 reached 59.57 billion Yuan, accounting for 22.48% of the total retailing volume.

4. Impact of supermarket development on dairy industry in China

The nearly exponential growth in supermarket sales described in the last section corresponded with a period of significant change in China's dairy sector, and the synergism between the two developments facilitated the explosive growth in China's dairy product consumption in the last decade. The major impacts of supermarket development on dairy industry in China can be summarized in the following 4 aspects: (1) supermarkets accelerated the demise of local monopolies on retail milk distribution, (2) supermarkets facilitate the transportation of milk from the pastoral region to coastal region, (3) supermarkets promoted the rapid formation of a unified dairy product market (4) supermarkets encouraged greater concentration and increased processing scale among firms in the dairy industry.

Table 3. Top 10 supermarket chains in order of gross sales in 2002[1].

		Number of chain stores			Total sales (billion Yuan)		
Order	Name of enterprise	2001	2002	Growth rate (%)	2001	2002	Growth rate (%)
1	Lianhua Supermarket	1224	1921	57	141.0	183.3	30
2	Hualian Supermarket	822	1200	46	85.1	150.6	77
3	Beijing Hualian	42	54	29	79.8	103	29
4	Shanghai Nonggongshang	324	702	117	74.6	87.3	17
5	Huarun Wanjia	328	397	21	62.7	85.9	37
6	Suguo Supermarket	662	940	42	52.8	70.2	33
7	Chongqing Trade Society	59	72	22	44.7	53.6	20
8	Jinjiang Maidelong	15	16	7	49.6	53.1	7
9	Beijing Wumei	201	355	77	25.2	50.6	101
10	Tianjin Home World	28	40	43	32.6	44	35
	Average	370	570	46	65	88	39

[1]Source: Analytical Report of operation of Interlocking enterprises in China in 2002

4.1 The demise of local retail milk monopolies

In section 2.2, the local monopolies in retail milk sales were cited as one of the factors that slowed the expansion of milk consumption in the early reform period. As supermarkets like Lianhua and Hualian expanded outside of their city of origin in the late 1990s, they brought their distribution networks, refrigerated supply chains, and dairy products from preferred suppliers. The dairy product suppliers from outside the region could bypass the established household and depot distribution systems operated by the local dairy monopoly. Supermarkets posed a competitive threat to local dairy companies by offering both an alternative source of fluid milk to consumers and an expanded selection of dairy products. Compared to the existing retail outlets for dairy products, supermarket had the advantages of greater customer traffic, a greater selection of dairy products, rapid information feedback systems, and refrigerated transfer and storage equipment (Hu *et al.*, 2003). These advantages have enabled supermarkets to become the dominant retail format in the dairy product retail sector.

A survey conducted in 2001 by the authors in Beijing, Guangzhou, and Shanghai asked consumers where they preferred to purchase dairy products. One hundred households were surveyed in each city, and the results for yogurt and fluid milk (both pasteurised and UHT) are shown in Table 4. Delivery systems are most important for milk in Shanghai, where 47% of the surveyed households purchased milk through household or depot deliveries. In Beijing and Guangzhou, however, delivery systems were used by less than 20% of the households. More than 40% of the households in all 3 cities purchased milk at supermarkets. The dominance of supermarkets is even more striking for yogurt. Nearly 80% of the households surveyed in Shanghai and Beijing purchase yogurt at supermarkets.

Why do consumers prefer to purchase dairy products from supermarket? Two important reasons are the quality and variety of dairy products offered by supermarkets. Using their buying power, supermarkets are able to demand quality products from their vendors. Consumers see purchases from reputable retailers as a means of reducing food safety risks and lowering search costs.

Table 4. Surveyed milk and yogurt purchases by retail outlet in Beijing, Shanghai and Guangzhou (%).

	Fresh & UHT milk				Yogurt			
	BJ	GZ	SH	Avg.	BJ	GZ	SH	Avg.
Household delivery	13.6	12.4	28.3	18.1	1.3	4.7	7.1	4.4
Dedicated delivery location	0.0	4.7	18.6	7.7	1.3	7.8	7.1	5.4
Supermarket	49.5	51.2	42.4	47.7	79.5	56.3	78.6	71.4
School	1.0	9.3	5.3	5.2	0.0	3.1	0.0	1.0
Grocery store/ Convenience store	20.4	14.7	1.9	12.3	12.8	20.3	0.0	11.0
Street vendor	5.8	4.7	1.8	4.1	3.9	3.1	0.0	2.3
Other	9.7	3.1	1.8	4.9	1.3	4.7	7.1	4.4
Total	100	100	100	100	100	100	100	100

Key: BJ = Beijing, GZ = Guangzhou, and SH = Shanghai

Consumers also value the broad selection of brands and products offered by supermarkets. A second survey conducted by the authors collected data about product offerings in supermarkets in Kunming, Shanghai and Beijing. Ten supermarkets were selected at random in each city, and the number of brands and the number of varieties of fresh milk, UHT milk, yogurt, and milk beverages offered in each store was collected. The results are displayed in Table 5. It is interesting to note that the number of product offerings tends to be larger in the more mature dairy markets of Shanghai and Beijing. Similarly, there tends to be greater variety in UHT milk, yogurt, and dairy beverages, products where supermarkets typically garner a greater share of household purchases. By stocking a greater variety of products and by offering price promotions on dairy products, supermarkets have been able to appeal to a broad cross-section of urban consumers.

4.2 Supermarkets facilitate inter-regional milk flows

The discussion in section 2.1 highlighted the importance of inter-regional milk movements to correct regional supply and demand imbalances. Opening the flow of milk from the pastoral region to the coast has also been important for the economic development of the western provinces. Since the 1980s, dairy industry participants and government officials have recognized that developing the capacity to ship milk from the pastoral region to other parts of China would improve production efficiency and milk supplies. The adoption of UHT technology was a critical component of realizing the region's potential for dairy production because of the extended life of UHT products. However, the interregional distribution networks established by supermarkets have greatly facilitated the rapid spread of UHT products. Companies like Yili and Mengniu, both with primary production bases in the pastoral region, have gained swift national recognition by marketing UHT milk and other long-life products in supermarkets in the major metropolitan areas in the coastal and agricultural regions. The success of these companies has prompted substantial investments in milk production capacity in the pastoral region in the last 5 years.

4.3 Supermarkets promoted the rapid formation of a unified dairy product market

Before the appearance of supermarkets, the dairy product market in China (with liquid milk as the major dairy product) was divided into several, relatively independent local markets dominated by the local urban dairy processing enterprise. As noted above, supermarkets played a key role in destroying the local retail monopolies by providing diverse product offerings, including products

Table 5. Number of dairy products offered at supermarkets in Kunming, Shanghai and Beijing.

Brand	Fresh milk			UHT milk			Yogurt			Dairy beverages		
	KM	SH	BJ	KM	SH	BJ	KM	SH	BJ	KM	SH	BJ
Sanyuan	0	2.5	17.7	0.2	1.0	9.0	0	4	15.7	0	3.8	5.3
Bright	0	11	4.3	4.6	6.5	3.3	4	7.5	8.0	4.4	12.8	3.0
Yili	0	0	2.0	6.8	3.8	6.7	0	4	13.3	9.6	3.5	10.3
Mengniu	0	0	1.0	5.6	3.3	6.3	0	4.5	10.3	4.4	3.5	4.0
Others	7.6	2	0.3	5.4	2.0	4.3	7.6	16.5	13.7	24.4	27.8	42.0
Total	7.6	15.5	25.3	22.6	16.5	29.7	11.6	36.5	61.0	42.8	51.3	64.7

Key: KM = Kunming, SH = Shanghai, and BJ = Beijing

from companies outside the region. Therefore, as supermarket chains expand their networks from the large cities to medium- and small-size cities, townships, and rural areas, they bring with them dairy products from companies throughout China. Though the transformation is not complete, in the last few years several brands have achieved sufficient distribution to become nationally recognized brands.

Table 6 displays the number of products marketed by 4 of the top brands in China in supermarkets in Kunming, Beijing, and Shanghai. This information comes from the survey data discussed in section 4.1. Of the 3 cities, Kunming has the shortest history of dairy product consumption and is a less mature market. The average number of liquid milk products sold in supermarkets in Kunming was 84.6, of which, Sanyuan brand accounted for 0.2%, Bright 15.4%, Yili 19.4%, Mengniu 11.8% and other brands 53.2%. A large number of the other products are marketed by Xuelan, the local dairy company in Kunming. What is interesting to note is that companies from Inner Mongolia and Shanghai have been able to gain significant penetration into the Kunming market, eroding the share of the local companies. In Shanghai, Bright, the local company, has just 31.5% of the product offering, and in Beijing, Sanyuan has less than 30%. As time passes, the Kunming market, and other emerging dairy markets like it, will become more diversified in terms of the brands offered in stores. Supermarkets play an active role in enabling quality branded products to proliferate new markets, facilitating the formation of a nationally unified dairy product market.

4.4 Supermarkets have promoted dairy industry concentration and growth in large-scale processing enterprises

As supermarkets have increased their market share of dairy retail sales, the dairy processing enterprises have become increasingly dependent on supermarkets. The supermarkets have used their growing market power to increase the cost of placing products on their shelves by charging slotting fees, demanding a minimum sales volume, or demanding other trade concessions. These barriers to entry are making it increasingly difficult for medium- and small-scale dairy processing enterprises to market their products through supermarkets. At the same time, the costs have created incentives for large-scale dairy processing enterprises to expand their markets and merge medium- and small-scale dairy processing enterprises to achieve the sales volume and economies of scale required to profitably sell through supermarkets.

Table 6. Average number of dairy products sold in supermarkets in Kunming, Shanghai and Beijing.

	Kunming		Shanghai		Beijing	
	Number of brands	%	Number of brands	%	Number of brands	%
Sanyuan	0.2	0.2	11.3	9.4	47.7	26.4
Bright	13	15.4	37.8	31.5	18.6	10.3
Yili	16.4	19.4	11.3	9.4	32.3	17.9
Mengniu	10	11.8	11.3	9.4	21.6	12.0
Others	45	53.2	48.3	40.3	60.3	33.4
Total	84.6	100	120	100	180.5	100

In October 2003, the People's Network published a copy of a schedule of fees charged by J supermarket on the entrance of goods from the suppliers. A portion of the fee schedule is reproduced in Table 7. From this table, we see that if a dairy processing enterprise wants to market its dairy products through the supermarket, the company will have to pay more than just the 0.4 million Yuan required by the chain's general office (National Day Fee). There are a variety of fees related to product placement and store maintenance that are paid to the individual outlets. In addition, the supermarket may deduct a considerable proportion of marketing margin in the form of concessions and penalties. Obviously, these costs represent a very heavy burden for the medium and small dairy enterprises that seek to place products in supermarkets.

Given the strong pressure placed on dairy firms at the retail end of the market, the competition among dairy processing enterprises has become fierce. A few large-scale dairy processing enterprises have rapidly merged many medium- and small-scale dairy processing enterprises through equity purchases to expand their scale. In 2002, China had a total number of over 700 dairy processing enterprises, of which, there were 304 dairy processing enterprises with a daily processing capacity of over 50 tons of milk. Table 8 shows the breakdown of these firms by capacity and location of production. Note that the large-scale dairy processing enterprises, which account for only 3% of total dairy processing enterprises, have concentrated over 30% dairy processing capacity. For example, the 6 largest dairy processing companies – such as Shanghai Bright, Beijing Sanyuan, Inner Mongolia Yili and Mengniu, Heilongjiang Wandashan and Shijiazhuang Sanlu – own 52 dairy processing enterprises, accounting for 17.1% of total number of plants with a daily processing capacity of over 50 tons of milk. These firms have a total daily processing capacity of 17,554 tons of milk, accounting for 29.9% of the total industry processing capacity. Bright Dairy and Shijiazhuang Sanlu alone

Table 7. Supermarket entrance fees charged by supermarket J.

Name of Payment	Unit	Amount (1000 yuan)	Name of Payment	Unit	Charge ratio (%)
National Day	Total for all stores	400	Ex-factory interest concessions	Gross sales	8%
Opening a business	Total for all stores	10-20	Service charge	Gross sales	1.5-2%
Old store renovation	Total for all stores	10-20	Consultancy fee	Gross sales	2%
DM bulletin fee	Total for all stores	2.3	Administration fee for arranging goods (排面)	Gross sales	2.5%
Endcap product display (端头)	Per store	2	Fees for delayed goods Delivery	Per day	0.3%
New products	Each new product per store	1	Difference of tax	Gross sales	5-6%
Employee administration	Per man per month	2	Unconditional return goods	Gross sales	3-5%
Stacking product display（堆头）	Per store	30-100	Undamaged fee		fitting

Source: People's Network, 2003.

have 19% of the nation's processing from plants with a minimum capacity of 50 tons. The top 6 companies had total assets of 9.68 billion Yuan in 2003, representing 38.3% of total industry processing assets. Their marketing revenue, at 16.95 billion Yuan, was 50.7% of total revenues for industry that year. Sales revenue for Shanghai Bright Dairy and Yili alone accounted for nearly one-third of the total revenues for the industry.

These top 6 firms are organized as dairy groups: the Bright Group in Shanghai, Sanyuan Group in Beijing, Sanlu Group in Hebei, Yili Group and Mengniu Group in Inner Mongolia and Wandashan Group in Heilongjiang. Since mid-1990s, these dairy groups have tried to reduce milk procurement costs, increase scale economies, and service regional markets with fresh products by purchasing or merging processing companies in China's 3 regions. The dairy group structure was created to strengthen and centralize the administration of the dairy processing enterprises distributed across various regions of the country.

The Shanghai Bright Dairy Group is one of the largest dairy enterprises in China, and we briefly recount the company's development to illustrate the rapid consolidation that has occurred in China's dairy sector since the mid 1990s. The Shanghai Bright Dairy Group was established as the Shanghai Dairy Company in 1956 during the planned economy period. The Shanghai Dairy Company monopolized the dairy product markets in Shanghai until the 1990s. Since the early 1990s, the Bright Dairy Group has been adapting to the changing market climate by actively using its advantages in technology, funds and markets to expand its production scale and its share in raw milk markets. In 1994, the Bright Dairy Group established a contract with the Shanghai Xinghuo Milk Company. In 1997, it merged the Shanghai Dairy No. 6 Factory and Shanghai Xin'an Dairy Factory. Meanwhile, it established the Chongqing Bright Dairy Factory, the Inner Mongolia Hulunbeier Prefecture Bright Co. Ltd. and the Heilongjiang Provincial Bright Song'e Dairy Co. Ltd. through joint operation agreements and share holding. Through this process, Bright acquired access to raw milk resources in Inner Mongolia and Heilongjiang Province, as well as a foothold in the Chongqing market. Since the early 1990s, the Bright Dairy Group has established dairy processing factories in Beijing, Tianjin, Jiangsu, Anhui, Jiangxi, Hubei, Hunan, Guangdong and Shaanxi through joint operation agreements and share holding. Meanwhile, it has established marketing bases in Heilongjiang, Inner Mongolia, Beijing, Tianjin, Xi'an, Wuhan, Qingdao, Dezhou, Nanjing, Wuxi, Hangzhou, Jinhua and Guangzhou (Bao, 2000; Chen, 1998; Wang, 1998). By the end of 2002, the Shanghai Bright Dairy Group had established 10 dairy

Table 8. Number of dairy enterprises by processing scale and location.

	Below 50-100		101-300		301-500		501-1000		Over 1001	
	Number	%	Number	%	Number	%	Number	%	Number	%
Coastal Region	44	38.3	49	42.6	12	10.4	8	7.0	2	1.7
Agricultural Region	60	42.3	64	45.1	10	7.0	4	2.8	4	2.8
Pastoral Region	16	34.0	16	34.0	7	14.9	5	10.6	3	6.4
Total in the Nation	120	39.5	129	42.4	29	9.5	17	5.6	9	3.0

processing enterprises in the coastal, 5 in the central agricultural region, and 1 pastoral region. Table 9 displays the distribution of dairy processing facilities for the top 6 dairy groups.

5. Conclusions

The emergence of supermarkets along with rapid development of the economy is quickly changing the traditional dairy industry in China. As supermarkets swiftly replace various traditional dairy product distribution channels, this change is promoting the formation of a national unified dairy product market. The development of supermarkets has created objectively favourable conditions for the movement of milk products from the low-cost production bases in the western pastoral provinces to the major consumption markets in the coastal region, and supermarkets have encouraged the consolidation of dairy firms into large-scale dairy-processing enterprises with centralized management. Along with the improvement of dairy product distribution, supermarkets have created opportunities for consumers in remote townships and rural areas to gain access to a greater variety of dairy products. This is particularly important in light of the discussion in section 2.3, because the gradual increase in the number of rural children consuming dairy products may lead to tremendous growth in dairy product demand by the next generation. China, with a population of 1.3 billion, may one day become the largest consumer of dairy products in the world.

Table 9. Distribution of Dairy Processing Enterprises of the 6 Dairy Groups.

	Bright	Sanyuan	Yili	Mengniu	Sanlu	Wandashan
East Coastal Region						
Beijing	4	6	1	1		
Tianjin	1	1				1
Shanghai	2	1				
Jiangsu	2					
Guangdong	1					
Total in East Coastal Region	10	8	1	1	0	1
Agricultural Region						
Hebei			2		11	
Henan				1	1	
Anhui	1					
Jiangxi	1					
Hubei	1					
Hunan	1					
Shanxi	1					
Total in Agricultural Region	5	0	2	1	12	0
Western Pastoral Region						
Heilongjiang	1		2			1
Inner Mongolia		2	2	2		
Xinjiang			1			
Total in West Pastoral Region	1	2	5	2	0	1
Grand total	16	10	8	4	12	2

With such tremendous potential growth in consumption, many in the industry are wondering who will supply this demand. While there is still much scope for future growth in milk production in China (Fuller *et al.*, 2005), limited feed resources and competition from other livestock industries will challenge the domestic industry's ability to keep pace with consumption. Some combination of importation of feeds and dairy products from abroad is inevitable in the long term. The recent appreciation of the Chinese Yuan relative to the U.S. dollar combined with the global acquisition system of trans-national supermarket groups has created conditions that enable imported UHT milk to be competitive with domestic products. Given the growing acceptance of UHT milk, this may have important consequences for the domestic industry in the short run.

While certain aspects of the development of supermarkets in China are unique, many of the effects supermarkets are having on the dairy industry in China are similar to impacts of supermarkets in other developing countries (Reardon and Berdegué, 2002). The emergence of supermarkets (including big stores and chained convenient stores) and the extension of their marketing networks to medium- and small-size cities and rural areas is causing these new marketing channels to replace the traditional retail sector. The small stores and private peddlers of agricultural and livestock products are greatly affected. The influence of supermarkets is also playing a roll in the fundamental change in the structure of production, distribution, and consumption of agricultural and livestock products in China.

Therefore, agricultural policy makers, implementers, as well as researchers, must be aware of and account for the influence of supermarket development in their analysis and policy formulation. There is a need for further research to better understand the impacts of supermarkets and procurement and distribution chains on rural households, which still constitute the bulk of China's population. Evidence from other countries suggest that small producers and processors can be shut out of the supply chain because of their inability to make the investments necessary to meet quality standards demanded by large retail chains. Moreover, as informal retail chains disappear, so do the alternatives for these producers (Farina, 2002; Faiguenbaum *et al.*, 2002). The dairy groups in China and various levels of government are actively promoting and facilitating adoption of improved dairy production technology by rural households. Nevertheless, research is greatly needed to understand how the rapid changes in dairy markets are affecting the welfare and future prospects of rural households in China.

References

Bao, M., 2000. Shàng hǎi niú nǎi (jí tuán) yǒu xiàn gōng sī de guāng míng zhī lù (Bright path of the Shanghai Milk Group (Co. Ltd.)). In: Wang, H. (ed.) Zhōng guó nǎi yè 50 nián (50 years of dairy industry in China). Beijing Ocean Press, pp. 157-159.

Chen, X., 1998. Shàng hǎi niú nǎi (jí tuán) yǒu xiàn gōng sī shí shī yī tǐ huà jīng yíng de zuò fǎ yú chéng xiào (Practice and achievement of the Shanghai Milk Group (Co. Ltd.) in implementing integrated operation). Zhōng guó nǎi niú (Dairy cattle in China) 1, 7-9.

China Chain Store & Franchise Association, 2003. China chain operation yearbook, 2003. Beijing, China Commercial Press.

China Dairy Development Strategic Study Research Group, 1990. Zhōng guó nǎi yè zhàn lù yán jiū (Studies on dairy industry development strategy). Hubei Science and Technology Press, pp. 18-37.

Farina, E., 2002. Consolidation, multinationalisation, and competition in Brazil: impacts on horticulture and dairy product systems. Dev. Pol. Rev. 20, 441-457.

Faiguenbaum, S., J. Berdegué and T. Reardon, 2002. The rapid rise of supermarkets in Chile: effects on dairy, vegetable, and beef chains. Dev. Pol. Rev. 20, 459-471.

Fuller, F., J. Huang, H. Ma and S. Rozelle, 2005. The rapid rise of China's dairy sector: Factors behind the growth in demand and supply. Working paper No. 05-WP 394, CARD, Iowa State University, 38 pp.

Garnaut, R. and G. Ma, 1993. How rich is China: evidence from the food economy. Aus. J. Chn. Aff. 30, 121-146.

Hu, D., H. Yu and T. Reardon, 2003. Zhōng guó xiāo fèi zhě chāo shì gòu mǎi shēng xiān nóng fù chǎn pǐn xiāo fèi xíng wéi yán jiū (Study on the consumption behavior of consumers in purchasing agrifood in supermarket). Chn. Rur. Econ. 224, 12-17.

Liang Y. and X. Zhou, 1999. Wǒ guó nǎi lèi shì chǎng xū qiú yú diào kòng duì cè yán jiū (Study on the demand of dairy product market in China and macro control countermeasures). In: Li, Y. (ed.) Nǎi yè chūn qiū (Chronicle of the dairy industry). China Agricultural Press, pp. 186-195.

Liu, Y., 2000. Chǎn gōng xiāo yī tǐ huà cù jìn le běi jīng nǎi yè de fā zhǎn (Integration of production, supply and marketing has promoted the development of dairy industry in Beijing). In: Wang, H. (editor) Zhōng guó nǎi yè 50 nián (50 years of dairy industry in China). Beijing Ocean Press, pp. 14-21.

People's Network, 2003. Partial entrance fees for J supermarket. October 30.

Reardon, T. and J. Berdegué, 2002. The rapid rise of supermarkets in Latin America: challenges and opportunities for development. Dev. Pol. Rev. 20, 371-388.

Song, G., 2000. Wèi le rén rén jiàn kāng mù biāo tiān tiān xū yào hē niú nǎi (For the objective of everyone's health, we need to drink milk everyday). In: Wang, H. (ed.) Zhōng guó nǎi yè 50 nián (50 years of dairy industry in China). Beijing Ocean Press, pp. 381-384.

Wang, J., 1998. Kuò zhǎn jí tuán, yùn zuò zī chǎn, zhèn xīng mín zú rǔ yè – zài 98 zhōng guó nǎi niú kē xué lùn tán de fā yán (Expand dairy groups, operate assets and rejuvenate national dairy industry – speech at the 1998 China dairy science forum). In: Proceedings of the 1998 zhōng guó nǎi niú kē xué lùn tán zī liào huì biān (1998 Chinese dairy science forum), pp. 94-97.

Wu, C., 2003. Dairy products industry overview. GAIN Report No. CH3814. USDA, FAS, 10 pp.

Zhou, Z.-Y., W-M.. Tian and J.-L. Zhou, 2002. The emerging dairy economy in China: production, consumption, and trade prospects. Aus. Agribus. Rev. 10, Paper 8.

Organizational learning in the context of an agri-food chain

A.J. Dunne

Abstract

This paper reports on the outcomes of the first stage of a longitudinal study that focused on the transformational change process being undertaken within the Supply Chain and Operations Area of a major Australian food manufacturing company. Organizational learning is an essential prerequisite for any successful change process and an organization's ability to learn is dependent on the existence of an environment within the organization that nurtures learning and the presence of key enablers that facilitate the learning process. An organization's capacity to learn can be enhanced through its ability to form and sustain collaborative relationships with its chain partners. The results show that an environment that supports organizational learning is being developed through consultative leadership and the empowerment of individuals within a culture that supports innovation and cross-functional teamwork but demands responsibility and accountability. The impact of these changes within the Supply Chain and Operations Area is evident in the significant improvement in the Area's productivity and efficiency levels over the past twelve months. The company's endeavours to engage its major supply chain partners in the learning process have been limited by the turmoil within the company. However the company has involved its supply chain partners in a series of mutually beneficial projects that have improved communication and built trust thereby laying the foundations for more collaborative chain relationships.

Keywords: organizational learning, change management, supply chain management

1. Introduction

For the past ten years, a consistent theme in the agribusiness literature has been the ever increasing rate of change in the competitive environment bought on by globalization and technological innovation (Boehlje *et al.*, 1995; Boehlje, 1999; Dunne, 2001; O'Keeffe and Mavondo, 2004). This era is characterised by discontinuity and uncertainty and has been described by Huyett and Viguerie (2005) as a 'world of extreme competition'.

Organizations that thrive in such times are what Senge (1990) describes as 'learning organizations' – organizations that have the ability to learn and adapt faster than their competitors. According to Limerick *et al.* (2000) such organizations are characterised by the presence of metastrategic management, flexible structures and empowered collaborative individuals. Further they claim that inter-organizational networks and strategic alliances are core business for such organizations.

The literature pertaining to organizational learning is predominantly theoretic or prescriptive in nature rather than reporting empirical studies that evaluated the causes and outcomes associated with organizational leaning initiatives (Argyris and Schön, 1996; Garavan, 1997; Senge *et al.*, 1999; Smith and Tosey, 1999; Lakomski, 2001).

Most empirical studies on organizational learning are limited in their scope and depth. For example, in most cases the source of data is restricted to one manager in each participating organization and usually the data is collected at one point in time (Beeby and Booth 2000; Spekman *et al.*, 2002; Bessant *et al.*, 2003; Murray, 2003; Lopez *et al.*, 2005; Preiss and Murray, 2005).

Such approaches ignore the fact that learning occurs at different levels of an organization, in different forms (adaptive and generative) and over time. In addition, such approaches ignore the role that inter-organizational relationships have on the transfer on tacit and explicit knowledge.

This paper reports on the transformational change process being undertaken within the Supply Chain and Operations Area of a major Australian food manufacturer, Golden Circle Limited and the impact that collaboration with their major customers and suppliers has had on organizational learning and its performance.

The decision to restrict this study to a single case, or value chain in this instance was made because a single case could be studied in more detail, access to relevant people was more feasible and data collection was more manageable. Any deficiency related to statistical generalization from the findings of the research was compensated for by the opportunity to gain a deeper understanding of the factors that influence the success of a change management process within a value chain context. The validity of single case research is well supported in the literature (Boehlje, 1999; Sterns *et al.*, 1998; Stake, 1994; Yin, 1994).

2. The research setting

2.1 The Australian food industry

The Australian food industry has experienced a similar degree of consolidation as that reported in the UK, Europe and North America at all levels of retail, manufacturing, packaging and raw ingredient supply (Fearne *et al.*, 2001; Dawson, 2004; Stock, 2004; Regmi and Gehlhar, 2005).

Supermarkets and grocery stores dominate food retailing in Australia accounting for sales revenues of $55 billion in 2003-04 (DAFF, 2005). In turn, food retailing is dominated by two major supermarket chains, Woolworths Limited and Coles Myer Limited who control between 60 and 70% of the market (Delforce *et al.*, 2005).

Food manufacturing is the largest sector of Australia's manufacturing industry with an annual turnover of $54 billion in 2003 (DAFF, 2005). Of the top 25 grocery sector food suppliers, 20 are owned or controlled by overseas interests (Retail World, 2005). As Hughes (2004) points out, foreign direct investment has become an increasing important strategy for multinational food companies to access and dominate local markets. Golden Circle is ranked 21 in the list of national grocery sector food suppliers.

The retail shelf stable fruit category is dominated by two firms SPC Ardmona (acquired by Coca-Cola Amatil in 2005) and Golden Circle. However this category is facing intense competition from imports (Dole, Del Monte and Heinz) and the increasing presence of supermarket private label products.

2.2 The case study firm

Golden Circle Limited is a grower-owned cooperative that manufactures in excess of 800 products including shelf stable fruits and vegetables, fruit juices, cordial, conserves and baby food. Total sales revenues were $AUD 414 million in 2004 of which 12 percent were attributed to shelf stable pineapple products. (Golden Circle, 2005)

In 2003 the company recorded an annual operating loss of $AUD 32 million which threaten its on-going viability and triggered radical change in the company's management and operations. Within 15 months, all but one of the Senior Management team, including the long serving Managing Director, had been replaced, the grower dominated Board had been reconstituted, external equity capital had been raised and a restructuring strategy implemented.

Central to this reconstruction strategy was improved operational planning to better align production and inventory with sales and the elimination of unnecessary costs from the company's operations and its material supply chain.

> *"The substantial loss in 2003 brought into sharp focus the need for total performance improvement across the company's operations – from management procedures and systems and sourcing and processing of raw materials, to the product made, when and where." (Golden Circle 2004: 4)*

A key component of this transformation process was the lean supply chain project for pineapple which was funded under the Australian government's National Food Industry Strategy (AFFA, 2000).

3. The research project

3.1 The research focus and its conceptual framework

This research set out to investigate three specific questions concerning the transformational process being undertaken with the Supply Chain and Operations Area of Golden Circle:

1. What was the impact of the change process on the Area's capacity to generate new knowledge.
2. What was impact of the change process on the Area's capacity to collaborate with its key chain partners to generate shared knowledge.
3. What was the impact of the change process on the Area's capacity to improve its performance through the utilization of new and shared knowledge?

These themes of investigation emerged from a review of the literature which indicated that:

- Information is the basis on which knowledge is built (Sveiby, 2001; Collins *et al.*, 2002).
- Knowledge within an organization is created by individuals and their interaction with colleagues both internal and external to the firm – the firm's collaborative capacity (Senge, 1990; Kanter, 1994; Argyris and Schön, 1996; O'Keeffe and Mavondo, 2005).
- The quality of information shared between an organization and its chain partners is influence by the strengths of their business relationships (Kanter, 1994; Morgan and Hunt, 1994; Barlow and Jashapara, 1998; Bessant and Tsekouras, 2001; O'Keeffe and Mavondo, 2005; Preiss and Murray, 2005).
- The absorptive capacity of an organization is influenced by its culture, leadership, structures and processes (Grant, 1996; Appelbaum and Reichart, 1997; Spekman et al., 2002; O'Keeffe and Mavondo, 2005).
- The capacity of an organization to execute effectively is a hallmark of a learning organization (Collins, 1999; Hubbard *et al.*, 2002; Liker, 2004).

Drawing on the literature a model of organizational learning in the context of a value chain has been constructed that identifies the key factors that nurture organizational learning (leadership, culture and structure & processes), the key enablers of learning (shared vision, willingness to act and the capability to act) which together with the nature of the organization's relationships with its key chain partners determine its ability to improve its competitive performance (Figure 1).

3.2 The research design

Participants from Golden Circle were drawn from Senior Management, Middle Management and Operations. In total 13 individuals were involved. Representatives from Golden Circle's chain partners were selected from their owner/suppliers and their major supplier of materials, Amcor Australasia through two of its Divisions – Amcor Fibre Packaging and Amcor Food Cans. Interviews were conducted with two representatives of each of these Divisions. In each case, one interviewee was at the General Manager level while the other was the company's Golden Circle Account Manager.

Interviews were sought with similar level representatives of a major retailer involved in the project but after an initial interview with the company's Planning and Projects Manager, the company declined to participate further in the project due to a lack of any perceived tangible benefit to them.

The primary source of data was collected though two rounds of semi-structured interviews that were conducted with these informants in February 2004 and April 2005. In total 44 semi-

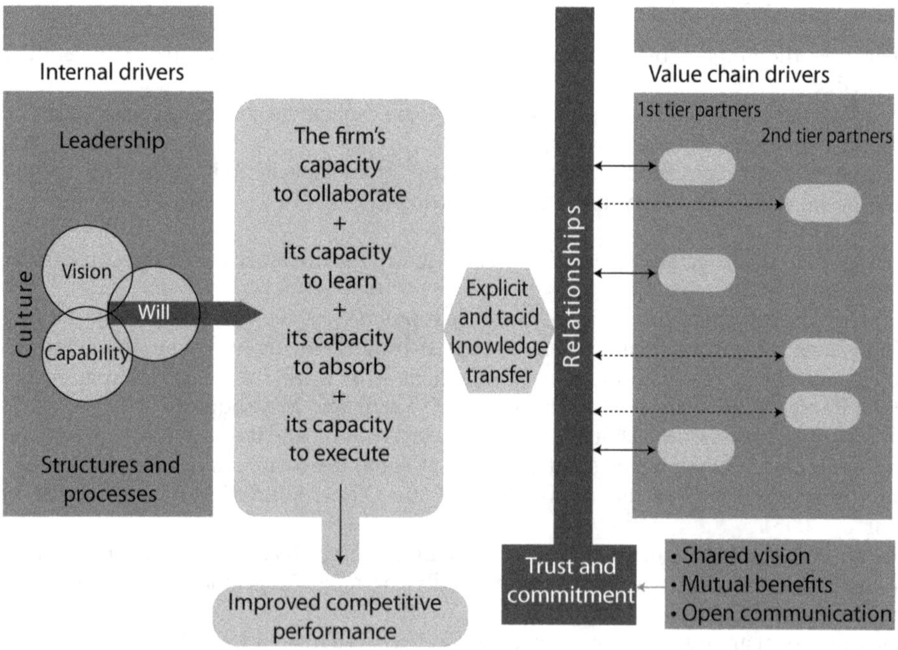

Figure 1. A conceptual model of organizational learning in a value chain context.

structured interviews were undertaken as part of this study. Other sources of data included my personal observations during workshops, presentations, informal conversations and relevant documents such as annual reports, workshop summaries and press releases.

The interview transcripts were analysed within a conceptual framework of organizational learning within a supply chain context constructed from the literature. The key dimensions of this conceptual framework within the focal organization were its learning environment (determined by the organizations leadership, culture, structure and processes) and its learning enablers (collective ownership of a vision for the future, collective motivation and the collective capabilities to achieve the vision). The type inter-organizational relationships that exist between Golden Circle and its major chain partners impact on both its learning environment and its learning enablers through the extent collaboration exists and the quality of the tacit and explicit knowledge transferred between them.

The first round of interviews focused on establishing a baseline measure of the focal company's leadership, culture, structure and performance at the time the transformational process began. In addition data was collected on the relationships and the extent of information transfer that existed between the supply chain partners. The second round of interviews was designed to monitor changes in these attributes, processes and relationships over the 12 month period and to probe the factors that promoted or hindered the change process.

4. Results and discussion

4.1 Golden Circle as it was (Pre February 2004)

The analysis of the transcripts from the initial round of interviews clearly demonstrated that Golden Circle at the commencement of the transformational process exhibited few of the characteristics of an organization that nurtured learning.

The leadership style was autocratic.

> *"One could describe it as probably benevolent dictatorship to a certain degree. I think the style's reasonably autocratic." (Senior Manager)*

> *"The staff were essentially told what to do and they did it and there wasn't I guess an environment whereby they were invited to participate in decision-making or even make their own decisions when needed on a daily basis." (Middle Manager)*

The organizational culture while supportive of the individual in a paternalistic fashion was conservative and reactionary.

> *"Golden Circle is not a harsh organization to work for. It does value its people. It probably doesn't communicate well to its people and do things effectively from time to time but I still think it does value its people and it does have a soul." (Senior Manager)*

> *"We've been in comfortable circumstances and there wasn't really seen to be the great need for a shake up or change." (Senior Manager)*

> *"Historically I would have to say they (senior management) don't accept change very well." (Middle Manager)*

Its structures and processes were inadequate and discouraged collaboration.

> *"The company didn't really know which of its products were profitable and which ones weren't.*
>
> *I mean, we never use to measure anything, well not accurately anyway." (Middle Manager)*
>
> *"It was like everybody wanted their own patch instead of pulling together." (Senior Manager)*

This description of Golden Circle fits that of what Limerick *et al.* (2000) refer to as a 'functional organization' that is appropriate under conditions of stability but inappropriate under conditions of change because of its inherent formality and the presence of functional silos that inhibit the organization's responsiveness to changes in its competitive environment.

Golden Circle's inability to respond to the rapid changes that had engulfed the company was compounded by the absence of the enabling factors of organizational learning – the presence of a shared corporate vision plus a motivated and capable workforce.

> *"There didn't seem to be any clear strategy in terms of what businesses the company wanted to be in and what ones it didn't want to be in." (Senior Manager)*
>
> *"We've been a very fragmented company whereby each department has had its own agendas and we haven't really aligned to each other's agendas as a complete strategy." (Middle Manager)*
>
> *"I think they've been frustrated in the past from not being able to contribute. They want to see Golden Circle succeed and they just need some direction in terms of how they can contribute." (Senior Manager)*
>
> *"We're currently doing skills assessment on a lot of our people. Doing manpower plans and seeing whether we've got gaps, to make sure that we have those skill sets in place for the future." (Middle Manager)*

A consistent theme that emerged from the interview data was that the strength of Golden Circle lay in its brand that had been built on being 'Australian Owned and Australian Grown'. However its declining financial performance indicated that the strength of its brand was no longer sufficient to sustain the company's competitive position in the market; its corporate culture had discouraged innovation and initiative, while its staff recruitment and development policies were inadequate to meet the challenges confronting the company. In addition, the insular nature of Golden Circle had also impacted on its ability to form collaborative relationships with its main supply chain partners.

Any analysis of the relationships between Golden Circle and its chain partners is complicated by the fact that its pineapple suppliers are the company's major shareholders and historically they had dominated its Board of Directors. This grower dominance of the Board impacted on the performance of Golden Circle in two important ways – they were conservative and reluctant to change and their power affected the profitability of the company through fruit payments and quality of the fruit intake.

> *"Grower politics is something that we have to deal with on an on-going basis and some of the pineapple growers don't want to see a whole lot of change. A lot of growers simply want to get paid money that would make them profitable without making any changes". (Middle Manager)*

> *"The quality variability in our fruit intake is just so out of control and it's gone into our operational costs which are too high." (Middle Manager)*

All major suppliers to Golden Circle, including growers, were bound by medium to long term supply contracts that meant that they all had a vested interest in the continuing viability of the company and as a result they were willing to participate in the change process.

> *"Golden Circle is probably one of Amcor's biggest customers in Australia. So we have got a huge financial interest in making sure it's successful and participate in the process." (Supplier)*

> *"We growers are there in the open market, the global market. So we have to become more efficient, hence we have to go through this process of change that we're talking about." (Grower)*

> *"It's very important to get the stakeholders involved and you get them to understand the value in the project for themselves. There's a lot of learning that they can take away and then apply to their own businesses and other business relationships." (Senior Manager)*

A major impediment to the change process in its initial stage was the absence a clear strategy for the company going forward which inhibited the commitment of stakeholders to the change process.

> *"We wondered initially if there had been enough homework done on their strategic direction. They embarked down the supply chain route without really having a strategic plan as to what they really wanted to achieve with that process." (Supplier)*

A second impediment to this process was the quality of the information exchanged between the chain partners. Historically while communication at the operations level was frequent and open, the information exchanged particularly in regards to forecasts was often inaccurate or incomplete. At the strategic level collaboration had been minimal between chain partners.

> *"There are long term relationships there but the real detailed understanding of how what you're doing impacts on the next guy and so on has been quite low. The quality of the information transfer has been very poor." (Senior Manager)*

> *"There was a reluctance I suppose even internally to Golden Circle to say 'well make sure you don't tell them too much'. Definitely in the early stages there was a little bit of playing cards close to the chest that needed to be worked through." (Senior Manager)*

As Spekman *et al.* (2002) explain the type of relationship that exists between supply chain partners is one of the key factors that affect the learning process. Golden Circle's inability to forge close relationships with its key chain partners restricted its access to their tacit and explicit knowledge.

In summary, Golden Circle was a 1950's style cooperative struggling to survive in the global competitive environment of the 21st century. Its relationships with its major chain partners while long standing and functional at the operational level, were normal buyer-seller relationships that were neither collaborative nor strategic in nature. As Golden Circle embarked on its transformational journey of change it was the antithesis of a learning organization.

> *"Historically we haven't been a strong learning organization and we haven't been open-minded about actually absorbing and learning from other organizations." (Senior Manager)*

4.2 Golden Circle 12 months on (April 2005)

By mid 2004 the Senior Management team had been revamped and a company wide restructuring plan had been put in place. Part of this plan was the recruitment of an external equity partner which resulted in the restructuring of the Board and the possibility of a public listing of the company in 2008.

The analysis of the interview data collected in April 2005 clearly showed that the internal environment of Golden Circle had changed significantly and these changes were beginning to deliver results. Although the company's annual results reported a further loss of $AUD 19 million for 2004, by the second half of the year it had reported earnings before interest and tax of $AUD 7.5 million and had moved rapidly to a position where it was trading profitably.

> *"This improvement was largely bought about by an increased focus on cost control and the benefits of sustained improvement programs in the Supply Chain and Operations Area which has resulted in marked improvements in gross margins." (Golden Circle, 2005: 2).*

The Senior Management team had stabilized with the appointment of a new Chief Executive Officer and was working more coherently. A three year business plan to rebuild profitability and shareholder value had been developed and was in the process of being implemented. While the focus of the business plan was company-wide performance improvement, the greatest impact was evident in the Supply Chain and Operations Area.

Under the leadership of the General Manager, the Supply Chain and Operations Area had embraced the principles of 'lean thinking' developed by Toyota (Liker, 2004) and applied them. Within 12 months the operational sections had been reconfigured into 7 value streams that represented the company's major product lines; cross-functional teams had been established to eliminate sources of inefficiency such as waste, non-conformance and duplication; communication channels had been unblocked and individuals had been empowered.

The crucial factors that emerged from the interview data that underpinned the transformation within the Supply Chain and Operations Area were:

1. the total commitment of the General Manager and his middle management team to a change process that was team based and disciplined;

   "I think everyone is working well together. There is a clear direction in terms of where we are going to go. We as a team have achieved quite a lot and everyone is feeling pretty good about that. We set the direction, said we want you on board and the majority of people got on board." (Senior Manager)

2. the empowerment of staff;

   "There has been a fair sort of transition in leadership with the Senior Management team very heavily focused on the team concept and the delegation of authority. I've found that my leadership style had to change fairly significantly. (Middle Manager)

   People want to be involved in teams, be involved in the process. I want to be involved so that I can be part of making a difference. This is a good place to work if you apply yourself and we are moving into exciting times here." (Operations Staff)

3. the implementation of an open communication system that reported on the progress of the change process and its impact on the company's performance;

   "All the production managers are finding that the better they can communicate with their people the better the results and the more that they can foster that communication back from their people then the better the change goes as well." (Middle Manager)

   "Being involved has made me want to improve because it makes sense to me I can see where they are coming from and I can see where they want to go." (Operations Staff)

4. the implementation of a process that identified, approved and monitored operational improvement projects and the incorporation of their outcomes into the Area's standard operating procedures;

   "All our Process Enhancement Projects are reviewed monthly to ensure that they are on track and delivering the on-going results as promised. We have a much better track on our performance which helps us with the prioritization – we are still refining that but we are definitely a long way ahead of last year." (Senior Manager)

5. the implementation of a staff development process that provided mentoring and opportunities for formal and informal skill development;

   "We are spending a lot of money and effort on developing the people we have here. We just had a session where the first lot of graduates got their certificates in front line management and we're going even bigger this year. I think that developing the people and their leadership skills has been where most of the gains have come from." (Middle Manager)

6. the implementation of a system that recognized and rewarded success on an on-going basis.

> *"We are now issuing recognition certificates signed by the CEO and the General Manager for a job well done. We've issued 20 of those in the last 2 months and we have had good feedback on that." (Middle Manager)*

> *"I am enjoying it because I am learning a lot more about how the company runs. Everybody appreciates the barbeques and the certificates because they know that management recognizes what we are doing. It's a pretty good way for us to meet them as well." (Operations Staff)*

The results confirmed that at least within the Supply Chain and Operations Area, Golden Circle was well on the way to establishing an environment that nurtured and enabled organizational learning to occur and to translate these learnings into improved financial performance.

Golden Circle commenced engaging its major suppliers in its change process in December 2003 when it initiated a workshop to inform its chain partners of its internal change initiative and to invite them to participate in a whole of chain approach to improve customer service while at the same time reduce their collective 'cost to serve'. Subsequently, the chain partners agreed on a series of joint chain improvement projects across four major areas – fruit intake, internal processing, customer service and packaging.

> *"We are saying to, particularly our larger suppliers that we want you to understand what our goals are and we want you to help us achieve those goals because if we achieve those goals it's going to be good for everyone." (Senior Manager)*

The analysis of the interview data provided indicated that Golden Circle's progress in developing more strategic relationships with its major suppliers had been minimal. There were several reasons given for this, the most important being the pressures confronting the Supply Chain and Operations management team to improve its internal performance had restricted their ability and the resources available to fully engage with their chain partners.

> *"Unfortunately there was a climate that we had when we embarked on this project that's not conducive to quickly getting into strategic partnering because of a 'them and us' type of supplier relationship - the degree of non-disclosure, a degree of win-lose rather than win-win. Another thing that's affected it over the last 12-18 months has been the other business priorities that we had to deal with which has made it difficult as a lot of the resources in the company have been stretched to the maximum." (Senior Manager)*

In spite of this failure to develop more strategic relationships there were instances of increased collaboration between Golden Circle and its chain partners that were delivering cost savings and better alignment. For example, a quality based payments system for pineapples had been introduced; packaging materials had been down-gauged and progress was being made in more accurately forecasting the demand for packaging material which impacted on inventory holdings.

> *"Growers are changing their practices as a result of the quality based payments system and we are hitting our targets. It's very positive stuff compared to last year where those things were costing us millions of dollars." (Middle Manager)*

"We were able to complete the down gauging project in 6 months or less which is a pretty quick time in this business to do that sort of change. The cost saving was about $200,000 per annum." (Supplier)

"Our systems are still not aligned but the exchange and the quality of the information exchanged has improved pretty dramatically over the past 12 months. There is less fluctuation in the forecast numbers." (Supplier)

By April 2005, the Supply Chain and Operations management team had indicated that developing closer relationships with their chain partners was now their priority:

"Getting our own house in order has been our primary focus. It's hard to go out and develop relationships with suppliers when you don't have the platform in place in your own organization. We have made progress on the grower side and while there has been a little work done with customers and packaging suppliers it hasn't been fast enough and we are poised to make a big push in those areas at the moment." (Senior Manager)

At the time of writing (December 2005) there were signs that a more collaborative relationship had been established with Amcor Fibre Packaging but not with other key suppliers.

5. What are the outcomes arising from this project so far?

The interim results clearly indicate that the transformational change process being undertaken within the Supply Chain and Operations Area of Golden Circle has been successful in terms of improving the efficiency and productivity of the Area. The reduction in operating costs that was reported in the final six months of 2004 had continued with further savings $2.2 million in the first half of 2005, overall consumer complaints had reduced and customer service levels had increased. These results were achieved through the commitment of employees to the change process.

"The other point, harder to quantify, is the great development in skills and cross functional teamwork that is behind our progress – 'the goose that lays the golden eggs'." (Senior Manager)

As Kotter (1996) has pointed out any transformational change within an organization is a difficult and drawn out process which requires commitment and perseverance. The Golden Circle Board had endorsed the change process and the General Manager had the skills and attributes to implement it.

"The Board took the decision that they were to give him (the General Manager, Supply Chain & Operations) the resources (to drive the process) because we knew we had to do something. But just as importantly, we had to make it clear to management at all levels that the Board was going to back this fellow to get the job done." (Board Member)

"We needed somebody in the supply chain area with that expertise. You know once he focuses on something he's tireless. It's people like him that will take the company forward." (Senior Manager)

Schein (1992) observed that organizations that were going out of business often discovered to their surprise, high levels of motivation and commitment among their employees. This was the situation within the Supply Chain and Operations Area – a situation that was exploited by the Area's General Manager to develop a culture of continuous improvement based on individual and group learning that would underpin their future performance.

The key to this cultural change was the empowerment of individuals within a framework that supported innovation and cross-functional teamwork but demanded responsibility and accountability. This framework was based on the Toyota problem solving process and Deming's 'plan, do, check, act which forms the basis of individual and organizational learning (Senge, 1990; Liker, 2004).

The motivation and commitment of employees has been sustained by the role model provided by the General Manager, a process that has delivered short term wins and the presence of an effective system that recognizes and rewards performance. In addition the involvement of management in a mentorship role when coupled with the formal and informal training opportunities presented to employees has helped overcome what Schein (2002) has described as 'the anxiety of failure' which is often associated with organizational change.

In summary, an environment that nurtures learning is evolving within the Supply Chain and Operations Area due to changes in leadership that has become more inclusive, in organizational culture that is supportive of innovation and in organizational structures that facilitate open communication and teamwork. Individual and group learning has been enabled by the presence of a clear focus to reduce costs, positive reinforcement and staff development opportunities.

The absence of significant progress in Golden Circle developing strategic partnerships with its chain partners is not surprising given the state of turmoil within the company. Cann (1998) identifies eight steps that are involved in building business-to-business relationships, five of which have an internal focus associated with developing the appropriate strategy, culture and capabilities that potential partners will find attractive. For the past 12-18 months Golden Circle has been following this process while engaging its supply chain partners in cost reduction projects that have improved communication and built trust. The foundations have been laid to take their supplier relationships to the next stage on supplier-customer relationship continuum.

The difficulties experience by Golden Circle in engaging its major supply chain partners in comprehensive whole of chain project is consistent with examples reported in the literature. Lambert and Cooper (2000) report that it would be counter productive for a firm to develop collaborative relationships with all members of a supply chain. Instead the firm should concentrate on allocating scarce relationship resources to those supply chain partners that are critical to the success of the company - a strategy Golden Circle is adopting with respect to its grower/suppliers and Amcor. This approach is similar to that described by Duffy and Fearne (2004) who assert that the supply chains in the UK retail food sector are managed as a series of dyadic relationships. Normally such relationships are what Cox (2004: 353) describes as *adversarial collaboration* where exchange partners provide *extensive operational linkages and relationship specific adaptations but seek the appropriation of value.* Spekman *et al.* (1998) argues that such relationships are coordinated rather than collaborative chain partnerships because of the limitations placed on the transfer of tacit and explicit knowledge between the exchange partners.

In summary, while the quality and quantity of operational information exchanged between the chain partners has improved, its contribution to learning and its improved operational performance within Golden Circle has been minimal at this stage. This is not unexpected given the absence of a shared vision between the chain partners and the nature of the historical relationships that had existed between the chain participants.

6. Conclusion

This paper reports the interim results of an empirical study of organizational learning in the context of an agri-food chain. The preliminary results of this study support the theoretical and prescriptive models of organizational learning found in the literature but highlight three important points: the development of the ideal learning environment and learning enablers even within a section of an organization requires commitment and perseverance, the necessity of getting the internal learning environment and learning enablers in place before engaging chain partners in the process and finally the difficulty in engaging multiple chain partners in the process simultaneously.

It was obvious from the interview data that sustaining the change process within the Supply Chain and Operations Area of Golden Circle was placing severe pressure on the General Manager and his middle management team. They have recognized this problem and have scaled back on the number of Process Improvement Projects initiated. It will be interesting to monitor the impact of this decision on the group's ability to prioritize projects and still maintaining the enthusiasm of the people whose projects have been deferred.

Another challenge facing the Senior Management team is the transfer the learnings from the change process being undertaken in the Supply Chain and Operations Areas to other Areas of the company such as Sales and Marketing, Finance and Corporate Services. This transfer of knowledge and its application is essential if Golden Circle as a whole is to become a learning organization.

The third major challenge is for Golden Circle to successfully engage its major supply chain partners in the collaborative learning process. This may prove the biggest challenge of all since its major material supplier is a global company and while Golden Circle is a major customer in Queensland, it is a minor player in Amcor's entire business. Similarly Golden Circle's major customers are the two major supermarket chains who have their own supply chain agenda that could marginalize the Golden Circle brand over time.

Golden Circle has been able to successfully implement a transformational change process within its Supply Chain and Operations Area that has resulted in the company reporting and operating profit for 2004. However embarking on a change process that produces short term wins and one which sustains a culture of continuous improvement are two different things. Given the rapidly evolving competitive environment of the Australian food industry there is a real possibility that time is against Golden Circle effecting the necessary changes in its corporate culture and performance to survive in its present form.

References

AFFA, 2000. National Food Industry Strategy. Canberra: Department of Agriculture, Fisheries and Forestry.

Appelbaum, S.H. and W. Reichart, 1997. How to measure an organization's learning ability: a learning orientation: part 1. Journal of Workplace Learning 9, 225-239.

Argyris, C. and D.A. Schön, 1996. Organizational Learning II. Reading: Addison-Wesley.
Barlow, J. and A. Jashapara, 1998. Organizational learning and inter-firm „partnering in the UK construction industry. The Learning Organization 5, 86-98.
Beeby, M. and C. Booth, 2000. Networks and inter-organizational learning: a critical review. The Learning Organization 7, 75-88.
Bessant, J., R. Kaplinsky and R. Lamming, 2003. Putting supply chain learning into practice. International Journal of Operations and Production 23.
Bessant, J. and G. Tsekouras, 2001. Developing learning networks. A.I. and Society 15, 82-98.
Boehlje, M., 1999. Structural changes in agricultural industries: how do we measure, analyse and understand them? American Journal of Agricultural Economics 81, 1028-1041.
Boehlje, M., J. Akridge and D. Downey, 1995. Restructuring Agribusiness for the 21st Century. Agribusiness 11, 493-500.
Cann, C.W., 1998. Eight steps to building a business-to-business relationship. Journal of Business and Industrial Marketing 13, 393-405.
Collins, J., 1999. Good to Great. Sydney, Australia: Random House.
Collins, R., T. Dunne and M. O'Keeffe, 2002. The 'locus of value': a hallmark of chains that learn. Supply Chain Management 7, 318-321.
Cox, A., 2004. The art of the possible: relationship management in power regimes and supply chains. Supply Chain Management 9, 346-356.
DAFF, 2005. Australian Food Statistics 2004. Canberra: Department of Agriculture, Fisheries and Forestry.
Dawson, J., 2004. Food Retailing, Wholesaling and catering. In: M.A. Bourlakis and P.W. Weightman (eds.), Food Supply Chain Management. Oxford, U.K.: Blackwell Publishing.
Delforce, R., A. Dickson and J. Hogan, 2005. Australia's Food Industry. Australian Commodities 12, 397-390.
Duffy, R., and A. Fearne, 2004. Partnerships and Alliances in UK Supermarkets Supply Networks. In: M.A. Bourlakis and P.W. Weightman (eds.), Food Supply Chain Management. Oxford, U.K.: Blackwell Publishing.
Dunne, A.J., 2001. Supply chain management: fad, panacea or opportunity. Perspectives, Paper 48, pp.1-38.
Fearne, A., D. Hughes and R. Duffy, 2001. Concepts of collaboration: supply chain management in a global food industry. In: J.F. Eastham, L. Sharples and S.D. Ball (eds.), Food Supply Chain Management. Oxford, U.K.: Butterworth-Heinemann.
Garavan, T. 1997. The Learning Organization: a review and evaluation. The Learning Organization 4, 18-29.
Golden Circle, 2004. Annual Report, 2003. Brisbane: Golden Circle Limited.
Golden Circle, 2005. Annual Report, 2004. Brisbane: Golden Circle Limited.
Grant, R.M., 1996. Toward a knowledge-based theory of the firm. Strategic Management Journal 17, 109-122.
Hubbard, G., D. Samuel, S. Heap and G. Cocks, 2002. The First XI: winning organizations in Australia. Brisbane, Australia: John Wiley & Sons.
Hughes, D., 2004. Food manufacturing. In: M.A. Bourlakis and P.W. Weightman (eds.), Food Supply Chain Management. Oxford, U.K.: Blackwell Publishing.
Huyett, W.I., and S.P. Viguerie, 2005. Extreme competition. The McKinsey Quarterly 1, 1-8.
Kanter, R.M., 1994. Collaborative advantage: the art of alliances. Harvard Business Review 69, 96-108.
Kotter, J.P., 1996. Leading change. Boston, Mass.: Harvard Business School Press.
Lakomski, G. 2001. Organizational change, leadership and learning: culture as cognitive process. The International Journal of Educational Management 15, 68-77.
Lambert, D.M., and M.C. Cooper, 2000. Issues in supply chain management. Industrial Marketing Management 29, 65-83.
Liker, J.K. 2004. The Toyota way. New York: McGraw- Hill.
Limerick, D., B. Cunnington and F. Crowther, 2000. Managing the new organization (2 ed.): Business and Professional Publishing.
Lopez, S.P., J.M. Peon and J.V. Ordas, 2005. Organizational Learning as a determining factor in business performance. The Learning Organization 12, 227-245.

Morgan, R.M. and S.D. Hunt, 1994. The Commitment -Trust Theory of Relationship Marketing. Journal of Marketing 58, 20-38.

Murray, P., 2003. Organizational learning, competencies, and firm performance: empirical observations. The Learning Organization 10, 305-316.

O'Keeffe, M. and F. Mavondo, 2005. Capabilities and competing: high performance in the food and beverage industries. Canberra: National Food Industry Strategy and the ANZ Bank.

Preiss, K.J. and P.A. Murray, 2005. Fashions of Learning: improving supply-chain relationships. Supply Chain Management 10, 18-25.

Regmi, A. and M. Gehlhar, 2005. New Directions in Global Food Markets (No. AIB794). Washington, D.C.: USDA, Economic Research Service.

Retail World, 2005. Australasian Grocery Guide. Sydney: Retail World Publications.

Schein, E.H., 1992. Organizational Culture and Leadership (2nd ed.). San Francisco: Jossey-Bass.

Schein, E H., 2002. The anxiety of learning. Harvard Business Review 80, 100-106.

Senge, P.M., 1990. The fifth discipline : the art and practice of the learning organization. Sydney: Random House Australia.

Senge, P.M., A. Kleiner, C. Roberts, R. Ross, G. Roth and B. Smith, 1999. The dance of change. London: Nicholas Brealey.

Smith, P.A. and P. Tosey, 1999. Assessing the learning organization: part 1 - theoretical foundations. The Learning Organization 6, 70-75.

Spekman, R.E., J.W. Kamauff and N. Myhr, 1998. An empirical investigation into supply chain management: a perspective on partnerships. Supply Chain Management 3, 53-67.

Spekman, P., J. Spear and J. Kamauff, 2002. Supply chain competency: learning as a key component. Supply Chain Management 7, 41-55.

Stake, R.E. 1994. Case studies. In: N.K. Denzin and Y.S. Lincoln (eds.), The handbook of quantitative research. Thousand Oaks, California: Sage Publications.

Sterns, J.A., D.B. Schweikhardt and H.C. Peterson, 1998. Using case studies as an approach for conducting agribusiness research. International Food and Agribusiness Management Review 1, 311-327.

Stock, J.R., 2004. The US food supply chain. In: M.A. Bourlakis and P.W. Weightman (eds.), Food supply chain management. Oxford, U.K.: Blackwell Publishing Ltd.

Sveiby, K., 2001. A knowledge-base theory of the firm to guide in strategy formulation. Journal of Intellectual Capital 2, 344-358.

Yin, R.K., 1994. Case study research (2nd Ed.). Thousand Oaks, California: Sage Publications.

Authors and affiliations

Adamowicz M.
Warsaw Agricultural University, Department of Agrarian Policy and Marketing Warsaw, Poland

Barcellos M.D de
Wageningen University, Marketing and Consumer Behaviour Group, Wageningen, The Netherlands

Bárdos K.
Hungarian Academy of Sciences, Institute of Economics, Budapest, Hungary

Beulens A.J.M.
Wageningen University, Information Technology Group, Wageningen, The Netherlands

Bijman J.
Wageningen University, Business Administration Group, Wageningen, The Netherlands

Bremmers H.J.
Wageningen University, Business Administration Group, Wageningen, The Netherlands

Campos E.M.
University of São Paulo, FEA-RP, Ribeirão Preto, São Paulo, Brazil

Consoli M.A.
University of São Paulo, FEA-RP, Ribeirão Preto, São Paulo, Brazil

Dunne A.J.
The University of Queensland, School of Natural and Rural Systems Management, Gatton, Queensland, Australia

Erve C. van
Product Board Animal Feed, The Hague, The Netherlands

Ferreira G.C.
Pontifícia Universidade Católica do Rio Grande do Sul (PUCRS), Mestrado em Administração e Negócios, Porto Alegre, Rio Grande do Sul, Brazil

Fritz M.
University of Bonn, Division for Business Management Organization and Information Management, Bonn, Germany

Fuller F.H.
Iowa State University, Center for Agricultural and Rural Development (CARD), Ames, Iowa, United States of America

Ghisi F.A.
University of São Paulo, Faculty of Economy, Administration and Accounting, Campus Ribeirão Preto, Department of Business, Ribeirão Preto, São Paulo, Brazil

Han J.
Nanjing Agricultural University, Nanjing, 210095, China

Haverkamp D.J.
Wageningen University, Business Administration Group, Wageningen, The Netherlands

Hausen T.
University of Bonn, Division for Business Management Organization and Information Management, Bonn, Germany

Hofstede G.J.
Wageningen University, Information Technology Group, Wageningen, The Netherlands

Hsiao H.I.
Wageningen University, Business Administration Group, Wageningen, The Netherlands

Hu D.
Professor Institute of Agricultural Economics and Development, Chinese Academy of Agricultural Sciences (CAAS), Beijing, China

Ingenbleek P.
Wageningen University, Marketing and Consumer Behaviour Group, Wageningen, The Netherlands

Kambewa E.
Wageningen University, Marketing and Consumer Behaviour Group, Wageningen, The Netherlands

Koot W.B.
University Utrecht, Utrecht School of Economics, Utrecht, The Netherlands

Kramer M.R.
Wageningen University, Information Technology Group, Wageningen, The Netherlands

Kristensen T.
Copenhagen Business School – Handelshøjskolen, Department of Marketing, Copenhagen, Denmark

Lans I. van der
Wageningen University, Marketing and Consumer Behaviour Group, Wageningen, The Netherlands

Lemanowicz M.
Warsaw Agricultural University, Department of Agrarian Policy and Marketing Warsaw, Poland

Li Y.
Wageningen University, Information Technology Group, Wageningen, The Netherlands

Liu X.
China Agriculture University, Department. of Agricultural Engineering, Beijing, P.R.China

Lopes F.F.
University of São Paulo, FEA-RP, Ribeirão Preto, São Paulo, Brazil

Lu H.
Wageningen University, Business Administration Group, Wageningen, The Netherlands

Martinelli D.P.
University of São Paulo, Faculty of Economy, Administration and Accounting, Campus Ribeirão Preto, Department of Business, Ribeirão Preto, São Paulo, Brazil

Martino G.
University of Perugia, Dipartimento di Scienze Economico-estimative e degli Alimenti, Perugia, Italy

Meuwissen M.P.M.
Wageningen University, Business Economics Group, Wageningen, The Netherlands

Meijer S.
Wageningen University, Information Technology Group, Wageningen, The Netherlands

Neves M. F.
University of São Paulo, FEA-RP, Ribeirão Preto, São Paulo, Brazil

Obersojer T.
Technical University Munchen, TUM Business School, Center of Life Sciences Weihenstephan, Freising-Weihenstephan, Germany

Omta S.W.F.
Wageningen University, Business Administration Group, Wageningen, The Netherlands

Paiva H.A.B.
University of São Paulo, FEA-RP, Ribeirão Preto, São Paulo, Brazil

Perugini C.
University of Perugia, Dipartimento di Economia, Finanza e Statistica, Perugia, Italy

Readron T.
Michigan State University, Department of Agricultural Economics, East Lansing, Michigan, United States of America

Revell B. J.
Harper Adams University College, Newport, Shropshire, United Kingdom

Rosenkranz S.
University Utrecht, Utrecht School of Economic, Utrecht, The Netherlands

Ruben R.
Radboud University Nijmegen, Centre for International Development Issues Nijmegen (CIDIN), Nijmegen, The Netherlands

Schulze B.
Georg-August-University Goettingen, Institute of Agricultural Economics, Goettingen, Germany

Sodano V.
University of Naples 'Federico II', Department of Agricultural Economics, Portici, Italy

Spiller A.
Georg-August-University Goettingen, Institute of Agricultural Economics, Goettingen, Germany

Szabó G.G.
Hungarian Academy of Sciences, Institute of Economics, Budapest, Hungary

Tan T.
Nanjing Agricultural University, Nanjing, 210095, China

Theuvsen L.
Georg-August-University Goettingen, Institute of Agricultural Economics, Goettingen, Germany

Tilburg A. van
Wageningen University, Marketing and Consumer Behaviour Group, Wageningen, The Netherlands

Ton G.
Agricultural Economics Research Institute, The Hague, The Netherlands

Trienekens J.H.
Wageningen University, Business Administration Group, Wageningen, The Netherlands

Velthuis A.G.J.
Wageningen University, Business Economics Group, Wageningen, The Netherlands

Vieira L.M.
UNISINOS, PPG em Administração, São Leopoldo, Rio Grande do Sul, Brazil

Vorst J.G.A.J. van der
Wageningen University, Logistics & Operations Research Group, Wageningen, The Netherlands

Wang K.
Nanjing Agricultural University, Nanjing, 210095, China

Weindlmaier H.
Technical University Munchen, TUM Business School, Center of Life Sciences Weihenstephan, Freising-Weihenstephan, Germany

Wijnands J.H.M.
Wageningen University, Business Economics Group, Wageningen, The Netherlands

Zúñiga-Arias G.
Wageningen University, Development Economics Group, Wageningen, The Netherlands

www.ingramcontent.com/pod-product-compliance
Lightning Source LLC
LaVergne TN
LVHW082053011026
849162LV00020B/1636
* 9 7 8 9 0 7 6 9 9 8 9 5 4 *